Active Power Line Conditioners

Dedication

To my wife, Carmen, and my daughters, Laura and Patricia; they are my life. My parents, Patricio and Mª Dolores, will live forever in the memory.

Patricio Salmerón Revuelta

Dedicated to my parents, Agustín and Carmen, my wife, Mary Carmen, my children, Salvador and Gonzalo.

Salvador Pérez Litrán

To my parents, Jaime and Clara, to my wife, Claudia, and my daughter, Emily, for the time "borrowed."

Jaime Prieto Thomas

Active Power Line Conditioners

Design, Simulation and Implementation for Improving Power Quality

Patricio Salmerón Revuelta

Salvador Pérez Litrán

Jaime Prieto Thomas

AMSTERDAM • BOSTON • HEIDELBERG • LONDON
NEW YORK • OXFORD • PARIS • SAN DIEGO
SAN FRANCISCO • SINGAPORE • SYDNEY • TOKYO

Academic Press is an Imprint of Elsevier

Academic Press is an imprint of Elsevier
125, London Wall, EC2Y 5AS, UK
525 B Street, Suite 1800, San Diego, CA 92101-4495, USA
225 Wyman Street, Waltham, MA 02451, USA
The Boulevard, Langford Lane, Kidlington, Oxford OX5 1GB, UK

British Library Cataloguing-in-Publication Data
A catalogue record for this book is available from the British Library

Library of Congress Cataloging-in-Publication Data
A catalog record for this book is available from the Library of Congress

ISBN: 978-0-12-803216-9

For information on all Academic Press publications
visit our website at http://store.elsevier.com

Publisher: Joe Hayton
Acquisition Editor: Lisa Reading
Editorial Project Manager: Natasha Welford
Production Project Manager: Melissa Read
Designer: Greg Harris

Printed and bound in the United States of America
16 17 18 19 20   10 9 8 7 6 5 4 3 2 1

Table of Contents

Preface

The purpose of this book is to present a modern content, updated and practiced on one of the main topics of interest within the electric power quality (EPQ): mitigation technologies to periodic disturbances with special emphasis on distributed generation systems. The topics covered in the theoretical foundations, control strategies, and practices related to active power line conditioners (APLCs) configurations are summarized – a text for researchers, professionals, and students interested in power engineering. The authors have tried to develop the content of the book in an understandable way, with a large number of practical cases that can be reproduced by the readership to implement the techniques discussed. We believe that all cases of simulation and laboratory implementations presented in the book will mean a journey that will allow the reader to learn more about the latest landmarks on the different APLC topologies for load active compensation.

In the beginning of the book, the PQ problem and the need for equipment to correct the lack of PQ in power grids are introduced. Power terms are defined according to the decomposition model, the apparent power collected within the IEEE Standard 1459, and PQ indexes derived there from. The instantaneous reactive power theory has been the theoretical framework that enabled the practical development of APLCs. The original and modified formulations, along with other proposals are set out in this text. Different APLC configurations are used in practice for load compensation. The shunt APF, series APF, hybrid APF, and shunt APF combined with series APF, known as UPQC, are treated in a separate chapter in this book. The different simulation examples and case studies with real APLCs are developed carefully. Finally, the new paradigm that has meant the emergence of distribution systems with dispersed generation is treated in a chapter where mitigation technologies are addressed in a distributed environment.

Specifically, we summarize the contents of each chapter.

Chapter 1: A problem of the EPQ or PQ, can be understood as a disturbance that causes the voltage and current in an electrical system deviate from a reference state considered ideal. This has led to specific studies to classify and characterize the types of disturbances that can occur in the mains. The acceptable limits for these electromagnetic compatibility phenomena, EMC, are established in the various standards, such as IEC 61000. Regarding the reduction of negative effects of PQ disturbances, different mitigation technologies have been proposed, including the active power filters (APFs), as most recent compensation systems. In this chapter, first a comprehensive description of different voltage waveform disturbances is performed: frequency variations, voltage fluctuations, voltage dips, voltage unbalances, and harmonics. Special attention to harmonics and limits of voltage and current harmonics contained in the standards is provided. The active power line conditioners (APLCs), as modern mitigation technologies to periodic disturbances are described. The different topologies and control principles are presented. The chapter, then, establishes the connection between the PQ and equipments to improve the lack of PQ, without decreasing the load operation. Classification herein established between nonlinear loads such as harmonic current sources (HCS) or harmonic voltage sources (HVS) is important, as this will dictate the choice of topology (series or parallel) most appropriate for compensation.

Chapter 2: Apparent power is one of the highest quantities of use and application in the field of electrical power. Its significance for the case of single phase and balanced three-phase sinusoidal system is clear, and its division in terms of active and reactive power. This is no longer true when dealing with unbalanced systems and/or distorted, where there is no definitive agreement to define the apparent power and where the partition in the most appropriate power terms is questionable. The IEEE 1459 Standard attempts to resolve this issue with the proposal of a definition based on the effective voltage and current; and a decomposition of the effective apparent power has three useful features: (1) distinguish between fundamental apparent power (and its active and reactive components) and the nonfundamental apparent power, (2) provides a useful measure of harmonic "pollution" by decomposition of the nonfundamental apparent power in different terms related to voltage and current THDs (total harmonic distortion), and (3) provides a measure of the degree of unbalance by the unbalance fundamental apparent power term. The Std 1459 is a framework to assess the operation of APLCs in situations of unbalance and distortion, so in this chapter, are defined terms and PQ indices greater international acceptance.

Chapter 3: This chapter presents the foundations of the instantaneous reactive power theory for three-phase, four-wire systems. The original theory was published in 1983 by Akagi, Kanazawa, and Nabae and it is they for whom it was named. A description model of three-phase systems is obtained through the introduction of two power quantities: the instantaneous real power and instantaneous imaginary power. Both the instantaneous active and instantaneous reactive currents explain the power flow between source and load. However, the initial goal of the theory was eminently practical: a basis to design static compensators without energy storage elements (active power filters, APFs) in your compensation strategy. They have subsequently proposed alternative formulations, but all with a clear link of inspiration in the original instantaneous reactive power theory. In this chapter, special attention is devoted to developments in phase coordinates established to prevent the transformation to coordinates 0, α, β. Here, the fundamentals of the instantaneous reactive power theory are presented, as well as those derived formulations have greater acceptance. Especially is incised between the similarities and differences between them, and their application to the determination of control strategies for active power conditioners for compensation of nonlinear three-phase loads.

Chapter 4: Nowadays, the APFs can be used as a practical solution to solve the problems caused by the lack of EPQ. The emerging technology of power electronic devices and the new developments in digital signal processing (DSP) have made possible its practical use. These power filters can compensate fully the nonlinear loads of electrical power systems: harmonics, reactive power, unbalances, etc. A shunt APF is an electronic power circuit (power inverter), which is connected in parallel with the load and behaves as a controlled current source. Until now, there have been many proposals for possible control strategies and in particular the methods used to obtain the reference waveforms. In this chapter, the usual power blocks, the control strategy, and the modulation control method will be presented. In particular, a shunt APF design will be detailed. In this case, the goal is to inject, in parallel with the load, compensation current to get a sinusoidal source current and/or power factor unit. At the end of the chapter, practical design considerations will be presented and the APF parameters will be justified. The proposed filter will be issued in a simulation platform to adjust the component values. Finally, a laboratory prototype will be probed to contrast the final design.

Chapter 5: The APFs of parallel connection inject a harmonic current with the same amplitude and opposite phase to the harmonic loads. This configuration has proven effective for the compensation of loads named current-source nonlinear loads (HCS) as in the case of rectifiers with a high inductance in the dc side. However, loads against the so-called type of voltage-source nonlinear loads (HVS) such as rectifiers with capacitor in the dc side, shunt APFs have an ineffective response. The series connection APFs have proven to be an alternative solution for clearing this type of load. Three compensation strategies are considered: control by detecting the source current, control by detecting the load voltage, and the hybrid control that combines the source current detection and the load voltage detection. These strategies are analyzed in steady state from the equivalent single-phase circuit. This analysis is done from the point of view of the distortion produced by the load and that this distortion in the system is due to nonlinearities other than those produced by the load itself. In addition, the state model is obtained; from the state equations, the behavior of the system for the three compensation strategies is analyzed. From the analysis in the state space, design rules for each control strategy are established. The theoretical analysis is compared with the results obtained using different simulation examples and experimental tests.

Chapter 6: Besides the single active filters, either a shunt APF or a series APF, other topologies consisting of a combination of active and passive filter were tested. Passive LC filters are branches tuned to the prevalent harmonic frequencies of the load. These configurations include associations of APF and passive filters; we refer to generically as hybrid filters. This chapter takes the study of a hybrid filter configuration consisting of a combination of series APF and passive filters for the harmonic compensation in systems with nonlinear loads. We analyzed, from a practical point of view, the application of such compensation equipment to loads of harmonic voltage source type, and against harmonic current source. Thus, in the same way as in Chapter 5, the three compensation strategies for both load types are presented. As is usual in this text, the analysis is performed from the point of view of the distortion produced by the load and distortion present in the system due to other nonlinearities than those produced by the load itself. In addition, the state model for the hybrid filter configuration SAPPF is obtained. The behaviors of the whole from their state equations for the three compensation strategies are analyzed. From the analysis in the space state, design rules for each control strategy is established. Different case studies are developed, both in the simulation platform based on MATLAB–Simulink, as in the experimental platform.

Chapter 7: A unified power quality conditioner (UPQC) consists of a combination of an APF connected in series and an APF connected in parallel with the load for voltage and current simultaneous compensation. The presence of loads that produce distortion can affect other sensitive loads (criticals) to the harmonic currents and connected to the same PCC. The UPQC can be controlled not only to compensate for harmonic currents and unbalances produced by nonlinear loads, but also for harmonics and unbalances in the supply voltage. Thus, the UPQC in one device combines the specific functions of the parallel current compensation and series voltage compensation. In this chapter, an overview of the UPQCs structure and corresponding control strategies is performed. In the last part, the functionality of UPQC on the quality of power, with voltage regulation and power flow at the fundamental frequency are expanded; that is, what has been named the universal active power line conditioner (UPLC). The chapter develops different simulation examples and experimental laboratory prototype.

Chapter 8: In previous chapters, the question of PQ was raised and associated the mitigation technologies for compensating voltages and currents in a particular location (node) of the network. This chapter considers the voltage profile of the distribution lines in systems with distributed generation (DG). Likewise, the impact on the PQ of a large number of distributed power injections to the grid is analyzed. This is of great interest as many of the generating units connected to the networks of the future are small power units such as those based on the gas technology or renewable energy sources, which require a power inverter stage for grid connection. Therefore, this chapter addresses the interaction of DG, and its impact on the PQ and the line compensation distribution. Different examples designed for this purpose will analyze the situation from a practical point of view.

Finally, we show that this book is about a specific and current topic, located at the frontier of knowledge. The systematic use of models developed in the environment of SimPowerSystems of Matlab–Simulink is one of its outstanding features. It is intended that the reader can reproduce the case studies raised, and that he can build his own practical examples. Likewise, special attention to work with real equipment and analysis of experimental results is provided. We used a system of acquisition and control signals of general application, developed by dSpace that can handle models tested with Matlab, which facilitates the reproduction of the reviewed case studies. These developments were performed in the power laboratory of research group in Electrical and Electronics of La Rábida, GEYER, at the University of Huelva. The files of simulation examples that have been developed throughout the book, for reader, it will be available at: http://booksite.elsevier.com/9780128032169/.

Patricio Salmerón Revuelta
Salvador Pérez Litrán
Jaime Prieto Thomas

About the Authors

Patricio Salmerón Revuelta was born in Huelva, Spain. He received PhD from the Electrical Engineering Department of the University of Seville, Spain, in 1993. In 1983, he joined the Seville University as an assistant and then Associate Professor in the Department of Electrical Engineering. Since 1993, he has been a Full Professor at Escuela Técnica Superior de Ingeniería, University of Huelva. He is head of the research group Electrical and Electronics of la Rábida, GEYER. He has directed several research projects related to electrical measurements in nonsinusoidal systems and equipment implementation to mitigate the lack of electric power quality. His research interests include electrical power quality, electrical power systems, active power filters, and distributed generation.

Salvador Pérez Litrán was born in Sanlúcar de Barrameda, Spain. He received the BSc degree in Automation and Industrial Electronics Engineering from the University of Seville, Spain, in 2002, and the PhD degree in Electrical Engineering from the University of Huelva, Spain, in 2011. He is currently Professor and Head at the Department of Electrical Engineering, University of Huelva. His research interests include application of power electronics in distribution systems, power quality analysis, active power filters, hybrid power filters, and renewable energy.

Jaime Prieto Thomas was born in Madrid, Spain, in 1969. He received the MSc degree in Electrical Engineering from the University of Seville, Spain, in 1994. After 3 years of engineering practice, he joined the Electrical Engineering Department of the University of Huelva in 1997 as Assistant Professor. He is currently Associate Professor in the same Department of the University of Huelva.

His main research interests are active power conditioning; power quality; and power converter analysis, design, and control.

1

Introduction to Power Quality from Power Conditioning

This chapter introduces the general objectives of this book. To do this, it begins with an introduction to the concept of power quality, which has become a discipline for defining the reference parameters for assessing the suitability of the waveforms of voltage and current of an electrical system. This allows to achieve compatible operation of all equipment that constitute it. This makes it possible to set limits on the disturbances in the voltage and current source which are reflected in the different standards.

An important source of the disturbances that occur in an electrical network is the presence of nonlinear loads. A model based on the Norton equivalent of a load is presented, in order to develop a theoretical analysis of the system when this load type is included.

Furthermore, an update of compensation equipment configurations most common in the technical literature that includes active power filters is also performed. This type of equipment has proven to be effective in the dynamic compensation of nonlinear loads.

P. Salmerón Revuelta, S.P. Litrán, and J.P. Thomas: Active Power Line Conditioners. http://dx.doi.org/10.1016/B978-0-12-803216-9.00001-8

1

Different topologies are presented as an introduction. The following chapters will analyze these in depth, and will also introduce some control strategies that allow the quality of electrical power to be improved.

1.1 Introduction

In recent decades, the concept of power quality (PQ) has become more important within the field of electrical engineering, such that currently it has become a matter of great interest for producer/supplier electricity companies, equipment manufacturers and end consumers [1].

A power quality problem can be understood as a disturbance that causes the system voltage or current to differ from an ideal reference value [2]. This definition has led to specific studies that aim to obtain a detailed account of the phenomena of electromagnetic compatibility (EMC) that cause disturbances in PQ [3]. This cataloging of limiting values are detailed in various national and international standards.

A range of ways of improving power quality have been proposed that are based on passive filters, which can be understood as devices which change impedance versus frequency. Other solutions include active filters, which are able to inject harmonics that counter to the network harmonics using power-electronic converters. These filters can be connected in parallel or in series depending on the type of load that needs to be compensated. A combination of both filter types (passive and active) can also be used, and this is known as a hybrid filter. In this case the active filter can improve the frequency response of the passive filter.

In this chapter, Section 1.2 discusses the various disturbances that may be present in the voltage waveform and the limits on the distortion levels of electric current. To do this, a load classification in different classes as provided in the standard as well as its current harmonic limits are presented. Nonlinear loads generate harmonic currents that can cause disturbances and malfunction of the facilities. To limit its effects, standards set limits to the harmonic levels that loads can inject to the electric network. Section 1.3 presents these standards and summarizes these limits. In Section 1.4, a nonlinear load model that allows the analysis of systems including these loads is proposed. They are also classified according to the harmonic type generated in the system. In Section 1.5, the filter configurations most common in the literature are summarized and their main features are outlined.

1.2 Power Quality

Many social and economic activities depend on the quality and efficiency of an electrical power supply. Both industrial and commercial users are interested in guaranteeing the electrical waveform quality that supplies their different systems. The proliferation of electronic equipment increases the nonlinearity of the load and so worsens the quality of the power in the system. Harmonic current drawn from a supply by the nonlinear load results in the distortion of the supply voltage waveform at the point of common coupling (PCC) due to the source impedance. Both distorted current and voltage may cause end-user equipment to malfunction, conductors to overheat and may reduce the efficiency and life expectancy of the equipment connected at the PCC.

Nowadays, problems with power quality can be very expensive due to bad operation of sensitive load. Hence, International Standards have established limits for harmonic current emissions, power quality measurement conditions and testing techniques to apply. A summary of the disturbances and limits set by the main standards are presented here.

1.2.1 Voltage Disturbances

Electricity supplied to customers has many features that can vary and affect how it can be used. From the point of view of the consumer, it is desirable that the supply voltage has a frequency and amplitude that does not vary and that the sinusoidal waveform is not distorted. In practice, there are numerous factors that prevent this aim being achieved. Of these, the principal factor is due to using the same users make the electrical wave that disturbs their characteristics with respect to the ideal situation [4].

Electrical current demanded by customers flowing through the distribution network produces voltage drops. This means that the supply voltage is continuously affected by these voltage drops, which in turn depend on the existing power demand at a given time. On the other hand, the system components may be subject to faults that may affect the supply voltage which could interrupt the supply to one or many consumers.

To maintain the frequency at a constant value, it is necessary to have a production capacity that can continuously adapt to simultaneous demand by all customers, although both production and demand are likely to vary in different ways. Specifically, in the case of production loss and damage to the transmission or distribution networks, the risk of an increase or decrease in frequency that will be corrected with the secondary regulation or tertiary regulation of the power system.

There are many phenomena that can disturb the normal functioning of consumer's equipment. Some them are associated with inevitable transients or are caused by defects: maneuvers or atmospheric phenomena. Others are due to the use that is now made electrical energy, since equipment that modifies the waveform of the system voltage is connected directly to the network. This is due to the proliferation of loads that produce these effects. On the other hand, these loads include control circuits that are sensitive to these disturbances.

Therefore, the power quality in relation to some of its features depends more on client than distributor or producer, so if the goal is to obtain a certain level of quality, both client and supplier must work together to achieve this.

Either way, the standards seek to ensure on the one hand that the supply voltage presents values that guarantee the PQ within limits, [5] and on the other try to limit disturbances that are produced by customer loads [6] and so minimizing any effect on the voltage.

The standard of the voltage wave is characterized [5] by the following parameters: frequency, amplitude, shape, and symmetry. From the point of view of generation, the voltage waveform produced in large power plants can be considered to be sinusoidal. However, in the process of transmission and distribution of energy, the aforementioned parameters undergo alterations that may affect users. The origin of these abnormalities is often found in the electrical system itself as a result of maneuvers, breakdowns, natural phenomena (lightning) or in the normal operation of certain receptors such as electronic converters [7]. To check the quality of power supply in IEC 61000-4-30 [8] methods and measurement conditions are established.

1.2.1.1 Frequency Variations

In general, the network frequency is usually very stable, barring exceptional cases, because there is a high degree of interconnection of electric power systems. The frequency variation in an electrical system occurs when the balance between load and generation is altered. In an interconnected system such as that in Spain a load change of 12,000 MW is needed to vary the frequency by 0.1 Hz.

Generally, frequency variations affect the speed of rotating machines, clocks synchronized to the network, and in general any type of electronic regulation that uses the frequency as a time reference.

As established in the standard EN 50160 [5], the nominal frequency is 50 Hz under normal operating conditions, and the average value of the fundamental frequency measured in periods of 10 s should be within the following ranges:

- To networks coupled by synchronous connection to an interconnected system:
  - 50 Hz ± 1% (from 49.5 Hz to 50.5 Hz) for 99.5% a year.
  - 50 Hz + 4%/−6% (from 47 Hz to 52 Hz) for 100% of the time.
- For networks which have no synchronous connection to an interconnected system (existing power networks in island areas):
  - 50 Hz ± 2% (from 49 Hz to 51 Hz) for 95% of the time over 1 week.
  - 50 Hz ± 15% (42.5 Hz to 57.5 Hz) for 100% of the time.

1.2.1.2 Slow Voltage Variations

An important factor in determining the quality of the power supply is the amplitude of the voltage waveform and hence its root mean square (rms) value (Figure 1.1). In an electrical system the rms value of the voltage may vary with respect to its nominal value. It is

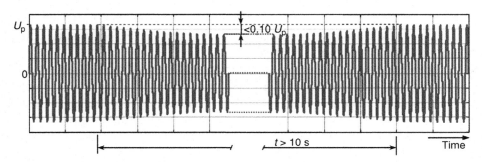

FIGURE 1.1 Slow voltage variations.

considered a slow voltage variation when its duration exceeds 10 s. Many factors can cause such variations, from supply failures, mostly due to atmospheric phenomena, to variations in the impedance of the receiver (no constant energy consumption, uneven distribution areas).

Standard EN 50160 states that the normalized voltage for low voltage networks is:

- For four-wire, three-phase systems
   U_n = 230 V between phase and neutral
- For three-wire, three-phase systems
   U_n = 230 V between phases

With respect to voltage variations, regardless of the situations that causes the defects or interruptions, the Standard establishes that the following normal operating conditions must be in place:

- For each period of 1 week, 95% of the rms values of the supply voltage averaged over 10 min must be in the range U_n ± 10%.
- For all periods of 10 min, the averaged rms values of voltage must be in the range U_n + 10% / −15%.

1.2.1.3 Voltage Fluctuations

Voltage fluctuations (Figure 1.2) occur when there are variations in supply voltage with a duration from several milliseconds up to 10 s, and an amplitude that does not exceed ±10% of the nominal value (if the variation is less than 90% (Figure 1.3) of the rated voltage, the event is called a sag or dip). Voltage fluctuations are mainly due to load variations in customer installations or maneuvers in the network (connecting arc furnaces, welding equipment, large motors, frequency converters, compressors, laser printers, etc.).

Such fluctuations may affect large numbers of consumers, but because the amplitude does not exceed ±10% of the nominal voltage, its effects do not usually affect the normal operation of electrical equipment. These phenomena can affect humans, however, because they cause lamps to flicker in brightness, which is generally uncomfortable for the consumer.

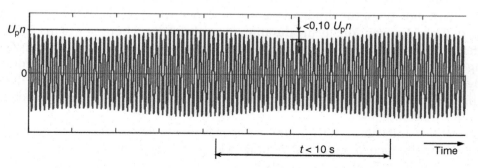

FIGURE 1.2 Voltage fluctuations.

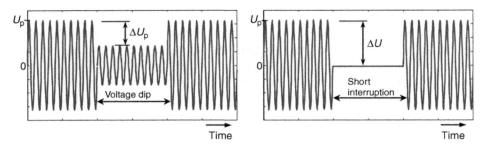

FIGURE 1.3 Voltage dips definition ($0.10U_p < \Delta U < 0.99U_p$) and short interruptions ($\Delta U > 0.99U_p$).

To quantify this phenomenon of flickering lamps, the Flicker concept has been defined. This is based on a statistical calculation which uses measurements of rapid variations in voltage:

- P_{st} (short-term flicker) assesses the severity of Flicker over a short time period at intervals of 10 min. This interval is valid for estimating disturbances caused by individual sources such as rolling, heat pumps or appliances. Values are typically given per unit, so that rates above 1 are considered to be visible and affect vision.
- P_{lt} (long-term flicker) considers a measurement duration of 2 h, which is considered to be appropriate to the operating cycle of the load over which an observer can be sensitive to long term flicker. Is calculated from twelve consecutive values of P_{st} according to the expression:

$$P_{lt} = \frac{\sqrt[3]{\sum_{i=1}^{12} P_{st,i}^3}}{12} \tag{1.1}$$

1.2.1.4 Voltage Sags or Dips and Short Interruptions

"Sags" or "dips" are occasional voltage drops in electric power systems. According to EN 50160, a voltage dip is defined as: "a sudden reduction of the supply voltage to a value between 90% and 1% of the declared voltage followed by a voltage recovery after a short period of time." Conventionally the duration of a voltage dip is between 10 ms and 1 min. The depth of a voltage dip is defined as the difference between the minimum rms voltage during the voltage dip and the declared voltage. Voltage changes that do not reduce the supply voltage to less than 90% of the declared voltage are not considered to be dips. According to some classifications, a dip in the declared voltage of 1% is considered to be a short interruption.

Voltage dips and short interruptions are usually due to defects that occur in the facilities of customers, such as sudden switching of large loads, or they can occur as short circuits or malfunctions in the general network.

1.2.1.5 Voltage Imbalances

In a three-phase system a voltage imbalance occurs when the rms values of the phase voltages or the phase angles between consecutive phases are not equal. EN 50160 sets the limit on the imbalance as follows: under normal operating conditions, for each period of 1 week, 95% of the rms values averaged over 10 min from the negative sequence voltage supply must be between 0% and 2% of the positive sequence. In some areas with partly single-phase or two-phase lines, imbalances may reach 3% at supply points.

These asymmetries in the supply voltage are mostly due to the circulation of an imbalanced current system by the network.

Voltage imbalances can cause an increase of temperature in induction motors and a decrease in use level in transformers.

1.2.1.6 Harmonic Voltage

These disturbances are caused by connecting nonlinear loads to the mains, such as equipment that include a part of power electronics.

According to EN 50160, under normal operating conditions, during each period of 1 week, 95% of the rms values of each harmonic voltage averaged over 10 min intervals should not exceed the values given in Table 1.1. In addition, the rate of total harmonic distortion (THD) of the supply voltage (including all harmonics up to 40) must not exceed 8%. The THD is defined in this standard by the following relationship:

$$THD = \sqrt{\sum_{h=2}^{40}(u_h)^2} \qquad (1.2)$$

Where u_h represents the relative amplitude of the harmonic, h, relative to the fundamental harmonic voltage.

Table 1.1 Values of Individual Harmonic Voltages at the Supply Terminals for Orders up to 25 Given in Percent of U_n

Odd Harmonics				Even Harmonics	
Nonmultiples of 3		Multiples of 3			
Order *h*	Relative Voltage (%)	Order *h*	Relative Voltage (%)	Order *h*	Relative Voltage (%)
5	6	3	5	2	2
7	5	9	1.5	4	1
11	3.5	15	0.5	6...24	0.5
13	3	21	0.5		
17	2				
19	1.5				
23	1.5				
25	1.5				

Note: No values are given for harmonics of order higher than 25, as they are usually small, but largely unpredictable due to resonance effects.

Voltage harmonics can cause the malfunction of certain electronic equipment, nuisance tripping of the protection, etc. In the long term it can even reduce the lifetime of rotating machines, capacitors, and power transformers.

1.2.2 Harmonics

The presence of harmonics in the voltage and current signals from the mains supply is not a new phenomenon [9]. Associated problems have been a constant concern for engineers since the beginning of the electrical industry [10], meaning that harmonics have been linked to a wide variety of problems for several decades.

All nonlinear loads produce harmonics in either voltage or current. IEEE Std 519-2014 [11] establishes recommended practice for the design of electrical systems that include both linear and nonlinear loads. The voltage and current waveforms that may exist throughout the system are described, and waveform distortion goals are established for the system designer. The interface between sources and loads is described as the point of common coupling and observance of the design goals will minimize interference between different items of electrical equipment.

Furthermore, the IEC standards do not establish limits to the circulation of current harmonics through to distribution network, however, they establish harmonic levels that loads can inject to the network. Thus, IEC 61000-3-2 [6] and IEC 61000-3-4 [12] deal with the limitations of current harmonic that can be injected into the public network.

In IEC 61000-3-2 [6] the limits of harmonic current components that can be produced by equipment tested under specific conditions are specified. This standard is limited to devices with lower input current 16 A per phase.

According to the cited standards, loads are classified into three types, as is shown in Table 1.2. Based on this classification the limits shown in Tables 1.3–1.5 are established. In the case of Class B equipment, the harmonics of the input current must not exceed the

Table 1.2 Equipment Classification According to Standard EN 61000-3-2

	Equipment
Class A	• Balanced three-phase equipment • Household appliances, excluding equipment identified by Class D • Tools excluding portable tools • Dimmers for incandescent lamps • Audio equipment • Everything else that is not classified as B, C, or D
Class B	• Portable tools • Arc welding equipment which is not professional equipment
Class C	• Lighting equipment.
Class D	• Personal computers and personal computer monitors • Television receivers • Note: Equipment must have power level 75W up to and not exceeding 600W

Table 1.3   Harmonics Limit for Class A Equipment

Order Harmonic, *n*	Maximum Permissible Harmonic Current (*A*)
Odd Harmonics	
3	2.30
5	1.14
7	0.77
9	0.40
11	0.33
13	0.21
$15 \leq n \leq 39$	$0.15 \dfrac{15}{n}$
Even Harmonics	
2	1.08
4	0.43
6	0.30
$8 \leq n \leq 40$	$0.23 \dfrac{8}{n}$

Table 1.4   Harmonics Limit for Class D Equipment

Order Harmonic, *n*	Maximum Permissible Harmonic Current per Watt (mA/W)	Maximum Permissible Harmonic Current (A)
3	3.4	2.30
5	1.9	1.14
7	1.0	0.77
9	0.5	0.40
11	0.35	0.33
$15 \leq n \leq 39$ Only odd harmonics	$\dfrac{3.85}{n}$	See the table corresponding to Class A

Table 1.5   Harmonics Limit for Class C Equipment

Order Harmonic, *n*	Maximum Permissible Harmonic Current (%) of the Input Current at the Fundamental Frequency
2	2
3	$30 \cdot \lambda*$
5	10
7	7
9	5
$15 \leq n \leq 39$ Only odd harmonics	3

*λ is the power factor.

absolute values given in Table 1.3, corresponding to the limits for Class A equipment multiplied by a factor of 1.5.

1.3 Nonlinear Loads Model

In the harmonic analysis of a power system, there are techniques that can be applied to linear circuits at steady state. These techniques vary depending on the required data, model complexity, the problem formulation and the computational algorithms in use. Nonlinear loads are considered as sources that inject harmonics into a linear network [13]. Depending on the type of harmonics and their behavior in the system, the nonlinear loads can be considered as current source or voltage source type loads [14–16].

Thus, in the repertoire of nonlinear loads, there will be a load type whose input current is almost invariant to changes in the source impedance, so that it can be considered as a nonlinear load that generates harmonic current sources and it is named HCS (harmonic current source). The harmonic voltage distortion at the point of common coupling that produces this type of load is usually relatively low, typically less than 5%. Figure 1.4 shows a model for this type of load, which consists of an ideal current source, representing the harmonics injected by the load.

One example of load with this type of behavior is a rectifier with a high enough inductance to obtain a practically constant dc side current. This inductance is much larger than the source impedance, so that its variations do not significantly affect the load current. Figure 1.5 shows the voltage and current for a load of this type. The figure shows how the harmonic distortion of the current is much higher than the voltage distortion at the load.

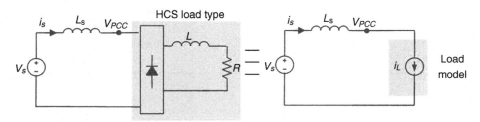

FIGURE 1.4 HCS type nonlinear load model.

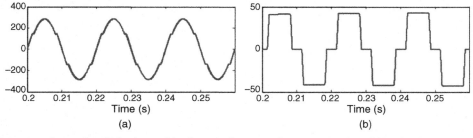

FIGURE 1.5 Waveforms of a HCS load type: (a) voltage at the point of common coupling; (b) load current.

For a steady state analysis this simple model is usually sufficient for most applications. However, in certain situations this model is not adequate, and it is necessary to perform a frequency analysis or a transient analysis [17,18]. In such cases, a model in which the nonlinear load is represented by a Norton or Thevenin equivalent can be used [19,20]; that is, by a real current source or a real voltage source, as shown in Figure 1.6. In such cases the value of the equivalent impedance should be determined using field measurements or detailed simulation models of the nonlinear load.

The parameter values of the elements that constitute the mentioned models can be determined by measuring the voltage and current harmonics of the load under two different operating conditions, [21,22]. Thus, in the circuit of Figure 1.7, it can be seen that connecting or disconnecting the switch k_1 causes the load voltage V_h, the harmonic current I_h and the current $I_{ZN,h}$ to have different values. The values of the current $I_{h,1}$ and the voltage $V_{h,1}$ before closing k_1, along with the values of voltage and current, $I_{h,2}$, $V_{h,2}$, once the switch is closed allows us to solve the equations:

$$Z_{N,h} = \frac{V_{h,1} - V_{h,2}}{I_{h,2} - I_{h,1}} \tag{1.3}$$

$$I_{N,h} = I_{h,1} + \frac{V_{h,1}}{Z_{N,h}} \tag{1.4}$$

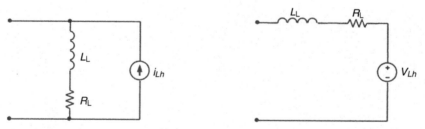

FIGURE 1.6 Norton and Thevenin models of a nonlinear load.

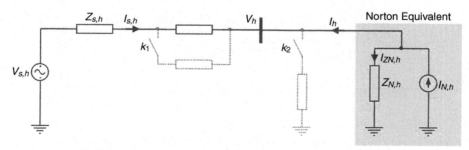

FIGURE 1.7 Determination of the Norton equivalent of a nonlinear load.

These equations can be used to calculate the Norton current source and the equivalent impedance for each harmonic. We must keep in mind that equations (1.3) and (1.4) are complex, so that not only measures of the rms values of voltage and current but also measurements of phase angles are necessary. It is also important that the measurements carried out for the two sets of operating conditions must be referred to a phase angle of a common variable that does not change with the condition of the system. In Figure 1.7 this common reference is the voltage $V_{S,h}$.

Similar equations can be used when considering the connection and disconnection of the switch k_2 in the circuit of Figure 1.7.

This load model allows better accuracy in a wider range of operating conditions than the ideal model with current source [9]. Another important feature of the proposed method to obtain the Norton and Thevenin equivalent parameters is that it is not necessary to know the system fully [23].

There are others type of loads where the current drawn by them is strongly affected by the value of the source impedance. However, the voltage at the PCC remains virtually unchanged with reasonable changes in the impedance of the supply side. It can be considered that this load behaves like a harmonic voltage source (HVS) connected to the network, Figure 1.8.

A typical example of a HVS load is a rectifier with a large capacitor to eliminate the ripple and to achieve a substantially constant voltage at the dc side. Figure 1.9 shows the current and voltage waveforms at the PCC for this rectifier type load. The current and voltage

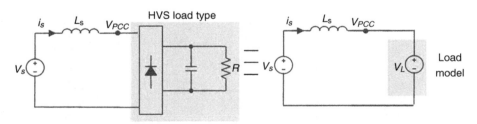

FIGURE 1.8 HVS type nonlinear load model.

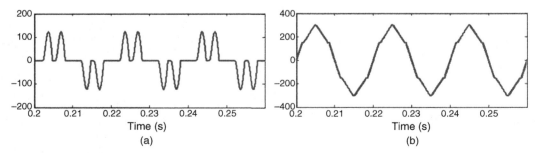

FIGURE 1.9 Waveforms of a HVS type load: (a) current; (b) voltage at the point of common coupling.

THD at the point of common coupling are 35.74% and 6.66%, respectively. Variations in the source impedance maintain the voltage at the PCC substantially constant, so in this case the ideal voltage source model in Figure 1.8 could be used.

Like with the HCS type loads, a more complete model can be obtained, formed by a real voltage or current source as shown in Figure 1.7.

EXAMPLE 1.1

The purpose of this case study is to obtain the Norton model of a nonlinear load. Figure 1.10 shows the circuit diagram consisting of an uncontrolled three-phase rectifier with a resistor and 50/3 Ω and an inductance of 55 mH connected in series at its dc side. This load is connected to a three-phase sinusoidal source with a phase voltage of 400 V and frequency of 50 Hz, with a source impedance modeled by an inductance of 2.8 mH and a resistance of 1.8 Ω. Another RL resistance and inductance branch with and inductance of 13 mH and a resistance of 50 Ω has been included to modify the network conditions through the switch k.

Voltage and current harmonics v_h, i_h are measured before and after closing the switch. Table 1.6 shows the rms values and phase angles for the most significant harmonics.

The application of expressions (3) and (4) enable to calculate $Z_{N,h}$ e I_{Nh}. Table 1.7 shows the values of the Norton currents and impedances for the regarded harmonics from the data of Table 1.6.

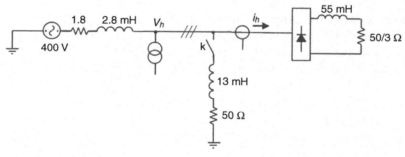

FIGURE 1.10 Scheme to model a HCS type load.

Table 1.6   Voltages and Currents Measured at Two Different Operating Conditions

	k, on				k, off			
	Current		Voltage		Current		Voltage	
Harmonic	rms (A)	fase	rms (V)	fase	rms (A)	fase	rms (V)	fase
1°	20.47	−12.55	190.4	−2.99	19.88	−12.64	184.6	−3.67
5°	3.52	114.6	16.75	2.32	3.46	114.3	15.44	−1.36
7°	2.01	90.95	12.89	−15.34	1.99	90.74	11.84	−19.64
11°	0.76	−152.6	7.44	106.9	0.78	−151.2	6.86	103.7
13°	0.46	172.5	5.29	73.54	0.48	175.2	4.89	71.75
17°	0.21	−118.90	3.15	144.30	0.2	−111.6	2.58	147.3

Table 1.7 Impedance and Current Source Values of the Norton Model for Example 2.1

Harmonic	5	7	11	13	17
$Z_{N,h}$	26.59∠−92.60°	65.75∠−86.98°	25.64∠68.50	14.42∠52.55°	1.78∠−74.72°
$I_{N,h}$	2.99∠−120.56°	2.19∠89.25°	0.48∠−159.20°	0.22∠120.70°	1.96∠−138.67°

1.4 Active Power Line Conditioners

Various compensation methods and circuit topologies have been proposed for harmonic mitigation and load compensation. So, in the 1970s [17,19] basic compensation principles of active power filters (APFs) were proposed. APFs are inverter circuits, comprising of active devices, i.e., semiconductor switches that can be controlled to act as harmonic current o voltage generators. However, there was almost no advance in active power line conditioners beyond a laboratory testing stage, because circuit technology was too poor to practically implement the compensation principle in the 1970s. Over the last 5–10 years, remarkable progress of fast switching devices such as bipolar junction transistors and static induction thyristors has spurred interest in the study of shunt and series active power filter for reactive power and harmonic compensation. In 1982, a shunt active conditioner of 800 kVA, which consisted of current source pulse width modulation (PWM) inverters using gate turn-off (GTO) thyristors, was put into practical use for harmonic compensation for the first time in the world.

Active Power filters, or APFs, can be used as a practical solution to solve the problems caused by the lack of electric power quality. The merging technology of power-electronic devices and the new developments in digital processing (DSP) have made their use practical [24,25]. These power filters can fully compensate the nonlinear loads of electrical power systems: harmonics, reactive power, unbalances, etc. So, they can be called active power line conditioners (APLCs).

Since there is a large number of possible APLC configurations, this section will present in a necessarily summarized form, the characteristics of those commonly used, in view of the number of published papers where they are referred as well as to their industrial use.

1.4.1 Shunt Active Power Filters

The operation principle of the shunt active filter (Figure 1.11) is based on injecting into the network the harmonic current consumed by the load but in opposite phase, so that the current source presents a pure sinusoidal waveform [26,27]. Unlike passive filters, the impedance of the system rarely affects in the filtering features, so they avoid one of the problems presented by passive filters. This configuration provides good compensation characteristics although their practical application has some drawbacks among which can be summarized as follows:

- It is difficult to build a high power PWM converter with fast response and low losses.
- Active filters have a high initial cost compared with passive filters.
- The injected current to the network by the active filter can flow through other passive filters and capacitors connected in the system.

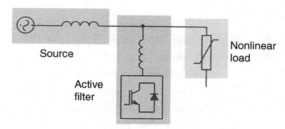

FIGURE 1.11 Basic scheme of an active filter with shunt connection.

1.4.2 Series Active Filters

In this configuration the active filter is connected in series with the load (see Figure 1.12). This arrangement allows disturbances in the voltage signal (unbalances, harmonics, etc.) to be eliminated [28–30]. According to the type of control to be established by the active filter it is possible on the one hand to eliminate the distortion in the waveform voltage produced by the load and on the other hand, to regulate the voltage at the terminals of the load regardless of the voltage drop, overvoltage, distortion, or unbalances in the supply voltages. This is carried out with a configuration that is known as a dynamic voltage restorer or DVR [31].

When the series active filter is designed to compensate the voltage harmonics generated by the load, it generates the appropriate waveform to neutralize the voltage distortion caused by the load, so that this is not transmitted to the supply voltage. The operating principle of DVR [32] is based on generating a voltage waveform of the appropriate frequency and magnitude to neutralize imbalances or distortions in the supply voltage so that the voltage at the terminals of the load has the appropriated waveform.

In industrial environments, DVR has been proposed as a way of reducing the impact that voltage disturbances produce in some loads. However, most of the time these devices are simply waiting for some disturbance to occur, and are usually operating well below their capacity [29,33,34]. In addition, these devices introduce an extra impedance in the network (because it will be connected to the network by coupling transformers), with the consequent losses. Therefore, it is usual to include additional functionality in these devices in addition to their operation as a DVR.

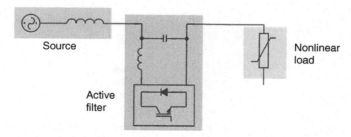

FIGURE 1.12 Basic scheme of an active filter with series connection.

1.4.3 Hybrid Active Filters

In order to take advantage of the various strengths of each of the filtering methods, different combinations of filter topologies have been proposed, [14,35]. Twenty-two of these have so far been proposed. This section briefly presents some of those most cited in the technical literature.

Figure 1.13 shows the combination of a passive filter and an active filter connected in parallel with the load [36,37]. The passive filter removes the most significant harmonics, reducing the power of the active filter. This topology has the same drawbacks as those associated with passive filters with parallel connection.

Figure 1.14 shows a hybrid topology composed of an active filter in series with the source and a passive filter in parallel with the load [38,39]. Various control strategies on the active filter have been tested in order to improve the filtering characteristics of passive filter [40,41]. Other strategies have also included controlling the voltage at the load [42]. For the latter purpose a common strategy is to generate a voltage in the active filter proportional to the harmonic of the supply current. Here, the active filter behaves as a high impedance barrier to the harmonics of the load current. This improves the performance of the passive filter, first by avoiding possible resonances and second by making the filtering feature of the passive filter independent of the source impedance. Furthermore, this can be achieved with a reduced nominal power active filter.

Finally, Figure 1.15 shows another hybrid filter topology used for eliminating current harmonics [43]. An active and a passive filter are connected in series, and the set

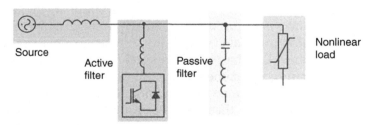

FIGURE 1.13 Parallel combination of an active and a passive filter.

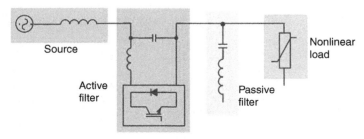

FIGURE 1.14 Hybrid topology with series active filter and passive filter in parallel with the load.

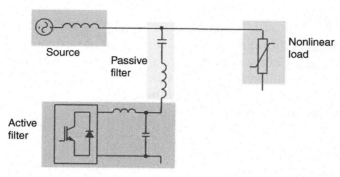

FIGURE 1.15 Combined topology of series active and passive filters, in parallel with the load.

is connected in parallel with the load. The control strategy for the active filter is based on the generation of a voltage proportional to the harmonics of the source current. This configuration improves the performance of the passive filter, with an active filter of reduced power, compared to the power required for an active filter connected in parallel with the load. This configuration has proved an enhanced behavior for the mitigation of harmonic currents and reactive power compensation of the load.

Other configurations have been investigated [44–46] in attempts to optimize performance parameters, rated power or dynamic response of the APF. A comparative analysis from the functional point of view can be found in [47].

1.4.4 Unified Power Quality Conditioners

Another combination is shown in Figure 1.16, which shows a series active filter and a parallel active filter [48]. Thus the active filter connected in series eliminates the voltage perturbances and the shunt active filter acts to eliminate disturbances in current. From the functional point of view this configuration can act dynamically both on voltage and current, so it has been called universal active filter or unified power quality conditioner [36,49,50]. The main drawback of this configuration is its comparatively high cost.

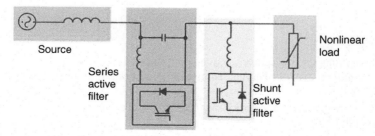

FIGURE 1.16 Combined topology of a series active filter and a parallel active filter.

1.5 Summary

This chapter provides an introduction to the concept of electric power quality. The disturbances that can appear in the voltage and current of the network are presented. With respect to the voltage, the different types of disturbances are defined, and the conditions for complying with the standards are given. In this case the standard that is referred is EN 50160. This establishes limits for:

- Frequency variations.
- Slow voltage variations.
- Voltage fluctuations.
- Voltage dips and short interruptions.
- Voltage imbalances.
- Voltage harmonics.

Standard IEEE Std. 519-2014 covers current distortions , and establishes recommendations in the design of electrical systems that include nonlinear loads. On the other hand, the IEC 61000 standard does not set limits to the circulation of harmonic currents in the distribution network, however it sets the harmonics levels that the loads can inject to the network.

Suitable models are necessary for the analysis of circuits where nonlinear loads are present. Simple or complex load models can be used, depending on the accuracy required in the analysis. In this chapter, a load model based on the Norton equivalent is proposed, and this will be used in later chapters.

A range of different topologies of active conditioners has been presented, and a brief introduction has been given of these conditioners and their usefulness from the point of view of improving the power quality. Thus, shunt active conditioners, which are connected in parallel with the load, have shown to be useful for compensating for current distortion. Series active filters, connected between supply and load, eliminate the disturbances caused by the supply voltages. Hybrid filters include LC (inductance and capacitor) branches tuned to specific frequencies and they can also improve the power quality with active filters of reduced power. Finally, the unified power quality conditioners (UPQC) are presented. They include a series active filter and a shunt active filter, allowing them to act simultaneously against voltage and current disturbances.

References

[1] Dell'Aquila A, Marinelli M, Monopoli VG, Zanchetta P. New power quality assessment criteria for supply systems under unbalanced and nonsinusoidal conditions. IEEE Trans. Power Delivery 2004;19(3):1284–90.

[2] Dugan RC, McGranaghan MF, Santoso S, Beaty HW. Electrical power systems quality. 2nd ed. New York: McGraw Hill; 2002.

[3] Sharon D, Montano JC, Lopez A, Castilla M, Borras D, Gutierrez J. Power quality factor for networks supplying unbalanced nonlinear loads. IEEE Trans Instrum Meas 2008;57(6):1268–74.

[4] Xu W, Liu Y. A method for determining customer and utility harmonic contributions at the point of common coupling. IEEE Trans Power Delivery 2000;15(2):804–11.

[5] European Standard UNE-EN 50160, Voltage characteristics of electricity supplied by public distribution systems; 2010.

[6] Standard IEC 61000-3-2. Electromagnetic compatibility (EMC) - Part 3-2: Limits - Limits for harmonic current emissions (equipment input current ≤ 16 A per phase), 2005.

[7] Alcántara FJ, Vázquez JR, Salmerón P, Litrán SP, Arteaga Orozco MI. On-line detection of voltage transient disturbances using ANNs. ICREPQ 09. Valencia: In: Proceedings of the International Conference on Renewable Energy and Power Quality; 2009.

[8] Standard IEC 61000-4-30. Electromagnetic compatibility (EMC) – Part 4–30: Testing and measurement techniques – Power quality measurement methods; 2008.

[9] Mack Grady W, Santoso S. Understanding power system harmonics. IEEE Power Eng Rev 2001;21(11):8–11.

[10] Owen EL. A history of hamonics in power system. IEEE Ind Appl Mag 1998;4(1):6–12.

[11] IEEE Std. 519-2014: IEEE recommended practice and requirements for harmonic control in electric power systems; 2014.

[12] Standard IEC 61000-3-4. Electromagnetic compatibility (EMC) – Part 3–4: Limits- Limitation of emission of harmonic currents in low voltage power supply systems for equipment with rated current greater than 16 A; 1998.

[13] Herrera RS, Pérez A, Salmerón P, Vázquez JR, Litrán SP. Distortion sources identification in electronic power systems. Zaragoza. In: Proc. Spanish Portuguese Congress on Electrical Engineering; 2009.

[14] Peng FZ, Adams DJ. Harmonics sources and filtering approaches. In: Proceedings of the Industry Aplications Conference 1999;1:448–55.

[15] IEEE Task Force on Harmonics Modeling and Simulation. Modeling and simulation of the propagation of harmonics in electric power networks. Part I: concepts, models, and simulation techniques. IEEE Trans. Power Delivery 1996;11(1):452–74.

[16] IEEE, Task Force on Harmonics Modeling and Simulation. Test systems for harmonics modeling and, simulation. IEEE Trans. Power Delivery 1999;14(2):579–87.

[17] Almeida CFM, Kagan N. Harmonic coupled Norton equivalent model for modeling harmonic producing loads. In: Proceedings of the fourteenth International Conference on Harmonics and Quality of Power; 2010(ICHQP).

[18] Fauri M. Harmonic modelling of nonlinear load by means of crossed frequency admittance matriz. IEEE Trans. Power Systems 1997;12(4):1632–8.

[19] Abdelkader S, Abdel-Rahman, MH, Osman MG. A Norton equivalent model for nonlinear loads. In: Proc. Conference on Large Engineering Systems, Power Engineering, LESCOPE '01, pp. 63–67, July 11–13, 2001.

[20] Thunberg E, Soder L. A Norton approach to distribution network modeling for harmonic studies. IEEE Trans. Power Delivery 1999;14(1):272–7.

[21] Nassif AB, Yong J, Mazin H, Wang X, Xu W. An impedance-based approach for identifying interharmonic sources. IEEE Trans. Power Delivery 2011;26(1):333–40.

[22] Thunberg E, Soder L. Influence of the network impedance on distribution system harmonic Norton models. In: Proc. International Conference on Power System Technology 2000;3(4–7):1143–8. PowerCon 2000.

[23] Balci ME, Ozturk D, Karacasu O, Hocaoglu MH. Experimental verification of harmonic load models. In: Proc. 43rd International Universities Power Engineering Conference 2008;1–4:1–4. UPEC 2008.

[24] Akagi H. New trends in active filters for improving power quality. In: Proc. IEEE International Conference on Power Electronics, Drives and Energy Systems, PEDES 1996;96:410–6.

[25] Akagi H. Active harmonic filters. Proceedings of the IEEE 2005;93(12):2128–41.

[26] Gyugyi L, Strycula EC. Active ac power filters. In: Proceedings of the IEEE Industry Applications Society Annual Meeting 1976;19-c:529–35.

[27] Miret J, Castilla M, Matas J, Guerrero JM, Vasquez JC. Selective harmonic-compensation control for single-phase active power filter with high harmonic rejection. IEEE Trans. Ind Electr 2009;56(8): 3117–27.

[28] Campos A, Joos G, Ziogas P, Lindsay J. Analysis and design of a series voltage unbalance compensator based on a three-phase VSI operating with unbalanced switching functions. IEEE Trans. Power Electr 1994;9(3):269–74.

[29] Dixon JW, Venegas G, Moran LA. A series active power filter based on a sinusoidal current-controlled voltage source inverter. IEEE Trans. Ind Electr 1997;44(5):612–20.

[30] Litrán SP, Salmerón P, Vázquez JR, Herrera RS. Different control strategies applied to series active filters. In: Proc. International Conference on Renewable Energy and Power Quality; 2007 ICREPQ 07. Sevilla.

[31] Woodley NH, Morgan L, Sundaram A. Experience with an inverter-based dynamic voltage restorer. IEEE Trans. Power Delivery 1999;14(3):1181–6.

[32] Newman MJ, Holmes DG, Nielsen jG, Blaabjerg F. A dynamic voltage restorer (DVR) with selective harmonic compensation at medium voltage level. In: Proceedings of the Industry Applications Conference, 2003. 38th IAS Annual Meeting 2003;2:1228–35.

[33] Chen Chin Lin, Lin Chen E, Huang CL. An active filter for unbalanced three-phase system using synchronous detection method. In: Proceedings of the Power Electronics Specialists Conference 1994;2:1451–5. PESC '94, 25th Annual IEEE.

[34] Ribeiro ER, Barbi I. Harmonic voltage reduction using a series active filter under different load conditions. IEEE Trans Power Electr 2006;1(5):1394–402.

[35] Peng FZ, Su GJ. A series LC filter for harmonic compensation of ac drives. In: Proceedings of the IEEE Power Elect Specilist Conf 1999;213–8. PESC'99.

[36] Fujita H, Akagi H. The unified power quality conditioner: the integration of series- and shunt active filters. IEEE Trans Power Electr 1998;13:315–22.

[37] Peng FZ, Akagi H, Nabae A. A new approach to harmonic compensation in power systems-a combined system of shunt passive and series active filters. IEEE Trans Ind Appl 1990;26(6):983–90.

[38] Peng FZ, Akagi H, Nabae A. Compensation characteristics of the combined system of shunt passive and series-active filters. In: Proceedings of the IEEE Ind Appl Soc Ann Meeting 1989;959–66.

[39] Peng FZ, Akagi H, Nabae A. A novel harmonic power filter. In: Proc. IEEE Power Electronics Specialists Conference 1988;1151–9.

[40] Herrera RS, Salmerón P, Vázquez JR, Litrán SP. A new control for a combined system of shunt passive and series active filters. In: Proceedings of the International Symposium on Industrial Electronics; 2007 ISIE'2007. Vigo.

[41] Fujita H, Yamasaki T, Akagi H. A hybrid active filter for damping of harmonic resonance in industrial power systems. IEEE Trans. Power Electr 2000;15(2):215–22.

[42] Litrán SP, Salmerón P, Prieto J, Herrera RS. Improvement of the power quality with series active filters according to the IEC 61000. In: Proceedings of the IEEE Mediterranean Electrotechnical Conference; 2006MELECON. Málaga.

[43] Fujita H, Akagi H. A Practical Approach to Harmonic Compensation in Power Systems. Series Connection of Passive and Active Filters. In: Conference Record of the 1990 IEEE Industry Applications Society Annual Meeting 1990;2:1107–12.

[44] Corasaniti VF, Barbieri MB, Arnera PL, Valla MI. Hybrid active filter for reactive and harmonics compensation in a distribution network. IEEE Trans Ind Electr 2009;56(3):670–7.

[45] Barrero F, Martinez S, Yeves F, Mur F, Martinez PM. Universal and reconfigurable to UPS active power filter for line conditioning. IEEE Trans Power Delivery 2003;18:283–90.

[46] Milanes Montero MI, Romero Cadaval E, Barrero F. Hybrid multiconverter conditioner topology for high power applications. IEEE Trans Ind Electr 2010;(99).

[47] Singh B, Al-Haddad K, Chandra A. A review of active filters for power quality improvement. IEEE Trans Ind Electr 1999;46(5):960–71.

[48] Prieto J, Salmerón P, Herrera RS, Litrán SP. Load compensation active conditioner for power quality. In: Proceedings of the IEEE Mediterranean Electrotechnical Conference; 2006 MELECON. Málaga.

[49] Monteiro LFC, Aredes M, Moor Neto JA. A control strategy for unified power quality conditioner. In: Proceedings of the IEEE Int. Symposium Industrial Electronics 2003;1:391–6.

[50] Rastogi M, Mohan N, Edris AA. Hybrid-active filtering of harmonic currents in power systems. IEEE Trans Power Delivery 1995;10:1994–2000.

2

Electrical Power Terms in the IEEE Std 1459 Framework

CHAPTER OUTLINE

Apparent power is one of the parameters commonly used and applied in the field of electrical power. Its significance for the case of single phase and sinusoidally-balanced three-phase systems is clear, as are the terms of power in which the partition is possible, namely, active power and reactive power. This is no longer true when it comes to unbalanced and/or distorted systems, where there is no unanimous agreement to define the apparent power and where the power terms most appropriate for your partition are even more questionable. IEEE Standard 1459 attempts to resolve this issue by proposing a definition based on the effective voltage and current, and a partition of the effective apparent power that has three useful features: (a) the fundamental apparent power, and active and reactive components, are conveniently separated from the nonfundamental apparent power; (b) provides a useful measure of harmonic "pollution" by the nonfundamental apparent power; and (c) provides a measure of the imbalance from the introduction of unbalanced fundamental apparent power. In this chapter the development of power in single-phase/three-phase systems in the most general conditions of asymmetry and distortion within IEEE Std. 1459 will be exposed.

P. Salmerón Revuelta, S.P. Litrán, and J.P. Thomas: Active Power Line Conditioners. http://dx.doi.org/10.1016/B978-0-12-803216-9.00002-X

2.1 Introduction

Since the beginning of the use of alternating current, the "engineers of electricity" were aware of the importance of the phase angle between the voltage and current, and the effects of this on the transmission of electricity. Further developments led to the concept of power factor as a figure of merit representing the efficiency of an electrical supply system, and defined as the ratio of the average power (real power) to the apparent power. The notion of power factor (PF) gained early acceptance as an important parameter in engineering economics of power systems. Its calculation and determination had no problems as it was applied to single-phase and balanced three-phase systems [1]. However, early on, when applied in three-phase systems, it became clear that the definition of PF had serious drawbacks when applied to unbalanced systems. Since the early 1920s, one or two definitions of apparent power emerged as extensions of the definitions found in single-phase circuits used [2,3].

Since the late 1920s, widespread controversy over the meaning of the various terms of power used by power engineers arose, particularly when they were raised with respect to nonsinusoidal waveforms and/or the presence of imbalances in the case of three-phase systems. This has led to an ambiguity in the basic definitions that remained unchanged for over 50 years; this has been the case with the definitions found in the IEEE Dictionary of as wide influence. Thus, in particular, in the IEEE Dictionary there was confusion about the "correct" definition of apparent power for unbalanced three-phase systems, which remained, and even the sixth edition of 1993 was still using two different definitions leading to two different factors power [3].

In 1996, the results of work carried out over 4 years by the IEEE working group on nonsinusoidal situations chaired by A. Emanuel and composed of 20 members from diverse backgrounds, (e.g., R. Arseneau, Y. Baghzouz, M. Cox, P. Filipski, A. Girgis, K. Srinivasan, or the same A. Emanuel, who had made previous contributions) [4], were published. In this paper, a number of terms of power that were intended to be practically used in situations where the voltage and/or current is distorted and/or unbalanced were proposed. Those terms could even be used to measure the level of distortion and imbalance of a system.

The IEEE working group expressly says that it does not attempt to solve theoretical controversies about the division of the apparent power in active components, reactive, distortion, and other nonactive components, but instead gives guidance to engineers and manufacturers of equipment for measurement, with respect to the quantities to be measured for billing purposes. The final product of the work of IEEE working group was the issue of IEEE Standard 1459-2000 (and the subsequent edition of 2010) reflecting largely the work that developed [5,6].

In this chapter the topics included in the IEEE Trial-Use Standard Definitions for the Measurement of Electric Power Quantities under Sinusoidal, Nonsinusoidal, Balanced, or Imbalanced conditions, IEEE Std 1459, will be presented. This development will be followed chronologically, starting from the drawbacks in the original definitions in the so-called IEEE Standard Dictionary of Electrical and Electronics Terms, ANSI / IEEE Std 100, followed by proposals by Emanuel and others, performed within the IEEE working group, and outside it, until the set of definitions and terms of power today largely accepted by the community of American engineers is finally arrived at.

2.2 Apparent Power and Power Factor in Distorted and Unbalanced Systems: The Background

Since its first edition, IEEE Std Dictionary has used two kinds of apparent power for three-phase systems: the arithmetically apparent power S_A and the vector-apparent power S_V. S_A is not explicitly defined in the Dictionary although it is mentioned and is used normally. S_A is the arithmetic sum of the apparent power of the individual phases,

$$S_A = \sum_{\forall j=a,b,c} S_j = \sum_{\forall j=a,b,c} V_j I_j = V_a I_a + V_b I_b + V_c I_c = \sum_{\forall j=a,b,c} \sqrt{P_j^2 + Q_j^2 + D_j^2}$$
$$= \sqrt{P_a^2 + Q_a^2 + D_a^2} + \sqrt{P_b^2 + Q_b^2 + D_b^2} + \sqrt{P_c^2 + Q_c^2 + D_c^2} \tag{2.1}$$

where V_j and I_j represent the rms values (root mean square) voltage and current of phase "j." In nonsinusoidal situations, the square of the apparent power of each phase is decomposed according to the Budeanu model distinguishing active power P_j, Budeanu reactive power Q_j and distortion power, D_j.

Conversely, S_V is defined as a vector sum,

$$S_V = \sqrt{\left(\sum_{\forall j=a,b,c} P_j\right)^2 + \left(\sum_{\forall j=a,b,c} Q_j\right)^2 + \left(\sum_{\forall j=a,b,c} D_j\right)^2}$$
$$= \sqrt{(P_a + P_b + P_c)^2 + (Q_a + Q_b + Q_c)^2 + (D_a + D_b + D_c)^2} \tag{2.2}$$

In view of the definitions, it can be verified that $S_A > S_V$. Both lead to two definitions of power factor, PF_A and PF_V. In sinusoidal balanced systems the two values are equal. However, outside the ideal conditions the results can be quite different [2]. The situation is illustrated with the help of an example.

EXAMPLE 2.1

A three-phase unbalanced load is constituted by two equal linear inductive impedances in phases "a" and "b" with a power factor $\sqrt{2}/2$, and a capacitive linear impedance in the phase "c" with a power factor of same value but active power consumption doubled. Determine the vector apparent power and arithmetic apparent power.

For active power consumption of P W per phase in the a and b phases, the complex power of each phase are given by (2.3),

$$S_a = S_b = P(1+j); S_c = 2P(1-j)$$
$$S_V = \sqrt{(P+P+2P)^2} = 4P$$
$$S_A = \sqrt{(P^2+P^2)} + \sqrt{(P^2+P^2)} + \sqrt{\left((2P)^2 + (2P)^2\right)} \tag{2.3}$$
$$= 4P\sqrt{2}$$

Figure 2.1 shows the situation. The definition of PF that uses the arithmetic sum for the apparent power produces a $PF_A = 0.707$. The definition used by the vector sum for the apparent power gives a $PF_V = 1$.

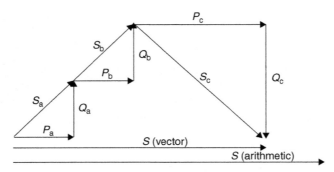

FIGURE 2.1 Vector apparent power, arithmetic apparent power, and power triangles at single-phase.

Although controversy has continued to the present, IEEE kept these definitions until the appearance of Std 1459, remaining oblivious to other proposals that had been the subject of various scientific publications over the years [7].

One of the properties of the single-phase apparent power has been considered most important: the apparent power is numerically equal to the maximum active power that exists in the input terminals of a system for an effective current and voltage value given. This is therefore directly related to the size of the equipment and generation and transmission losses. This concept can be extended to sinusoidal balanced circuits, in a limited sense to nonsinusoidal balanced circuits, but not directly to unbalanced circuits. This is mainly due to the nontrivial concerns that arise when choosing the voltage reference for the phase voltages for the line conductors. However, this does not stop the engineer from trying to find a definition that has practical utility for certain specific situations [2]. Thus, an important application of the apparent power is its use in determining the power factor, a measure of the energy consumption efficiency of an arbitrary load supplied by a three-phase source. The following discussion is limited to this. To determine the most efficient way to deliver energy to the load, the line currents can be divided on the one hand into an "indispensable" active component to transport the active power load, and a second component, the nonactive current, which is considered an "unused" component. While the active components are delivered by the supply source, nonactive components may be delivered to the load by a theoretically lossless compensator located close to the load, thereby causing no loss in the generation and transportation. The maximum active power, the apparent power, which can be supplied by the source to a perfectly compensated load, can be considered a reference quantity for determining the PF of uncompensated loads. If the approach of "system" is adopted, the apparent power of the system is the maximum power output of the system as a whole, that is, the maximum of the sum of phase active powers, provided that generation and transport losses proportional to ΣI_j^2 and ΣV_j^2 are constant.

The significance of the apparent power can be obtained by successive application of the Schwarz inequality,

$$|P| = \left| \frac{1}{T} \int_0^T \sum_{\forall j} v_j i_j \, dt \right| \leq \sum_{\forall j} \left| \frac{1}{T} \int_0^T v_j i_j \, dt \right| \leq \sum_{\forall j} \left(\sqrt{\frac{1}{T} \int_0^T v_j^2 \, dt} \sqrt{\frac{1}{T} \int_0^T i_j^2 \, dt} \right) = \sum_{\forall j} V_j I_j \tag{2.4}$$

But otherwise,

$$\sum_{\forall j} V_j I_j \leq \sqrt{\sum_{\forall j} V_j^2} \sqrt{\sum_{\forall j} I_j^2} \tag{2.5}$$

This allowed Buchholz to define apparent power, the maximum active power as possible for a given effective current and given a set of voltages in the system as,

$$S_{\text{sist}} = 3 \sqrt{\frac{1}{3} \sum_{\forall j} V_j^2} \sqrt{\frac{1}{3} \sum_{\forall j} I_j^2} \tag{2.6}$$

This definition has one of the significant attributes that was expected, as the apparent power is the power loss produced in the transport of energy and it is proportional to the square of the power factor derived from it [8]. The definition was used by the IEEE working group and subsequently extended to include the neutral conductor, to be adopted in the IEEE Standard 1459 as discussed later.

2.3 IEEE Working Group on Nonsinusoidal Situations

As has been established, the only consensus that originally existed was that the definitions used to evaluate the flow of electrical energy in power networks were inadequate in non-sinusoidal regimes or when unbalanced loads caused asymmetric regimes in three-phase systems. However, it was particularly necessary to reach consensus on definitions of power since at that time new measuring equipment based on digital sampling was becoming available. In essence it was to establish the measurement strategy most appropriate to be adopted by manufacturers and technical users when growing numbers of nonlinear loads in Points of Common Connection, PCC, were appearing, and needed to be measured for billing purposes.

Moreover, the energy generated by power plants is nearly sinusoidal of positive-sequence, therefore voltages are free of harmonics and negative and zero-sequence components. In the user side, synchronous and asynchronous motors develop torques that result from the interaction between the rotor currents and the positive-sequence rotating magnetic field. Distortion and voltage imbalance cause parasitic torques and additional losses. It therefore seems logical to separate the fundamental powers (60/50 Hz) from the rest.

This approach has the advantage of treating the fundamental power terms in the classic way. At the same time, it provides acceptable ways to measure the level of "pollution" harmonics present in the network. The IEEE working group considered "fundamental current" the current flowing to the same frequency as the fundamental voltage. Thus, the philosophy followed was to separate the main "product", terms of fundamental power, from the "pollution" nonfundamental components and their cross terms [4].

2.3.1 Single-Phase Nonsinusoidal

The instantaneous voltage $v(t)$ versus time is given by (2.7),

$$v(t) = V_0 + \sqrt{2} \sum_{h \neq 0}^{\infty} V_h \sin(h\omega t + \alpha_h) \tag{2.7}$$

where V_0 is average value, V_h is the rms value of the harmonic h of voltage and α_h is the phase angle of the voltage harmonic of order h.

Likewise there is the instantaneous current $i(t)$

$$i(t) = I_0 + \sqrt{2} \sum_{h \neq 0}^{\infty} I_h \sin(h\omega t + \beta_h) \tag{2.8}$$

where I_0 is average value or dc component, I_h is the rms value of harmonic h of current and β_h is the phase angle of the current harmonic of order h.

The effective or rms values of voltage and current are:

$$V = \sqrt{\sum_{h=0}^{\infty} V_h^2} \quad ; \quad I = \sqrt{\sum_{h=0}^{\infty} I_h^2} \tag{2.9}$$

Then the fundamental components are separated V_1, I_1 of the harmonic components V_H, I_H,

$$V^2 = V_1^2 + V_H^2 \quad ; \quad I^2 = I_1^2 + I_H^2 \tag{2.10}$$

where

$$V_H^2 = \sum_{\forall h \neq 1} V_h^2 \quad ; \quad I_H^2 = \sum_{\forall h \neq 1} I_h^2 \tag{2.11}$$

From (2.10) the following decomposition for the apparent power S is obtained,

$$S^2 = (VI)^2 = (V_1 I_1)^2 + (V_1 I_H)^2 + (V_H I_1)^2 + (V_H I_H)^2 \tag{2.12}$$

Thus, distinguished in (2.12), two components for the apparent power:

$$S^2 = S_1^2 + S_N^2 \tag{2.13}$$

where:

- S_1 is the fundamental apparent power, which in turn is decomposed into a fundamental active power P_1 and fundamental reactive power Q_1.

$$S_1^2 = (V_1 I_1)^2 = P_1^2 + Q_1^2 \tag{2.14}$$

with

$$P_1 = V_1 I_1 \cos\theta_1 \quad ; \quad Q_1 = V_1 I_1 \sin\theta_1 \quad ; \quad \theta_1 = \alpha_1 - \beta_1 \tag{2.15}$$

- S_N is the nonfundamental apparent power and comprises current distortion power, $V_1 I_H$, voltage distortion power, $V_H I_1$, and harmonic apparent power, $V_H I_H$,

$$S_N^2 = D_I^2 + D_V^2 + S_H^2 \tag{2.16}$$

The harmonic apparent power S_H can be further divided as (2.17),

$$S_H^2 = (V_H I_H)^2 = P_H^2 + N_H^2 \tag{2.17}$$

where P_H is the total harmonic active power,

$$P_H = \sum_{h \neq 1} V_h I_h \cos\theta_h \quad ; \quad \theta_h = \alpha_h - \beta_h \tag{2.18}$$

and N_H the total harmonic nonactive power.

It is recognized that while it is possible to assign a direction of flow to both P_1 and Q_1, this is not possible with the three components set for S_N. These are really only formal products, and unlike the active power have no physical meaning. However, such formal components can serve as useful indicators for the operation of a network. In fact, (2.16) is used to obtain (2.19),

$$S_N^2 = \left(\frac{I_H}{I_1}\right)^2 S_1^2 + \left(\frac{V_H}{V_1}\right)^2 S_1^2 + \left(\frac{V_H I_H}{V_1 I_1}\right)^2 S_1^2 \tag{2.19}$$

This can be rewritten in terms of total harmonic distortion indexes, THD, of voltage and current. The nonfundamental apparent power normalized to the fundamental apparent power is,

$$\left(\frac{S_N}{S_1}\right)^2 = (I\,\text{THD})^2 + (V\,\text{THD})^2 + (I\,\text{THD}.V\,\text{THD})^2 \tag{2.20}$$

In practice $S_N/S_1 \approx \text{ITHD}$. For VTHD < 5% and ITHD > 40%, the error incurred when using the above approach is less than 1%.

In general, both the value of S_N and its normalized form S_N/S_1 are better indicators of the level of harmonic "pollution" than the P_H value. Thus a well-balanced nonlinear load can be characterized by a low value of S_N/S_1. An increase in current distortion does not necessarily mean an increase in the value of P_H but always involves an increase in normalized value S_N/S_1.

Moreover, the figure of more convenient to quantify the effectiveness of power flow in a system merit is the "total power factor",

$$\text{PF} = \frac{P}{S} = \frac{P_1 + P_H}{S} \tag{2.21}$$

However, isolating P_1, Q_1, and S_1 in a nonfundamental power facilitates monitoring the flow of unpolluted fundamental power of electrical supply, and allows the application of standard techniques of compensating as power factor correction capacitors. Hence the value of the displacement power factor dPF,

$$\mathrm{d\,PF} = \frac{P_1}{S_1} = \cos\theta_1 \tag{2.22}$$

that retains significant value in cases where fundamental powers are monitored separately from nonfundamental powers.

2.3.2 Nonsinusoidal and Unbalanced Three-Phase Systems

Balanced three-phase systems are analyzed according to the suggested scheme in Section 2.3.1. Unbalanced systems require some additional considerations.

The IEEE working group on defining the apparent power follows the proposals suggested by Filipski [2], Arseneau, [9] and Emanuel, [1], among others. So the system apparent power or equivalent apparent power S_e originally suggested by F. Buchholz in 1922 and clarified by WM Goodhue in 1933 [10] was adopted,

$$S_e = 3V_e I_e \tag{2.23}$$

where

$$V_e = \sqrt{\frac{V_a^2 + V_b^2 + V_c^2}{3}} \quad ; \quad I_e = \sqrt{\frac{I_a^2 + I_b^2 + I_c^2}{3}} \tag{2.24}$$

The IEEE working group provides the following information to calculate V_e:

- For a four-wire system, V_a, V_b, V_c, are the rms values of the phase-neutral voltages.
- For a three-wire system, the equivalent voltage can be calculated with the same expression (2.24) where now V_a, V_b, V_c are the voltages measured from each conductor to an artificial neutral point (the star point of three equal resistors), or by (2.25),

$$V_e = \sqrt{\frac{V_{ab}^2 + V_{bc}^2 + V_{ca}^2}{9}} \tag{2.25}$$

where rms values correspond to line-to-line voltages. The equivalent current I_e is obtained from the rms values of the currents of the conductors I_a, I_b, and I_c.

Each of the rms values of phase to neutral voltage and current is determined in the manner,

$$V_a^2 = V_{a1}^2 + \sum_{\forall h \neq 1} V_{ah}^2$$

$$\cdot$$
$$\cdot \tag{2.26}$$
$$\cdot$$

$$I_c^2 = I_{c1}^2 + \sum_{\forall h \neq 1} I_{ch}^2$$

In the same way as in the single-phase case, the resolution of effective voltage and current (preferably as called the Std 1459) is convenient in the fundamental effective component and the harmonic effective component,

$$V_e^2 = V_{e1}^2 + V_{eH}^2 \quad ; \quad I_e^2 = I_{e1}^2 + I_{eH}^2 \tag{2.27}$$

where subscript "1" refers to the rms values of the fundamental component,

$$V_{e1} = \sqrt{\frac{V_{a1}^2 + V_{b1}^2 + V_{c1}^2}{3}} \quad ; \quad I_{e1} = \sqrt{\frac{I_{a1}^2 + I_{b1}^2 + I_{c1}^2}{3}} \tag{2.28}$$

and the subscript "H" refers the totalized nonfundamental rms components in the form,

$$V_{eH} = \sqrt{\sum_{\forall h \neq 1} \left(\frac{V_{ah}^2 + V_{bh}^2 + V_{ch}^2}{3} \right)} \quad ; \quad I_{eH} = \sqrt{\sum_{\forall h \neq 1} \left(\frac{I_{ah}^2 + I_{bh}^2 + I_{ch}^2}{3} \right)} \tag{2.29}$$

If the same approach as the one followed in Section 2.3.1 is used, it is separated from the effective apparent power S_e, the fundamental apparent power of the terms of nonfundamental apparent power [6]:

$$S_e^2 = (3V_e I_e)^2 = (3V_{e1} I_{e1})^2 + (3V_{e1} I_{eH})^2 + (3V_{eH} I_{e1})^2 + (3V_{eH} I_{eH})^2 \tag{2.30}$$

The first term is the fundamental effective apparent power,

$$S_{e1} = 3V_{e1} I_{e1} \tag{2.31}$$

This power can be separated into the fundamental positive-sequence apparent power S_1^+ and a complementary component, S_{U1}, attributed to the unbalance of the system. S_1^+ is conveniently resolved in the fundamental active and reactive power of positive-sequence,

$$\left(S_1^+\right)^2 = \left(P_1^+\right)^2 + \left(Q_1^+\right)^2 \tag{2.32}$$

with

$$P_1^+ = 3V_1^+ I_1^+ \cos\theta_1^+ \quad ; \quad Q_1^+ = 3V_1^+ I_1^+ \sin\theta_1^+ \tag{2.33}$$

The next three terms are the nonfundamental effective apparent power

$$S_{eN} = \sqrt{S_e^2 - S_{e1}^2} = \sqrt{D_{eI}^2 + D_{eV}^2 + S_{eH}^2} \tag{2.34}$$

where

$$D_{eI} = 3V_{e1} I_{eH} = 3S_{e1} (\text{THD}_{eI}) \tag{2.35}$$

is the current distortion power and,

$$\text{THD}_{eI} = \frac{I_{eH}}{I_{e1}} \tag{2.36}$$

is the equivalent total harmonic distortion of the effective current.

$$D_{eV} = 3V_{eH}I_{e1} = 3S_{e1}(\text{THD}_{eV}) \qquad (2.37)$$

is the voltage distortion power, and

$$\text{THD}_{eV} = \frac{V_{eH}}{V_{e1}} \qquad (2.38)$$

is the equivalent total harmonic distortion of the effective voltage.

$$S_{eH} = 3V_{eH}I_{eH} = 3S_{e1}(\text{THD}_{eI})(\text{THD}_{eV}) \qquad (2.39)$$

is the harmonic apparent power, containing harmonic active power,

$$P_H = \sum_{\substack{h \neq 1 \\ i=a,b,c}} V_{ih}I_{ih}\cos\theta_{ih} = P - P_1 \qquad (2.40)$$

and finally, the remaining nonactive term (total harmonic nonactive power),

$$D_{eH} = \sqrt{S_{eH}^2 - P_H^2} \qquad (2.41)$$

The expression (2.42) describes the decomposition analyzed according to seven terms of power for $S\,S_e$ and helps to summarize the situation,

$$S_e^2 = \left(P_1^+\right)^2 + \left(Q_1^+\right)^2 + S_{U1}^2 + D_{eI}^2 + D_{eV}^2 + P_H^2 + D_{eH}^2 \qquad (2.42)$$

According to the IEEE working group, the following approach has the following advantages:

- The "main product" P_1^+ is separated from the remaining components of active power. Nonlinear loads convert a small part of P_1^+ in P_H, P_1^-, and P_1^0 that are generated into the power grid and dissipated. Normally, P_H, P_1^-, $P_1^0 < 0$, and compared to P_1^+ have very small values, which makes its measurement problematic. A typical value is $P_H/P_1 < 0.02$.
- Positive sequence fundamental reactive power, Q_1^+, is also separated and helps estimate capacitor banks needed to correct the positive-sequence power factor, $\cos\theta_1^+$.
- The nonfundamental apparent power, S_{eN}, is a quantity useful to assess at a glance the severity of the distortion. It indicates whether a particular load or group of loads, are operating under conditions of harmonic, low, moderate, or excessive pollution. S_{eN} helps determine power compensation equipment. It can also serve to detect and protect equipment that are behaving like harmonics sinks, behaving similarly to harmonic filters.
- The effective apparent power S_e and its components reproduce the format of the classical sinusoidal definitions that were the norm for over a century and have always been well understood.

2.3.3 Depenbrock's Comments

One of the "discussions" that originated from the paper of the IEEE working group was signed by Depenbrock and Staud [11]. The German Professor M. Depenbrock has been working for quite a few years on such matters, although his works were little known outside his country of origin. In his comments, a number of examples are presented, as well as an alternative proposal for the definition of apparent power for four-wire systems that largely were subsequently adopted by the Standard [10,12].

In the first example, Depenbrock–Staud have two circuits, one of three conductors and other of four conductors. Both consume the same amount of active power, and have the same rms line current, the same apparent power, and the same power factor according to the IEEE working group. However, the second circuit circulates neutral current 1565 times the rms value of the line currents. Depenbrock reasons that if the power factor is a more convenient merit figure to quantify the efficiency of the energy flow on a system, both circuits should not have the same power factor. Moreover, any current flowing in the neutral conductor is a zero sequence; as the zero-sequence voltage is normally considered negligible in distribution networks, the zero-sequence current carries no active power. The costs of an additional neutral conductor could be saved if the zero sequence is null. Therefore, it is justified to consider the always nonactive zero-sequence currents as an even quantity pernicious on other nonactive currents requiring no neutral conductor.

The Depenbrock–Staud commentary includes a further example of a symmetrical nonlinear load (three single-phase rectifiers connected in wye with accessible neutral) content a high zero sequence current. In this case, the rms value of the current flowing in the neutral conductor is $\sqrt{3}$ times the rms value of each of the line currents, while the IEEE working group does not consider this current in the calculation of apparent power. Therefore, Depenbrock concludes that the neutral current, I_n, should be accounted for in the calculation of the apparent power of a four-wire three-phase system. Moreover, the intensity I_n to be measured specifically, since it is impossible to calculate the rms value of I_n from the rms values of the three-phase currents.

While the above examples are part of a practical case, it is also possible to support this claim with the following theoretical considerations: apparent power can be defined as being equal to the maximum active power that may be achieved without varying averages over a period of the sum of the instantaneous squared values of the voltages and currents without the limitation that any conductor can assume a special conductor to gather and return all currents of other conductors. Of course, the (total) collective power does not depend on the choice of the designated return conductor. If the apparent power is defined as the limiting value of the active power, apparent power must also be independent of the choice of the return conductor.

As has been indicated, the rms value of the current in the return conductor cannot be expressed from the rms values of the other conductors, therefore, the currents of all conductors have to contribute to the collective current (or equivalent) in the same way, no current can be omitted. This is the only way to avoid the problem that the determination of the value of the apparent power would otherwise depend on which conductor was chosen as the return conductor. Thus, for a system of "n" conductors and an outer reference "O," the proposed Depenbrock adopts the following expression,

$$S_\Sigma = U_\Sigma I_\Sigma \qquad (2.43)$$

Where rms values of voltage U_Σ and current I_Σ are determined as (2.44),

$$U_\Sigma = \sqrt{\sum_{r=1}^{n} U_{rO}^2} \quad ; \quad I_\Sigma = \sqrt{\sum_{r=1}^{n} I_r^2} \tag{2.44}$$

The "O" reference point may be the point of a star consisting of "n" equal resistors connected to terminals 1, 2, ..., n of the system. Furthermore, the rms value of the voltage can also be determined through the values of line-to-line voltage,

$$U_\Sigma = \sqrt{\frac{1}{2n} \sum_{r=1}^{n} \sum_{s=1}^{n} U_{rs}^2} \tag{2.45}$$

All the above expressions can always be expressed as notation IEEE working group [11].

2.4 Standard IEEE 1459

In 1998 and 1999 two papers of Emanuel appear [10,12] on the subject. In this work, the author adopts the proposal by Depenbrock and redefines the results of the IEEE working group according to the vision of a three-phase system with neutral conductor as a four-conductor system. Thus, the Standard established a generic definition of apparent power, defined the so-called effective voltage and effective current, and thereafter assumed the decomposition of the effective apparent power in terms of power terms proposed by the IEEE working group [5,6].

The definition of apparent power that handles Std 1459 is as follows: the apparent power is the maximum power that can be transmitted under ideal conditions (sinusoidal single-phase or sinusoidal balanced three-phase system) with the same voltage impact (on the insulation and on no-load losses) and the same current impact (or line losses) of system over the network.

For this definition, the rms voltage and rms current value characterize the impact of the load on the power system, i.e. what current value is carried by the line, and what isolation is necessary and what load losses are expected. The explicit expression of apparent power depends on the impact that cause voltage and current to the load. In summary, to determine the apparent power as defined above, we must determine an equivalent current and equivalent voltage of a balanced system of positive sequence producing the same impact on the network as actual system currents and voltages.

Figure 2.2 outlines a general three-phase system, where an unbalanced load is supplied by a four-wire system.

Power losses in the lines are,

$$\Delta P = r\left(I_a^2 + I_b^2 + I_c^2 + I_n^2 \right) \tag{2.46}$$

The ideal system should dissipate exactly the same power in the lines, Figure 2.3,

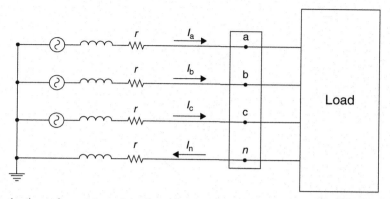

FIGURE 2.2 Four-wire three-phase system where the resistance of the line conductor and neutral conductor are equal.

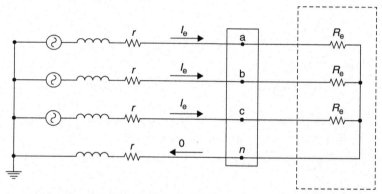

FIGURE 2.3 Equivalent three-phase circuit for the definition of I_e.

The load is formed by three equal resistors and lines operate three equal currents I_e, then,

$$\Delta P = 3rI_e^2 \tag{2.47}$$

Equality of expressions (2.46) and (2.47), the equivalent or effective current is obtained,

$$I_e = \sqrt{\frac{1}{3}\left(I_a^2 + I_b^2 + I_c^2 + I_n^2\right)} \tag{2.48}$$

The next step is to find an equivalent voltage V_e. This takes into account the no-load losses in magnetic cores of transformers and isolates "upstream" of the load [6,13,14].

In the first version of the Standard it is assumed that voltage dependent (no-load) losses, P_Y, which are due to the phase to neutral voltages, and voltage dependent (no-load) losses, P_Δ, which are due to line-to-line voltages are equal, Figure 2.4.

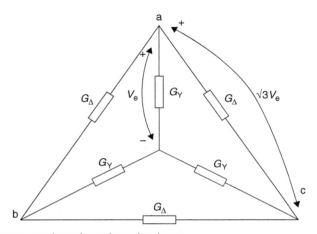

FIGURE 2.4 Conductances representing voltage-dependent losses.

This is,

$$P_\Delta = 3G_\Delta \left(\sqrt{3}V_e\right)^2 = 3G_Y V_e^2 = P_Y \quad \Rightarrow \quad G_\Delta = \frac{1}{3}G_Y \tag{2.49}$$

where the conductances G_Δ and G_Y are introduced to represent the losses which are voltage dependent, P_Δ and P_Y, respectively. The equivalent voltage is given by (2.50):

$$G_\Delta \left(V_{ab}^2 + V_{bc}^2 + V_{ca}^2\right) + G_Y \left(V_{an}^2 + V_{bn}^2 + V_{cn}^2\right) = 9G_\Delta V_e^2 + 3G_Y V_e^2 \tag{2.50}$$

If considered as has been said that both powers are equal, or what is the same as saying $G_\Delta = (1/3)G_Y$, then,

$$G_\Delta \left(V_{ab}^2 + V_{bc}^2 + V_{ca}^2\right) + 3G_\Delta \left(V_{an}^2 + V_{bn}^2 + V_{cn}^2\right) = 18G_\Delta V_e^2 \tag{2.51}$$

from which follows (2.52),

$$V_e = \sqrt{\frac{1}{18}\left\{3\left(V_{an}^2 + V_{bn}^2 + V_{cn}^2\right) + \left(V_{ab}^2 + V_{bc}^2 + V_{ca}^2\right)\right\}} \tag{2.52}$$

For a three-wire three-phase system where $I_n = 0$, the Standard recommends the simplified expressions,

$$V_e = \sqrt{\frac{V_{ab}^2 + V_{bc}^2 + V_{ca}^2}{9}} \quad ; \quad I_e = \sqrt{\frac{I_a^2 + I_b^2 + I_c^2}{3}} \tag{2.53}$$

referred to as Buchholz–Goodhue, and proposals for original works of the IEEE working group, (2.14), (2.15).

Finally, the partition of the effective apparent power such is included and as published in the IEEE standard 1459 as a summary,

$$S_e^2 = \left(P_1^+\right)^2 + \left(Q_1^+\right)^2 + \left(S_{U1}\right)^2 + D_{eI}^2 + D_{eV}^2 + P_H^2 + D_{eH}^2 \tag{2.54}$$

These are the seven terms of power, which have been introduced in the previous sections.

Just as a reminder to state that these power terms are defined from the fundamental and harmonic effective components of voltage and current, and therefore, in the environment of IEEE Standard, they now take the form,

$$I_{e1}^2 = \frac{1}{3}\left(I_{a1}^2 + I_{b1}^2 + I_{c1}^2 + I_{n1}^2\right) \quad ; \quad I_{eH}^2 = I_e^2 + I_{e1}^2 \tag{2.55}$$

for current, and,

$$V_{e1}^2 = \frac{1}{18}\left\{3\left(V_{a1}^2 + V_{b1}^2 + V_{c1}^2\right) + \left(V_{ab1}^2 + V_{bc1}^2 + V_{ca1}^2\right)\right\} \quad ; \quad V_{eH}^2 = V_e^2 - V_{e1}^2 \tag{2.56}$$

for voltage.

2.4.1 After the Standard: Std 1459-2010

Unquestionably, the influence of Emanuel dominated the development and dissemination of Standard. It was he who chaired the working sessions of the IEEE working group and as recognized in [15], took the moral responsibility of the theoretical and technical Standard content. The Standard reflects developments regarding the acceptance of the values of equivalent voltage and current followed by Emanuel alone. In case of the equivalent current it is clear that Depenbrock proposals were accepted, not in case the equivalent voltage that is less inclined, and relates to the losses that voltage depends on. Once the first edition of the Standard was published, Emanuel redefined the equivalent voltage and current as a result of collaboration with the School of European power, this time involving Willems [14]. In what follows we will discuss the new definitions of I_e and V_e according to the same scheme in the previous section.

To determine the equivalent current, a three-phase system consisting of an unbalanced load that is fed by a four-wire system where each of the lines having resistance r, and the neutral conductor is assumed resistance, r_n. Power dissipation of the line are,

$$\Delta P = r\left(I_a^2 + I_b^2 + I_c^2\right) + r_n I_n^2 \tag{2.57}$$

The ideal system consisting of three equal resistors fed a balanced load voltage system verifies that $I_a = I_b = I_c = I_e$, $I_n = 0$, and therefore must dissipate exactly the same power in the line that the original system,

$$\Delta P = 3r I_e^2 \tag{2.58}$$

Equality of two expressions for ΔP is the value of equivalent or effective current, I_e,

$$I_e = \sqrt{\frac{1}{3}\left(I_a^2 + I_b^2 + I_c^2 + \rho I_n^2\right)} \quad ; \quad \rho = \frac{r_n}{r} \tag{2.59}$$

The Std 1459 supposed $\rho = 1$. In installations for typical medium and low voltage $\rho = 0.2 - 4$. Today, digital instrumentation can develop equipment to adjust ρ for any given value.

To determine the equivalent voltage, V_e, assumes that the load consists of a group of resistors connected in Y and a residual group connected in Δ. An equivalent resistance R_Y and R_Δ, respectively, characterizes each group. The equivalence criterion is based on identical electrothermal effects, that is,

$$\frac{V_{an}^2 + V_{bn}^2 + V_{cn}^2}{R_Y} + \frac{V_{ab}^2 + V_{bc}^2 + V_{ca}^2}{R_\Delta} = \frac{3V_e^2}{R_Y} + \frac{9V_e^2}{R_\Delta} \tag{2.60}$$

If you define the ratio of powers (2.61),

$$\xi = \frac{P_\Delta}{P_Y} = \frac{(9V_e^2 / R_\Delta)}{(3V_e^2 / R_Y)} = \frac{3R_Y}{R_\Delta} \tag{2.61}$$

and is substituted in expression (2.60), it (2.62),

$$\frac{V_{an}^2 + V_{bn}^2 + V_{cn}^2}{R_Y} + \frac{V_{ab}^2 + V_{bc}^2 + V_{ca}^2}{3R_Y}\xi = \frac{3V_e^2}{R_Y} + \frac{9V_e^2}{3R_Y}\xi \tag{2.62}$$

finally produced (2.63),

$$V_e = \sqrt{\frac{3\left(V_{an}^2 + V_{bn}^2 + V_{cn}^2\right) + \xi\left(V_{ab}^2 + V_{bc}^2 + V_{ca}^2\right)}{9(1+\xi)}} \tag{2.63}$$

This new definition of V_e represents an important conceptual change of Emanuel. In fact, as stated explicitly in [15], the concept of voltage dependent losses was abandoned in favor of an approach of load that transfers equivalent active power; that is how it was reflected in the 2010 edition of the standard [5].

Electrical meaning associated with these concepts is summarized in the following statement; the system allows for maximum power transfer is a perfectly balanced and symmetrical with a line current I_e and phase to neutral voltage V_e. Figures 2.5 and 2.6 show the compensated three-phase system and its equivalent circuit, respectively.

The apparent power and equivalent voltage and current can also be expressed by means of symmetrical components. If denoted with subscripts +, −, and 0, the positive, negative, and zero sequence components, respectively, have

$$I_e = \sqrt{I_+^2 + I_-^2 + (1+3\rho)I_0^2} \tag{2.64}$$

$$V_e = \sqrt{V_+^2 + V_-^2 + \frac{1}{1+\xi}V_0^2} \tag{2.65}$$

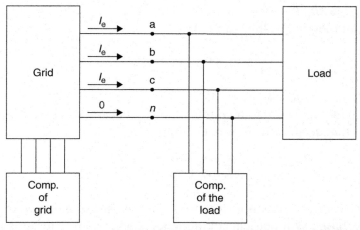

FIGURE 2.5 Current compensation for a three phase system according to the scheme of IEEE Standard.

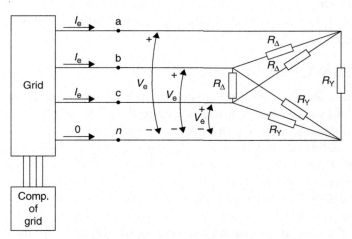

FIGURE 2.6 Equivalent circuit for a three phase system according to the approach of IEEE.

The value ratios of powers, ξ, lead to different situations; four of them are of interest:

1. $P_\Delta = 0$, $\xi = 0$, $R_\Delta \rightarrow \infty$, $R_e = R_Y$;
Loads do not connected in delta:

$$V_e = \sqrt{\frac{1}{3}\left(V_{an}^2 + V_{bn}^2 + V_{cn}^2\right)} \tag{2.66}$$

2. $P_Y = 0$, $\xi \rightarrow \infty$, $R_Y \rightarrow \infty$, $R_e = R_\Delta/3$;
Load of three wires:

$$V_e = \sqrt{\frac{V_{ab}^2 + V_{bc}^2 + V_{ca}^2}{9}} \tag{2.67}$$

3. $P_\Delta = P_Y$, $\xi = 1$, $R_\Delta = 3R_Y$, $R_e = R_Y/2 = R_\Delta/6$;

Situation recommended by the Std 1459:

$$V_e = \sqrt{\frac{1}{18}\left\{3\left(V_{an}^2 + V_{bn}^2 + V_{cn}^2\right) + \left(V_{ab}^2 + V_{bc}^2 + V_{ca}^2\right)\right\}} \tag{2.68}$$

4. $P_\Delta = 3P_Y$, $\xi = 3$, $R_\Delta = R_Y$, $R_e = R_Y/4 = R_\Delta/4$;

Case where all resistances of four conductor are equal and produce the same expression as that used by the European approach:

$$V_e = \sqrt{\frac{1}{12}\left(V_{an}^2 + V_{bn}^2 + V_{cn}^2 + V_{ab}^2 + V_{bc}^2 + V_{ca}^2\right)} \tag{2.69}$$

The differences between the four situations can be easily understood from the expressions of V_e and I_e based symmetrical components (2.55), (2.56). Thus, it is found that the main reason for the discrepancy between the four cases is due to the zero sequence voltage, V_0, and ratio ξ. Only when $\xi \rightarrow \infty$, has no effect on the zero-sequence voltage. In power systems V_0 is usually very small, so that ξ has a very small influence on the value of V_e.

2.5 A Practical Case

To illustrate numerically the developments made so far, an example given as a case study has been included in this section. It is a four-wire three-phase system with the general conditions of unbalance and distortion shown in Figure 2.7.

Before that a brief digression on digital signal processing applied to the electric power quality is needed. Experience gathered through this research has shown that the measurement equipment must have built-in digital techniques to ensure suitable resolution and accuracy. Therefore, an efficient digital algorithm to estimate the terms of current/power system constitutes a key part of the measurement process. Here, a procedure for the calculation of power and current components in unbalanced and distorted three-phase systems is proposed. This method is based on the Fast Fourier Transform (FFT) applied to some complex instantaneous time vectors of voltage and current, known as Park vectors. The FFT of Park vectors shows that their harmonic components are related to the steady state harmonic symmetrical components of positive and negative sequences. The positive sequence harmonic components are placed in the left hand side of the spectrum, and the negative sequence harmonic components are placed in the right hand side of the spectrum. Thus, just a glance at a voltage/current instantaneous time vector spectrum allows one to identify a degree of distortion and unbalance of three-phase waveforms.

The analysis of distorted signals is resolved satisfactorily from the method of Fourier series. Sampling waveforms of voltage and current, and further handling is used by the discrete Fourier transform (DFT); this is performed very quickly by means of a Fast Fourier

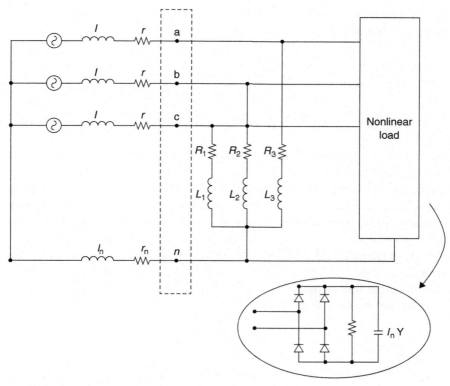

FIGURE 2.7 Circuit for the practical example 2.2.

Fast Transform algorithm (FFT). Thus, given a nonsinusoidal voltage signal $v(t)$ expressed in the form

$$v(t)=\sqrt{2}\sum_{\forall k}V_k\cos\left(2\pi kf_1t+\alpha_k\right)\qquad(2.70)$$

where f_1 is the frequency of the fundamental harmonic, V_k is the rms value of the voltage harmonic of order k and $\boldsymbol{V_k}=V_k\llcorner\alpha_k$ is the phasor corresponding to the harmonic of order k. Sampling $v(t)$ every Δt seconds (sampling frequency $f_s=1/\Delta t$), and a frequency resolution according to $\Delta f=1/N\Delta t$, can obtain a set of N samples $v(n\Delta t)$ for all $n=0,1,2,\dots,N-1$. DFT for the sequence of samples is given by (2.71)

$$V\left(h\Delta f\right)=\frac{1}{N}\sum_{n=0}^{N-1}v(n\Delta t)W_N^{-hn}\qquad(2.71)$$

for $h=0,1,2,\dots,N-1$, where h corresponds to the rate of change in the frequency domain, and $W_N=\exp(j2\pi/N)$. If the ratio $f_1/\Delta f$ is an integer, that is, when the sampling interval $N\Delta t$ is a multiple of the fundamental period $T_1=1/f_1$, the sampling process is synchronized with the signal, and there is no error dispersion [16]. The DFT of the signal (2.70) in such a case results in (2.72) and (2.73);

$$V(h\Delta f) = \frac{V_k}{\sqrt{2}} \quad \text{for} \quad h = \frac{kf_1}{\Delta f} < \frac{N}{2} - 1 \tag{2.72}$$

$$V(h\Delta f) = \frac{V_k^*}{\sqrt{2}} \quad \text{for} \quad h = N - \frac{kf_1}{\Delta f} > \frac{N}{2} + 1 \tag{2.73}$$

The expression (2.72) allows the spectrum of $v(t)$ for $h < N/2 - 1$, and the expression (2.73) of conjugate phasors for each harmonic k for $h > N/2 + 1$. A similar analysis is possible for a current distorted waveform [16]. The definition of the DFT used by Matlab (fft command) differs slightly from that in (2.71), which requires multiplying the results by fft of Matlab by the number of samples N; on the other hand, further multiplication by $\sqrt{2}$ gives the phasors of each harmonic of the distorted signal directly.

The analysis of unbalanced and distorted three-phase circuits is enhanced by the Discrete Fourier Transform (DFT) applied to a complex samples sequence [17]. In fact, it is usual in electrical machinery to define a space vector or Park vector as follows:

The Park voltage vector is defined as:

$$\boldsymbol{v}_p = \sqrt{\frac{2}{3}} \left(v_a + \gamma v_b + \gamma^2 v_c \right) \tag{2.74}$$

where $\gamma = \exp(j\,2\pi/3)$ and v_a, v_b, v_c are the phase voltages. An identical relationship is possible for the three-phase line currents:

$$\boldsymbol{i}_p = \sqrt{\frac{2}{3}} \left(i_a + \gamma i_b + \gamma^2 i_c \right) \tag{2.75}$$

In a three-wire, three-phase circuit with the voltage Park vector $\boldsymbol{v}_p$ and current Park vector $\boldsymbol{i}_p$, it is possible to determine only two line voltages and two line currents:

$$\boldsymbol{v}_p = \left(\frac{v_{ab}}{\sqrt{6}} - \frac{v_{cb}}{2\sqrt{6}} \right) - j\frac{v_{cb}}{2\sqrt{2}} \tag{2.76}$$

$$\boldsymbol{i}_p = \frac{\sqrt{3}\,i_a}{2\sqrt{2}} - j\left(\frac{i_a}{2\sqrt{2}} + i_c\sqrt{\frac{3}{2}} \right) \tag{2.77}$$

The DFT of N complex samples sequence $\{\boldsymbol{v}_p(n\Delta t)\}$ where Δt is the sampling interval and n varies in 0, 1, 2,. ... ., $N-1$, is given by:

$$V_p(h\Delta f) = \frac{1}{N}\sum_{n=0}^{N-1} v_p(n\Delta t) W_N^{-hn} \tag{2.78}$$

thus establishing

$$V_p\left(h\Delta f\right)=\frac{\sqrt{2/3}}{N}\left(\sum_{n=0}^{N-1}v_a(n\Delta t)W_N^{-hn}+\gamma\sum_{n=0}^{N-1}v_b(n\Delta t)W_N^{-hn}+\gamma^2\sum_{n=0}^{N-1}v_c(n\Delta t)W_N^{-hn}\right) \tag{2.79}$$

For each of the summands within brackets in (2.79) we have (2.80)

$$\sum_{n=0}^{N-1}v_{a,b,c}(n\Delta t)W_N^{-hn}=\begin{cases}\dfrac{V_{a,b,c;k}}{\sqrt{2}} & \text{for } h=\dfrac{kf_1}{\Delta f}<\dfrac{N}{2}-1\\[2ex]\dfrac{V_{a,b,c;k}^*}{\sqrt{2}} & \text{for } h=N-\dfrac{kf_1}{\Delta f}>\dfrac{N}{2}+1\end{cases} \tag{2.80}$$

Thus, the DFT vector Park adopts the expression

$$V_p\left(h\Delta f\right)=\frac{1}{\sqrt{3}}\left(V_{ak}+\gamma V_{bk}+\gamma^2 V_{ck}\right)\quad\text{for }\ h=\frac{kf_1}{\Delta f}<\frac{N}{2}-1 \tag{2.81}$$

and

$$V_p\left(h\Delta f\right)=\frac{1}{\sqrt{3}}\left(V_{ak}^*+\gamma V_{bk}^*+\gamma^2 V_{ck}^*\right)\quad\text{for }\ h=N-\frac{kf_1}{\Delta f}>\frac{N}{2}-1 \tag{2.82}$$

Each of the above expressions is identified by positive sequence, (2.81), and the conjugate of the negative sequence, (2.82), components in steady state,

$$V_p(h\Delta f)=\begin{cases}V_k^+ & \text{for } h=\dfrac{kf_1}{\Delta f}<\dfrac{N}{2}-1\\[2ex]\left(V_k^-\right)^* & \text{for } h=N-\dfrac{kf_1}{\Delta f}>\dfrac{N}{2}-1\end{cases} \tag{2.83}$$

Expression (2.83) confirms that it is possible to determine the steady state symmetrical components through the DFT of the complex samples (2.78). In fact, the first half of the spectrum (if $f_1 = \Delta f$) $h = k < (N/2)-1$, $V_k + = V_p(h)$ comprise positive sequence symmetrical components; the second half of the spectrum $h = N-k > (N/2) + 1$, determines $V_k^- = V_p(N-h)$, the negative sequence symmetrical components. A similar analysis can be done to determine the positive and negative symmetrical components of each harmonic of the line current.

EXAMPLE 2.2

An unbalanced three-phase source supplies two loads of four-wire: an asymmetric but linear load and a symmetrical but nonlinear load consisting of three single-phase rectifiers connected in star. The source and the loads are connected via three-wire line that have the same impedance, and a neutral conductor with a different line conductors impedance; specifically $r_1 = r_2 = r_3 = r = 4.2$ mΩ, $r_n = 6$ mΩ. Figure 2.7, indicates the place of measurement is taken into the system.

The waveforms of the phase to neutral voltages and line currents are shown in Figure 2.8.

Figure 2.9a shows the waveform where the distortion of neutral current due to third harmonics are seen; Figure. 2.9b shows one of the line-to-line voltages. The waveforms of voltage and current in Figures 2.8 and 2.9 correspond to unbalanced currents and voltages with some distortion. Both the nonlinear loads, such as unbalanced load are primarily responsible for the voltage and current disturbances at the point of measurement.

Table 2.1 lists the rms values of the voltages and currents, as well as corresponding to the fundamental harmonic.

Loss of power dissipated by the three lines is

$$\Delta P_L = r\left(I_a^2 + I_b^2 + I_c^2\right) = 121.78 \text{ W} \tag{2.84}$$

Furthermore, losses in the neutral conductor is,

$$\Delta P_n = r_n I_n^2 = 15.86 \text{ W} \tag{2.85}$$

Table 2.2 shows the rms values of the line-to-line voltages and the fundamental harmonic; Table 2.3 shows the effective values of voltage and current of the system. Here, the approach of Std 1459-2000 has been adopted; in Section 2.5 the results are displayed according to the approach of Std 1459-2010.

Finally, Table 2.4 presents the values of the different terms of power.

The results highlight the following load characteristics:

The values of voltage and current distortion are, $THD_{eV} = 10.69\%$, $THD_{eI} = 36.16\%$. The voltage has a high distortion but that of the current is lower, and there is also a reduced presence of harmonics with different sequence of the positive sequence, this causes $S_e > S_{e1} > S_{e1}{}^+$, but with values of the same order. The contribution of voltage and current harmonics produced a level of nonfundamental apparent power medium, $S_{eN}/S_{e1} = 0.38$, Table 2.4.

- The fundamental active power is the dominant term. However, due to the fact that the harmonic power is injected from the load to the network, it is verified that $P_1 > P$, that is, $P_H < 0$, $P_H = -4.86$ W. Since the supply voltages are unbalanced, the fundamental negative-sequence active power ($P_{1-} = 17.55$ W) and the fundamental zero sequence active power ($P_{10} = 86.72$ W) are positive, dissipated in the linear loads.

- Fundamental positive-sequence reactive power normalized $Q_1 + /S_{e1} = 0.082$ gives the power of shunt capacitor banks will get a fundamental positive-sequence power factor equal to one. In this case, PF_{1+} is 0.996, therefore the use of capacitors can improve the power factor of the load by only a very small amount.

- For comparison, the power factor values obtained from the definition of vector apparent power and arithmetic apparent power included above are indicated IEEE Std 100: $PF_V = 0.967$, $PF_A = 0.963$. Both are greater than $PF_e = 0.923$, since the effective apparent power S_e is greater than S_V and S_A, mainly include neutral current; in any case, in this example, the values obtained are very close to a load with a relatively high value of fundamental power.

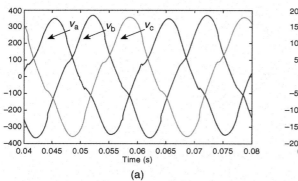
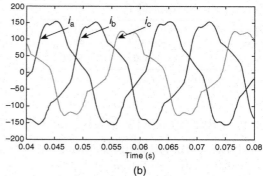

FIGURE 2.8 Waveforms of voltages (a) and currents (b) of the practical example 2.2.

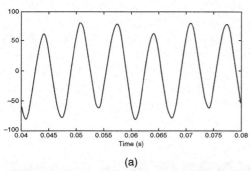
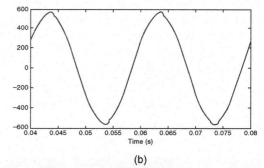

FIGURE 2.9 Waveforms of neutral current (a), and line to line voltage between a-b (b), are shown.

Table 2.1 Rms Values of the Phase Voltages (in V), and Line Currents (in A), and their Fundamental Harmonic

V_a	I_a	V_b	I_b	V_c	I_c	I_n
221.82	102.69	233.71	105.18	233.15	85.95	51.41
V_{al}	I_{al}	V_{bl}	I_{bl}	V_{cl}	I_{cl}	I_{nl}
219.25	101.04	231.20	103.26	230.66	83.77	9.53

Table 2.2 Rms Values of Line-to-Line Voltage (in V) and their Fundamental Harmonics

V_{ab}	V_{bc}	V_{ca}	V_{abl}	V_{bcl}	V_{cal}
393.30	390.80	395.30	393.23	390.66	395.22

Table 2.3 Effective Values of the Voltage (in V), the Current (in A), and Their Fundamental and Harmonic Components

V_e	I_e	V_{el}	I_{el}	V_{eH}	I_{eH}
228.30	102.69	227.01	96.57	24.26	34.92

Table 2.4   Results of the Different Terms of Power in V_A, W, and var

$S_e = 70338$	$S_{el} = 65772$		$S_{eN} = 24929$		
$P = 64929$	$S_1{}^+ = 65052$	$S_{U1} = 9706$	$D_{el} = 23784$	$D_{eV} = 7028$	$S_{eH} = 2541$
$P_l = 64934$	$P_1{}^+ = 64830$	$Q_1{}^+ = 5370$			$P_H = -4.86$ $D_{eH} = 2541$

Note: Terms from Equation 2.42 are shown in bold.

2.6 Discussions and Conclusions

As shown in the previous sections, after the publication of the first edition of Std 1459, his main inspiration A. Emanuel, supplemented and reinterpreted some of the concepts used, and these were then incorporated in the next issue, as outlined in Section 2.4.1. In this section we delve into the motivations of those corrections using an illustrative example.

In some situations the power factor determined according to the first edition of Std 1459 could be greater than 1. Section 2.4.1 presented, a new expression for V_e, in which the concept of voltage dependent losses is abandoned in favor of the approach of load active power. Now, in the definition of $\xi = P_\Delta/P_Y$, P_Δ is the power absorbed by loads in Δ and/or loads in Y without neutral wire, and P_Y is the active power absorbed by the loads connected in Y with neutral wire. The advantage of this approach is the fact that the load topology dictates the value of ξ and not the voltage-dependent losses. An example will help clarify the situation.

EXAMPLE 2.3

Consider a composite balanced three-phase load of resistors R connected in Y and supplied through four conductors with a set of asymmetrical voltages that phasors are,

$$V_a = V_+ + V_0 \quad ; \quad V_b = \gamma^2 V_+ + V_0 \quad ; \quad V_c = \gamma V_+ + V_0 \tag{2.86}$$

Determine the power factor of the load according to Std 1459.
The line and neutral currents are,

$$I_a = I_+ + I_0 \quad ; \quad I_b = \gamma^2 I_+ + I_0 \quad ; \quad I_c = \gamma I_+ + I_0 \quad ; \quad I_n = 3I_0 \tag{2.87}$$

where $I_+ = V_+/R$ and $I_0 = V_0/R$.

Considering the resistance of the neutral conductor null, $\rho = 0$. The effective current is then

$$I_e = \sqrt{I_+^2 + I_-^2 + I_0^2} = \frac{1}{R}\sqrt{V_+^2 + V_0^2} \tag{2.88}$$

If the effective voltage V_e is determined without considering the topology of the load and is considered $\xi = 1$ as required by Std 1459-2000, we have,

$$V_e = \sqrt{V_+^2 + V_-^2 + \frac{1}{2}V_0^2} = \sqrt{V_+^2 + \frac{1}{2}V_0^2} \tag{2.89}$$

Therefore,

$$S_e = \frac{3}{R}\sqrt{\left(V_+^2 + V_0^2\right)\left(V_+^2 + 1/2V_0^2\right)} \tag{2.90}$$

The active power is,

$$P = \frac{3}{R}\left(V_+^2 + V_0^2\right) \tag{2.91}$$

Then the power factor is,

$$PF_e = \frac{P}{S_e} = \frac{V_+^2 + V_0^2}{\sqrt{\left(V_+^2 + V_0^2\right)\left(V_+^2 + 0.5V_0^2\right)}} = \sqrt{\frac{1 + (V_0/V_+)^2}{1 + 0.5(V_0/V_+)^2}} \succ 1 \tag{2.92}$$

This unusual result is clear from the latter interpretation of Emanuel for V_e and described in Std 1459-2010. In fact, the load of example has a $P_\Delta = 0$, then $\xi = 0$ and not 1. In this case the value of V_e is,

$$V_e = \sqrt{V_+^2 + V_0^2} \tag{2.93}$$

and the apparent power,

$$S_e = \frac{3}{R}\left(V_+^2 + V_0^2\right) \tag{2.94}$$

resulting in the expected value for the power factor,

$$PF_e = \frac{P}{S_e} = 1 \tag{2.95}$$

In this same context, the calculations in Example 2.1 have been based on Std 1459-2000. As has been stated here, however there is a reinterpretation of Emanuel collected in Std 1459-2010 required to recalculate the example taking $\rho = r_n/r = 1.43$, and $\xi = 0$ since the load only includes elements of star connection with accessible neutral. Table 2.5 includes the most significant results.

From a global point of view, Std 1459 introduces a procedure to treat analysis in electrical power systems with asymmetry and distortion. This does keep some of the discrepancies in the European approach, but has its own positive characteristics.

The IEEE Standard uses a definition of effective apparent power, S_e, for three-phase systems due to Buchholz–Goodhue with an interpretation related to the efficiency of the

Table 2.5 Results of Practical Example for $r_0 = 1.43$ and $\xi = 0$

V_e	I_e	S_{1+}	S_{el}	S_e	PF_{el}	PF_e
229.62	104.52	65.052	65.844	72.004	0.986	0.901

transmission of power to the load, more in line with that associated to the apparent power under ideal conditions. Another one of its main features is the preferential treatment that assigns positive sequence terms of power at the fundamental frequency. The remaining nonfundamental powers are concentrated in a component S_{eN}, which helps the measurement of the amount of harmonics VA transferred at joint interest and their relative degree of distortion, S_{eN}/S_{e1}. Moreover, the calculation of S_{eN} can be used as a measure of warning in linear consumers against resonances as a complement in billing by utilities. Finally, it should be noted that modern digital instrumentation allows power terms included in the Standard to be configured easily.

A clear limitation in the Standard relates to the fact that it does not distinguish between the situation in which the harmonics flows into the load and the situation in which the harmonics flow from the load to the mains; however, no other Standard provides a better solution.

2.7 Summary

This chapter presents the development of the calculation of electric power for steady state sinusoidal/nonsinusoidal, balanced/unbalanced three-phase systems. For this we have taken the approach of IEEE Std 1459. The framework of the standard decomposes the apparent power in single-phase and three-phase systems. Thus, the apparent power has been partitioned into four terms: the fundamental effective apparent power, the current distortion power, the voltage distortion power, and the harmonic apparent power. Similarly, the fundamental effective apparent power supports a further partition into the fundamental positive-sequence active power, fundamental positive-sequence reactive power, and the fundamental unbalanced power, the latter being used in practice to assess the imbalance of an asymmetrical system. Of particular importance is the fact that the standard defines the apparent power in a three-phase system from the effective values of voltage and current accounting system losses differently to the traditional approach. Thus, to determine the effective voltage, the consumption of the load in star, and the consumption of the load corresponding to delta connections must be taken into account, and for calculating the effective current, losses from the neutral current must be taken into account.

The chapter has used a chronological approach with the aim of presenting the motivations that led to the definitions set out here. Likewise, for a register of magnitudes to be considered useful, it is important to know how it can be evaluated. A procedure based on the FFT, to determine efficiently the power terms and quality indices derivatives thereof, has been introduced and applied in a case study. Knowledge of the power terms in three-phase systems in the most general conditions of asymmetry and distortion is a useful tool for evaluating the effectiveness of the compensation equipments based in APFs to improve the electric power quality.

References

[1] Emanuel AE. On the definition of power factor and apparent power in unbalanced polyphase circuits with sinusoidal voltage and currents. IEEE Trans Power Deliv 1993;8(3):841–7.

[2] Filipski PS. Polyphase apparent power and power factor under distorted waveform conditions. IEEE Trans Power Deliv 1991;6(3):1161–5.

[3] Filipski PS, Bagzouz Y, Cox MD. Discussion of power definitions contained in the IEEE Dictionary. IEEE Trans Power Deliv 1994;9(3):1237–43.

[4] IEEE working group on non-sinusoidal situations. Practical definitions for powers in systems with non-sinusoidal waveforms and unbalanced loads: a discussion. IEEE Trans Power Deliv 1996;11(1):79–87.

[5] Definitions for the measurement of electric power quantities under sinusoidal, non-sinusoidal, balanced, or unbalanced conditions, IEEE Std 1459-2010; 2010.

[6] Emanuel AE. Summary of IEEE Standard 1459: definitions for the measurement of electric power quantities under sinusoidal, non-sinusoidal, balanced, or unbalanced conditions. IEEE Trans Ind Appl 2004;40(3):869–76.

[7] Czarnecki LS. Misinterpretations of some power properties of electric circuits. IEEE Trans Power Deliv 1994;9(4):1760–4.

[8] Emanuel AE. Power definitions and physical mechanism of power flow. IEEE Press-Wiley: Chichester, UK; 2010.

[9] Filipski, PS, Arseneau R. Definition and measurement of apparent power under distorted waveform conditions, IEEE Tutorial Course 90 EH-0327-7-PWR; 1990. pp. 37-42.

[10] Emanuel AE. The Buchholz–Goodhue apparent power definition: the practical approach for non-sinusoidal and unbalanced systems. IEEE Trans Power Deliv 1998;13(2):344–8.

[11] Depenbrock M, Staudt V. Discussion to Ref. [4].

[12] Emanuel AE. Apparent power definitions for three-phase systems. IEEE Trans Power Deliv 1999;14(3):767–72.

[13] Emanuel AE. Reflections on the effective voltage concept, Sixth International Workshop on Power definitions and Measurements under non-sinusoidal conditions, Milano; October 2003. pp. 1–7.

[14] Willems JL, Ghijselen JA, Emanuel AE. The apparent power concept and the IEEE standard 1459-2000. IEEE Trans Power Deliv 2005;20(2):876–84.

[15] Willems JL, Ghijselen JA, Emanuel AE. Addendum to Ref. [14].

[16] Arrillaga J, Watson NR, Chen S. Power system quality assessment. New Jersey: Wiley; 2000.

[17] Salmerón P, Alcántara FJ. A new technique for unbalanced current and voltage measurement with neural networks. IEEE Trans Power Deliv 2005;20(2):852–8.

3

Instantaneous Reactive Power Theory

CHAPTER OUTLINE

This chapter presents the fundamentals of the instantaneous reactive power theory for three-phase and three/four-wire systems. The original formulation was published in 1983 by Akagi–Kanazawa–Nabae and it is they for whom it was named. A description model of three-phase systems is obtained through the introduction of two power variables: the instantaneous real power and instantaneous imaginary power. Both the instantaneous active current and the instantaneous reactive current that achieve and explain the physical foundation to power flow between source and load are obtained. However, the initial objective of the formulation was eminently practical: to serve as a basis for designing static compensators (active power filters) that do not require energy storage elements for instantaneous

P. Salmerón Revuelta, S.P. Litrán, and J.P. Thomas: Active Power Line Conditioners. http://dx.doi.org/10.1016/B978-0-12-803216-9.00003-1
Copyright © 2016 Elsevier Inc. All rights reserved.

compensation. Following this, alternative formulations have been proposed, although they all have a clear nexus of inspiration in the original instantaneous reactive power formulation. In this chapter the basics of instantaneous reactive power theory are presented, as well as those derived formulations that have greater acceptance, especially the modified formulation of instantaneous reactive power. The chapter also addresses the similarities and differences from the original formulation, especially regarding its application to the determination of control strategies for active power filters for nonlinear three-phase loads compensation.

This chapter also includes two topics of interest. The first is a section dedicated to the formulations developed in a framework of rotating axes based on the so-called Park transformation that has led to the dq, i_d–i_q, and pqr methods. The second is a section for the dual formulation of the instantaneous reactive power directly applicable to the use of static compensators for serial connection. Several application examples throughout the chapter will help you understand the contents herein.

3.1 Introduction

The theory of instantaneous reactive power in three-phase circuits was presented in 1983 by Akagi, Kanazawa, and Nabae [1,2]. The formulation as originally developed is applicable to three-wire and four-wire three-phase circuits. It is further characterized by defining an instantaneous reactive power for each phase as a single expression for arbitrary waveforms of voltage and current and with an associated electric meaning. Ever since its appearance, the theory has remained one of the key concepts within the research and development of power engineering. In fact, the increasing penetration of active power filters for compensation of nonlinear three-phase loads, an activity that the formulation was linked from its beginning, has contributed to its use in different control strategies [3–5]. After the appearance of the instantaneous reactive power theory, different formulations have been presented intended to establish themselves as alternatives, so from now the Akagi, Kanazawa, and Nabae formulation will be called the "original formulation" of instantaneous reactive power [6–33]. The so-called modified formulation [22–25] developed both in phase coordinates and 0–α–β coordinates, the formulation dq rotating axes [26], the i_d–i_q method [27], derived from the above, or the latest pqr formulation [28,29] are some of the formulations that have become more widespread. However, all have their inspiration in the original formulation of instantaneous reactive power.

While the theory of instantaneous reactive power has its origin in the control design of shunt active power filters, a dual development is possible to design the control of series active power filters; a section of this chapter is dedicated to this issue.

In the following sections, the fundamentals of the instantaneous reactive power theory are presented, together with those derived formulations that have gained greater acceptance. The discussion also highlights the similarities and differences of the new formulations with respect to the original formulation and their application to the determination of control strategies for active power filters in order to facilitate industrial load compensation [18].

3.2 Original Instantaneous Reactive Power Formulation

The scope of what is named the original formulation of instantaneous reactive power [1–2], are the three-wire three-phase systems or four-wire three-phase systems that generally have to consider the zero sequence-phase voltages and currents, Figure 3.1. The instantaneous three-phase voltages and currents in 1–2–3 phase coordinates can be transformed to 0–α–β coordinates by the transformation equations (3.1), (3.2):

$$\begin{bmatrix} u_0 \\ u_\alpha \\ u_\beta \end{bmatrix} = \sqrt{\frac{2}{3}} \begin{bmatrix} \frac{1}{\sqrt{2}} & \frac{1}{\sqrt{2}} & \frac{1}{\sqrt{2}} \\ 1 & -\frac{1}{2} & -\frac{1}{2} \\ 0 & \frac{\sqrt{3}}{2} & -\frac{\sqrt{3}}{2} \end{bmatrix} \begin{bmatrix} u_1 \\ u_2 \\ u_3 \end{bmatrix} \tag{3.1}$$

$$\begin{bmatrix} i_0 \\ i_\alpha \\ i_\beta \end{bmatrix} = \sqrt{\frac{2}{3}} \begin{bmatrix} \frac{1}{\sqrt{2}} & \frac{1}{\sqrt{2}} & \frac{1}{\sqrt{2}} \\ 1 & -\frac{1}{2} & -\frac{1}{2} \\ 0 & \frac{\sqrt{3}}{2} & -\frac{\sqrt{3}}{2} \end{bmatrix} \begin{bmatrix} i_1 \\ i_2 \\ i_3 \end{bmatrix} \tag{3.2}$$

From (3.1) and (3.2), the line-to-neutral voltage and the line current variables are expressed in a new coordinate system 0–α–β. As a result, the voltage/current space vector can be placed in a system of mutually perpendicular axes in the directions of the unit vectors. **0, α, β**; Figure 3.2 helps to visualize the situation. The 0 axis perpendicular to the plane formed by the axes $\alpha\beta$, is the support of the zero-sequence phase component of the voltage/current.

The current flowing through the neutral of a four-wire three-phase system is related to the zero sequence by (3.4)

$$i_n = i_1 + i_2 + i_3 \tag{3.3}$$

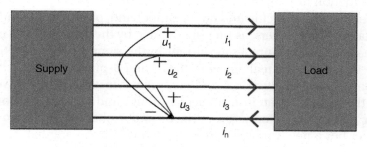

FIGURE 3.1 Reference voltage and current in a three-phase system.

FIGURE 3.2 Transformation of a phase coordinate system (123) to a 0–α–β coordinate system.

$$i_0 = \frac{1}{\sqrt{3}}(i_1 + i_2 + i_3) = \frac{1}{\sqrt{3}} i_n \qquad (3.4)$$

From (3.1) and (3.2), we can set the product between the voltage space vector and the current space vector for the instantaneous real power $p(t)$, that is,

$$p(t) = \boldsymbol{u.i} = u_1 i_1 + u_2 i_2 + u_3 i_3 = u_0 i_0 + u_\alpha i_\alpha + u_\beta i_\beta \qquad (3.5)$$

Since the transformation defined in (3.1), (3.2) (Clarke–Concordia transformation) is a transformation power invariant.

3.2.1 0, α, β Coordinates

In this subsection the formulation of the so-called original instantaneous reactive power will be developed within the coordinates 0, α, β. In four-wire three-phase systems, the original formulation defines two instantaneous real powers p_0 and $p_{\alpha\beta}$, and an instantaneous imaginary power $q_{\alpha\beta}$, in the manner prescribed in (3.6)

$$\begin{bmatrix} p_0 \\ p_{\alpha\beta} \\ q_{\alpha\beta} \end{bmatrix} = \begin{bmatrix} u_0 & 0 & 0 \\ 0 & u_\alpha & u_\beta \\ 0 & -u_\beta & u_\alpha \end{bmatrix} \begin{bmatrix} i_0 \\ i_\alpha \\ i_\beta \end{bmatrix} \qquad (3.6)$$

The matrix equation (3.6) suggests that p_0 ($=v_0 i_0$) involves an instantaneous real power in the zero-sequence phase circuit, and on the other hand, products $v_\alpha i_\alpha$ and $v_\beta i_\beta$ also correspond to instantaneous powers, as they are defined by the product of an instantaneous voltage in a phase by an instantaneous current in the same phase. Therefore, the $p_{\alpha\beta}$ is considered the instantaneous real power in the α- and β-phase circuit, in dimensions of Watts, W. By contrast, the products $v_\alpha i_\beta$ and $v_\beta i_\alpha$ are not instantaneous powers since they are defined by the product of an instantaneous voltage and an instantaneous phase current in another phase. Accordingly, $q_{\alpha\beta}$, at the α- and β-phase circuit is not an instantaneous real power but a new variable power defined under the original formulation, and the unit of measure is referred to [25] as imaginary watts, IW.

Since in (3.6) the three-phase voltages u_0, u_α, u_β are a set of three waveforms imposed by the supply, equation (3.6) can be interpreted as a geometric transformation ("mapping" is a term sometimes used) of a three-dimensional current space vector to a three-dimensional power space vector, and vice versa [24,25]. Although many of these "mapping" matrices are possible from a theoretical point of view, there are few that can offer a clear meaning from the practical point of view; indeed, the matrix given in (3.6) has proven useful in the control of active power filters.

The transformation matrix (3.6), first described in 1983, supports the inverse transformation,

$$
\begin{bmatrix} i_0 \\ i_\alpha \\ i_\beta \end{bmatrix} = \frac{1}{u_0 u_{\alpha\beta}^2} \begin{bmatrix} u_{\alpha\beta}^2 & 0 & 0 \\ 0 & u_0 u_\alpha & -u_0 u_\beta \\ 0 & u_0 u_\beta & u_0 u_\alpha \end{bmatrix} \begin{bmatrix} p_0 \\ p_{\alpha\beta} \\ q_{\alpha\beta} \end{bmatrix}
\tag{3.7}
$$

where

$$
u_{\alpha\beta}^2 = u_\alpha^2 + u_\beta^2.
\tag{3.8}
$$

From (3.7) the terms of the components of the instantaneous currents are obtained 0–α–β coordinates

$$
\begin{aligned}
i_0 &= \frac{1}{u_0} p_0 \\
i_\alpha &= \frac{1}{u_{\alpha\beta}^2} u_\alpha p_{\alpha\beta} + \frac{1}{u_{\alpha\beta}^2}\left(-u_\beta q_{\alpha\beta}\right) = i_{\alpha p} + i_{\alpha q} \\
i_\beta &= \frac{1}{u_{\alpha\beta}^2} u_\beta p_{\alpha\beta} + \frac{1}{u_{\alpha\beta}^2}\left(u_\alpha q_{\alpha\beta}\right) = i_{\beta p} + i_{\beta q}
\end{aligned}
\tag{3.9}
$$

where i_0 is the zero-sequence instantaneous current, $i_{\alpha p}$ is the instantaneous active current of the α-phase, $i_{\beta p}$ is the instantaneous active current of the β-phase, $i_{\alpha q}$ is the instantaneous reactive current of the α-phase, $i_{\alpha q}$ is the instantaneous reactive current of the β-phase.

The derivation of equation (3.9) from (3.7) is possible whenever $u_0 \neq 0$, since it would not be otherwise possible to calculate the inverse matrix. However, i_α and i_β in (3.9) do not depend on u_0 even when it is not zero. This means that the original formulation regards the zero-sequence circuit as a single-phase circuit independent of α- and β-phase circuits; therefore, it is acceptable to replace $u_0 = 0$ when the original formulation is applied to four-wire three-phase systems without zero-sequence voltage.

From equations (3.6) and (3.9), relations of power terms are followed, (3.10)–(3.11):

$$
\begin{aligned}
p(t) &= p_0(t) + p_\alpha(t) + p_\beta(t) = p_0(t) + p_{\alpha p}(t) + p_{\beta p}(t) + p_{\alpha q}(t) + p_{\beta q}(t) \\
&= u_0 i_0 + \frac{u_\alpha^2}{u_{\alpha\beta}^2} p_{\alpha\beta} + \frac{u_\beta^2}{u_{\alpha\beta}^2} p_{\alpha\beta} + \frac{-u_\alpha u_\beta}{u_{\alpha\beta}^2} q_{\alpha\beta} + \frac{u_\alpha u_\beta}{u_{\alpha\beta}^2} q_{\alpha\beta}.
\end{aligned}
\tag{3.10}
$$

$$
0 = u_\alpha i_{\alpha q} + u_\beta i_{\beta q} = u_\alpha \left\{ \frac{1}{u_{\alpha\beta}^2}\left(-u_\beta q_{\alpha\beta}\right) \right\} + u_\beta \left(\frac{1}{u_{\alpha\beta}^2} u_\alpha q_{\alpha\beta} \right) = p_{\alpha q} + p_{\beta q}
\tag{3.11}
$$

The instantaneous active and reactive power in each phase are named as follows,

$p_0 = u_0\, i_0$: zero-sequence instantaneous power
$p_{\alpha p} = u_\alpha\, i_{\alpha p}$: α-phase instantaneous active power
$p_{\beta p} = u_\beta\, i_{\beta p}$: β-phase instantaneous active power
$p_{\alpha q} = u_\alpha\, i_{\alpha q}$: α-phase instantaneous reactive power
$p_{\beta q} = u_\beta\, i_{\beta q}$: β-phase instantaneous reactive power

Figure 3.3a and b describe the power flow based on the original formulation for a four-wire three-phase system. In the original framework, the instantaneous reactive power zero-sequence circuit is considered as a separate single-phase circuit of the α- and β-phase circuit.

This approach is adopted from the method of symmetrical components that divides a four-wire three-phase circuit into a zero-sequence circuit, a positive-sequence and a negative-sequence circuit; the zero-sequence circuit is regarded as an independent single-phase circuit of the other phase-sequence circuits.

Equation (3.11) means that the sum of the α-phase instantaneous reactive power, $p_{\alpha q}$ and β-phase instantaneous reactive power, $p_{\beta q}$, is always zero. This assumes that both do not contribute to energy transfer between the source and load within the α–β-phase circuit. However, $p_{\alpha q}$ is involved in the transfer of energy within the α-phase circuit, in the same way as $p_{\beta q}$ in the β-phase circuit, increasing the value of the current flowing through each of the phases. Thus, the original formulation introduced by a side $q_{\alpha\beta}$ as a instantaneous imaginary power that determines $p_{\alpha q}$ and $p_{\beta q}$, and second, defines two independent instantaneous real powers p_0 and $p_{\alpha\beta}$; the three power variables form the three-dimensional power space.

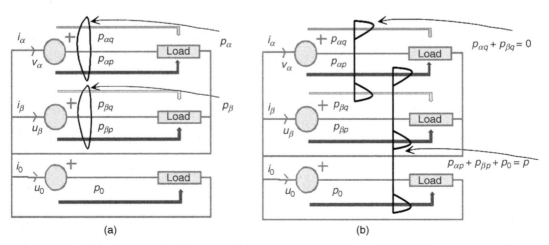

(a)　　　　　　　　　　　(b)

FIGURE 3.3 Power flow based on the formulation of the original instantaneous reactive power.

EXAMPLE 3.1

Determining the power variables for the reactive balanced three-phase load of Figure 3.4 that is supplied by the balanced three-phase voltage system (3.12).

$$u_1(t) = \sqrt{2}\, V_F \cos\omega t$$
$$u_2(t) = \sqrt{2}\, V_F \cos(\omega t - 120) \qquad (3.12)$$
$$u_3(t) = \sqrt{2}\, V_F \cos(\omega t + 120)$$

Line currents circulating in an inductive balanced three-phase load are of the form (3.13)

$$i_1(t) = \sqrt{2}\, I_F \cos(\omega t - \varphi)$$
$$i_2(t) = \sqrt{2}\, I_F \cos(\omega t - 120 - \varphi) \qquad (3.13)$$
$$i_3(t) = \sqrt{2}\, I_F \cos(\omega t + 120 - \varphi)$$

where φ for the load of the Figure 3.4 is 90°. From (3.1) the voltage components are obtained in the α, β plane.

$$u_\alpha = \sqrt{3}V_F \cos\omega t; \quad u_\beta = \sqrt{3}V_F \sin\omega t \qquad (3.14)$$

Similarly to (3.2) the current components obtained in the α, β plane are

$$i_\alpha = \sqrt{3}I_F \sin\omega t; \quad i_\beta = -\sqrt{3}I_F \cos\omega t \qquad (3.15)$$

From (3.6) the three power variables are obtained,

$$p_0 = 0$$
$$p_{\alpha\beta} = 0 \qquad (3.16)$$
$$q_{\alpha\beta} = -3V_F I_F \sin(90)$$

The balanced three-phase load of Figure 3.4 supplied by a balanced sinusoidal three-phase voltage of direct sequence phases absorbs an instantaneous real power equal to the active power (average power). The active power for purely reactive load is zero; in effect (3.16) becomes (3.17),

$$p(t) = p_0(t) + p_{\alpha\beta}(t) = P = 0 \qquad (3.17)$$

On the other hand, (3.16) shows that the instantaneous imaginary power for the load in Figure 3.4 for the conditions of this exercise is average reactive power value of opposite sign,

$$q_{\alpha\beta} = -3V_F I_F = -Q \qquad (3.18)$$

Two observations arise from this example. First, the description of the energy flow between source and load is not possible with only the instantaneous real power variable as happens in the case of single-phase systems. For a three-phase system treated as a global system it is necessary to define a new variable of power. The formulation of the original instantaneous reactive power introduces instantaneous imaginary power to complete the description of the process of energy transfer between source and load. Second, the definition given in (3.6) introduces an instantaneous imaginary power whose average value is the average reactive power of opposite sign. This situation will be overcome in Section 3.2.2, where an instantaneous imaginary power is introduced with the opposite sign; well, its average value takes a positive value for positive sequence of phases and a negative value for negative sequence of phases. This modification seems more consistent with the conventions of standard sign.

In this example, therefore, the line currents include only instantaneous reactive current component.

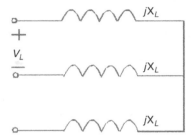

FIGURE 3.4 Balanced three-phase load in star which comprises three inductances.

The formulation of the instantaneous reactive power is established through the use of what we call a mapping matrix, as has been demonstrated in the development of this sub-section, however a vector development is possible. In fact, as has been stated in [17, 21], in a three-dimensional space defined by axes $0\alpha\beta$, three voltage space vector can be defined as

$$u_{\alpha\beta} = \begin{bmatrix} 0 \\ u_\alpha \\ u_\beta \end{bmatrix}; \quad u_0 = \begin{bmatrix} u_0 \\ 0 \\ 0 \end{bmatrix}; \quad u_{-\beta\alpha} = \begin{bmatrix} 0 \\ -u_\beta \\ u_\alpha \end{bmatrix} \tag{3.19}$$

The space vector $u_{\alpha\beta}$ is the projection of the voltage space vector $u_{0\alpha\beta}$ in the $\alpha\beta$ plane, the vector u_0 follows the direction of the axis 0 and the vector $u_{-\beta\alpha}$ called the orthogonal voltage vector, as $u_{\alpha\beta}$ is also located in the $\alpha\beta$ plane.

The three vectors are perpendicular to each other, and therefore, the dot product between any two of them is zero. Specifically, the following relationships are verified:

$$u_{0\alpha\beta} = \begin{bmatrix} u_0 & u_\alpha & u_\beta \end{bmatrix}^t = u_{\alpha\beta} + u_0 \tag{3.20}$$

$$u_{-\beta\alpha} \cdot u_{0\alpha\beta} = 0 \tag{3.21}$$

The current space vector i (3.22)

$$i = \begin{bmatrix} i_0 & i_\alpha & i_\beta \end{bmatrix}^t \tag{3.22}$$

can be split into three components that are the projections of the current vector on the three voltage vectors (3.19). In fact,

$$i = \frac{p_{\alpha\beta}(t)}{u_{\alpha\beta} \cdot u_{\alpha\beta}} u_{\alpha\beta} + \frac{q_{\alpha\beta}(t)}{u_{-\beta\alpha} \cdot u_{-\beta\alpha}} u_{-\beta\alpha} + \frac{p_0(t)}{u_0 \cdot u_0} u_0 \tag{3.23}$$

A power variable (3.6) appears in the numerator of each current term, the instantaneous real power in $\alpha\beta$-phase

$$p_{\alpha\beta}(t) = u_{\alpha\beta} \cdot i \tag{3.24}$$

the zero-sequence instantaneous real power

$$p_0(t) = u_0 \cdot i \tag{3.25}$$

and the instantaneous imaginary power in $\alpha\beta$-plane,

$$q_{\alpha\beta}(t) = u_{-\beta\alpha} \cdot i \tag{3.26}$$

In the denominators of each component of the instantaneous current, squared norms of each voltage vector appear,

$$u_{\alpha\beta} \cdot u_{\alpha\beta} = u_{\alpha\beta}^2; \quad u_0 \cdot u_0 = u_0^2; \quad u_{-\beta\alpha} \cdot u_{-\beta\alpha} = u_{-\beta\alpha}^2 \tag{3.27}$$

verifying the relations given in (3.28),

$$u_{0\alpha\beta}^2 = u_0^2 + u_{\alpha\beta}^2; \quad u_{\alpha\beta}^2 = u_{-\beta\alpha}^2 \tag{3.28}$$

In (3.23) the three current components are clearly identified in the same manner as (3.9),

$$\begin{bmatrix} i_0 \\ i_\alpha \\ i_\beta \end{bmatrix} = \frac{p_{\alpha\beta}}{u_{\alpha\beta}^2} \begin{bmatrix} 0 \\ u_\alpha \\ u_\beta \end{bmatrix} + \frac{p_0}{u_0^2} \begin{bmatrix} u_0 \\ 0 \\ 0 \end{bmatrix} + \frac{q_{\alpha\beta}}{u_{\alpha\beta}^2} \begin{bmatrix} 0 \\ u_{-\beta} \\ u_\alpha \end{bmatrix} \tag{3.29}$$

The inverse matrix of the transformation (3.1) or (3.2) is given in (3.30),

$$[T]^{-1} = \sqrt{\frac{2}{3}} \begin{bmatrix} \frac{1}{\sqrt{2}} & 1 & 0 \\ \frac{1}{\sqrt{2}} & -\frac{1}{2} & \frac{\sqrt{3}}{2} \\ \frac{1}{\sqrt{2}} & -\frac{1}{2} & -\frac{\sqrt{3}}{2} \end{bmatrix} \tag{3.30}$$

The matrix (3.30) to retrieve the components of phase 1, 2, 3 from 0–α–β components.

3.2.2 Phase Coordinates

In this section the formulation of original instantaneous reactive power takes place exclusively in phase coordinates, as was proposed in [13,19], based on the vector decomposition of the three-phase current in three orthogonal components. This formulation can be used to find the compensation current simply without resorting to a previous coordinate transformation.

For three-phase systems, the instantaneous currents and voltages can be represented in vector form as:

$$\mathbf{u} = \begin{bmatrix} u_1 & u_2 & u_3 \end{bmatrix}^t, \quad \mathbf{i} = \begin{bmatrix} i_1 & i_2 & i_3 \end{bmatrix}^t \tag{3.31}$$

where t, transpose.

Likewise an orthogonal voltage vector is defined,

$$\mathbf{u}_q = \begin{bmatrix} u_{q1} & u_{q2} & u_{q3} \end{bmatrix}^t = \frac{1}{\sqrt{3}} \begin{bmatrix} u_{23} & u_{31} & u_{12} \end{bmatrix}^t \tag{3.32}$$

verifying (3.33),

$$\mathbf{u}^t . \mathbf{u}_q = 0 \tag{3.33}$$

The instantaneous power is the dot product between the voltage vector and current vector

$$p(t) = \mathbf{u}^t . \mathbf{i} = u_1 i_1 + u_2 i_2 + u_3 i_3 \tag{3.34}$$

and the instantaneous imaginary power is defined as the scalar product between the orthogonal voltage vector and current vector

$$q(t) = \mathbf{u}_q^t . \mathbf{i} = \frac{1}{\sqrt{3}} \left(u_{23} i_1 + u_{31} i_2 + u_{12} i_3 \right) \tag{3.35}$$

Thus, one can define a yz plane determined by the voltage vectors as a basis of orthogonal unit vectors $\mathbf{y}_0$, $\mathbf{z}_0$, in the directions of the vectors $\mathbf{u}$ and $\mathbf{u}_q$, respectively, Figure 3.5,

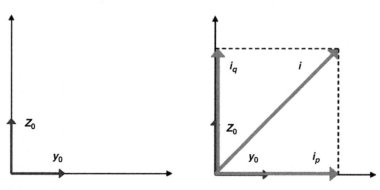

FIGURE 3.5 Vector diagram of current vector decomposition in systems that do not include zero-sequence components of voltage and current.

$$y_0 = \frac{u}{|u|} \qquad z_0 = \frac{u_q}{|u_q|} \tag{3.36}$$

where $|u| = \sqrt{(u^T u)}$ is the instantaneous norm of voltage space vector and $|u_q| = \sqrt{(u_q{}^t u_q)}$ is the instantaneous norm of orthogonal voltage space vector.

3.2.2.1 Without Zero-Sequence Components

For three-wire three-phase systems without zero-sequence components of voltage and current, the instantaneous current space vector i decomposes according to their orthogonal coordinate axes in Figure 3.5, that is,

$$i = i_p + i_q = \frac{p}{|u|} y_0 + \frac{q}{|u_q|} z_0 \tag{3.37}$$

i_p and i_q are mathematically determined by the projections of the vector i of the vector u and vector u_p, respectively,

$$i_p = \frac{u^t \cdot i}{|u|^2} u = \frac{p}{|u|} y_0 \tag{3.38}$$

$$i_q = \frac{u_q^t \cdot i}{|u_q|^2} u_q = \frac{q}{|u_q|} z_0 \tag{3.39}$$

In three-phase systems without zero-sequence components of voltage, equal instantaneous norms of voltage vector u and orthogonal voltage vector u_q can be verified. The current i_p is the instantaneous active current, also known as the instantaneous power current, and the current i_q is the instantaneous reactive current. Since these components are orthogonal, the following relationship can be set between the instantaneous norms,

$$i^2 = i_p^2 + i_q^2 \tag{3.40}$$

Electrically represent instantaneous currents minimum instantaneous norm transferring real power $p(t)$ to the load to voltage u, on one hand, and instantaneous imaginary power $q(t)$ to the voltage u_q, other, respectively. Thus, the scalar product of the orthogonal voltage vector and current vector, (3.35), is an alternative to the definition of instantaneous reactive power based on the cross product of voltage and current vectors included in [23] for three-wire three-phase systems where $u_1 + u_2 + u_3 = 0$. This new power variable (instantaneous imaginary power) has its origin in the line voltages (3.35) and does not contribute to the flow of instantaneous power; this was established in Figure 3.3 in terms of the $0–\alpha–\beta$ coordinates.

In Figure 3.6 current decomposition in the formulation of phase coordinates is presented graphically. It was considered the load current in one phase of a three-phase bridge rectifier with high inductance in the dc side. This current for phase 1 is the sum of instantaneous active current and instantaneous reactive current.

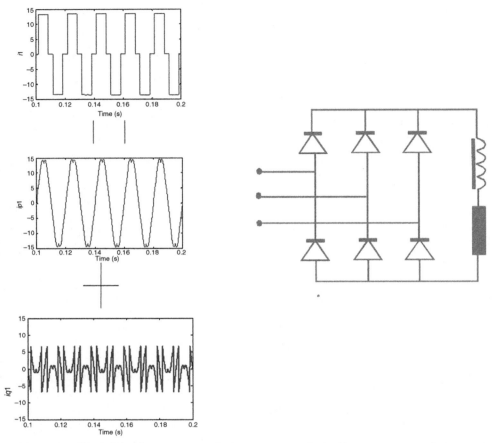

FIGURE 3.6 Decomposition of the phase 1 current at its instantaneous active and instantaneous reactive components for a bridge rectifier load type with high inductance on the dc side.

3.2.2.2 With Zero-Sequence Components

For four-wire three-phase system, the vector of instantaneous zero-sequence voltage $\boldsymbol{v}_0$, and the vector of instantaneous zero-sequence current, $\boldsymbol{i}_0$, is defined by (3.41)

$$\boldsymbol{v}_0 = \left[\frac{v_0}{\sqrt{3}} \ \frac{v_0}{\sqrt{3}} \ \frac{v_0}{\sqrt{3}} \right]^t = \frac{v_0}{\sqrt{3}} \mathbf{1}_3^t, \quad \boldsymbol{i}_0 = \frac{i_0}{\sqrt{3}} \mathbf{1}_3^t \tag{3.41}$$

where $\mathbf{1}_3$ is a vector whose elements are all 1 and

$$\sqrt{3}v_0 = u_1 + u_2 + u_3, \quad \sqrt{3}i_0 = i_1 + i_2 + i_3 \tag{3.42}$$

It is possible to decompose the voltage vector $\boldsymbol{u}$ according to equation (3.43)

$$\boldsymbol{u} = \boldsymbol{v} + \boldsymbol{v}_0 \tag{3.43}$$

where $\boldsymbol{v}$ represents the voltage vector excluding any zero-sequence components. The following relationship of instantaneous norms is verified

$$u^2 = v^2 + v_0^2 \tag{3.44}$$

where u refers to instantaneous norm of the vector $\boldsymbol{u}$, that is, $u^2 = \boldsymbol{u}^T \boldsymbol{u}$ (dispenses with the vertical bars to represent instantaneous norm for simplicity); similarly, v and v_0 denote the respective instantaneous norms of vectors $\boldsymbol{v}$ and $\boldsymbol{v}_0$.

In the same manner that (3.36) can define a coordinate system from the unit vectors $\boldsymbol{x}_0$, $\boldsymbol{y}_0$, $\boldsymbol{z}_0$ that follow the respective directions of the voltage vectors $\boldsymbol{v}_0$, $\boldsymbol{v}$, and $\boldsymbol{u}_q$, Figure 3.7:

$$\boldsymbol{x}_0 = \frac{\boldsymbol{v}_0}{|\boldsymbol{v}_0|} \qquad \boldsymbol{y}_0 = \frac{\boldsymbol{v}}{|\boldsymbol{v}|} \qquad \boldsymbol{z}_0 = \frac{\boldsymbol{u}_q}{|\boldsymbol{u}_q|} \tag{3.45}$$

The dot product of voltage and current space vectors is,

$$p = \boldsymbol{u}^t \cdot \boldsymbol{i} = \boldsymbol{v}^t \cdot \boldsymbol{i} + \boldsymbol{v}_0^t \cdot \boldsymbol{i} = p_v + p_0 \tag{3.46}$$

defining instantaneous real power delivered to the load (sum of the instantaneous powers of the three phases). The power form two variables:

$p_v = \boldsymbol{v}^t \cdot \boldsymbol{i}$, instantaneous real power exchanged between source and load when a zero-sequence (v_0 or i_0) or both (v_0 and i_0) are absent.

$p_0 = \boldsymbol{v}_0^t \cdot \boldsymbol{i}$, the zero-sequence instantaneous power exchanged between source and load when both zero-sequence components (v_0 and i_0) are present.

Regarding the power variables, the following identities contained in (3.47) are verified,

$$p_v \equiv p_{\alpha\beta}; \quad q \equiv -q_{\alpha\beta} \tag{3.47}$$

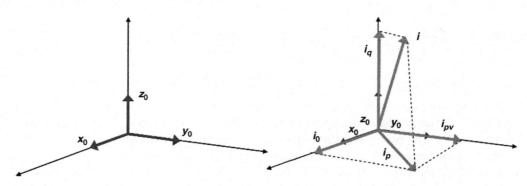

FIGURE 3.7 Vector diagram of current vector partition in a three-phase system including zero-sequence components.

The instantaneous current vector i is decomposed according to its orthogonal projections along the coordinate axes of Figure 3.7,

$$i = i_0 + i_{pv} + i_q = \frac{p_0}{|v_0|} x_0 + \frac{p_v}{|v|} y_0 + \frac{q}{|u_q|} z_0 \tag{3.48}$$

The physical meanings of i_0, i_{pv}, and i_q can be determined by the method of Lagrange multipliers [12],

$$i_0 = \frac{v_0^t \cdot i}{|v_0|^2} v_0 = \frac{p_0}{|v_0|^2} v_0 = \frac{p_0}{|v_0|} x_0 \tag{3.49}$$

$$i_p = \frac{v^t \cdot i}{|v|^2} v = \frac{p_v}{|v|^2} v = \frac{p_v}{|v|} y_0 \tag{3.50}$$

$$i_q = \frac{u_q^t \cdot i}{|u_q|^2} u_q = \frac{q}{|u_q|} z_0 \tag{3.51}$$

where it is verified that instantaneous norms v and u_q are equal. The current components (3.49), (3.50), and (3.51), are respectively, the currents of minimum instantaneous norm to transfer to the load, the power p_0 for voltage v_0, the power p_v for voltage v, and the power q for voltage u_q. The orthogonal projection of the vector i in the plane determined v and v_0 is the instantaneous active current i_p. This is responsible for the transfer of the instantaneous power $p + p_0$ between the generator and the load, and is obtained by adding up i_0 e i_{pv},

$$i_p = i_0 + i_{pv} = \frac{p_0}{|v_0|} x_0 + \frac{p_v}{|v|} y_0 \tag{3.52}$$

The current i_q is obtained by projecting the vector i on the orthogonal voltage vector u_q. Since i_q satisfies the orthogonality condition (3.53),

$$i_p^t \cdot i_q = 0 \tag{3.53}$$

the following relationship between the instantaneous norms of the current components is verified,

$$i^2 = i_p^2 + i_q^2 \tag{3.54}$$

Thus, it is clear that the proposed partition of the current in the terms given by (3.49)–(3.51) is not arbitrary. This is electrically imposed by those components necessary for the full description of the process of instantaneous energy transfer between source and load. The influence of the zero-sequence components in the instantaneous power flow is easily explained by the vector diagram of Figure 3.7. The current vector i_0 affects only the instantaneous active current. If, $i_0 \neq 0$ and $v_0 = 0$, then $u \equiv v$. The zero-sequence current is perpendicular to the voltage vector and does not contribute to the instantaneous real

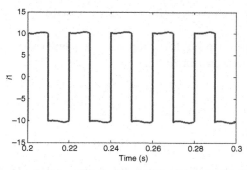

FIGURE 3.8 Current of phase 1 for a load of three single-phase rectifiers connected in star.

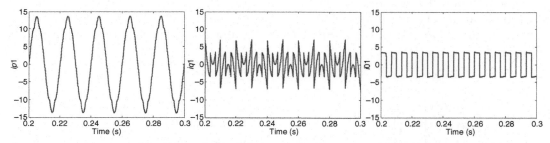

FIGURE 3.9 Decomposition of the current of phase 1 for a load of single-phase rectifiers connected in star (Figure 3.8) as the vector formulation (left to right): instantaneous active current, instantaneous reactive current, and zero-sequence instantaneous current.

power ($p \equiv p_v$). For $i_0 = 0$ and $v_0 \neq 0$, the instantaneous active current is perpendicular to v_0 ($i_p \equiv i_{pv}$); thus, the zero-sequence voltage does not contribute to the instantaneous real power. Finally, it can be concluded that only if both components are zero-sequence, $i_0 \neq 0$ y $v_0 \neq 0$, they contribute to the instantaneous transfer of real power ($p = p_v + p_0$).

Figure 3.9 shows a current decomposition for phase 1 of a system consisting of a load of four wires; it is a load constituted by three single-phase rectifiers in star, Figure 3.8. It can be noted that the zero-sequence current has a period one-third of that of the phase current.

The terms of power and current components obtained by the original formulation (3.22) when they are brought to the phase domain by the inverse transformation matrix (3.23) match the obtained expressions (3.49)–(3.51).

3.2.3 Applications to Active Compensation

Since its origin, the instantaneous power theory has been linked to seeking compensation strategies for the control of active power filters. One might even say that its origin was motivated primarily by the needs that the design of control of those required in the case of nonlinear phase loads [1–6].

In the framework of the original formulation, the power variables exchanged between the load are divided into two components,

$$p_{L0} = p_{L0dc} + p_{L0ac} \tag{3.55}$$

$$p_{L\alpha\beta} = p_{Ldc} + p_{Lac} \tag{3.56}$$

$$q_{L\alpha\beta} = q_{Ldc} + q_{Lac} \tag{3.57}$$

where p_{L0dc}, p_{Ldc}, q_{Ldc}, components of constant values; p_{L0ac}, p_{Lac}, q_{Lac}, time varying components or oscillatory. A low-pass filter is capable of removing a power component of the other.

An active power line conditioner, APLC (active power filter) connected in parallel with the load injects a compensation current determined by the currents of reference control circuit, Figure 3.10. For a four-wire three-phase system, the compensator currents are,

$$\begin{bmatrix} i_{ca}^* \\ i_{cb}^* \\ i_{cc}^* \end{bmatrix} = \sqrt{\frac{2}{3}} \begin{bmatrix} \dfrac{1}{\sqrt{2}} & 1 & 0 \\ \dfrac{1}{\sqrt{2}} & -\dfrac{1}{2} & \dfrac{\sqrt{3}}{2} \\ \dfrac{1}{\sqrt{2}} & -\dfrac{1}{2} & -\dfrac{\sqrt{3}}{2} \end{bmatrix} \frac{1}{u_0 u_{\alpha\beta}^2} \begin{bmatrix} u_{\alpha\beta}^2 & 0 & 0 \\ 0 & u_0 u_\alpha & -u_0 u_\beta \\ 0 & u_0 u_\beta & u_0 u_\alpha \end{bmatrix} \begin{bmatrix} p_{C0}^* \\ p_{C\alpha\beta}^* \\ q_{C\alpha\beta}^* \end{bmatrix} \tag{3.58}$$

where the subscript C refers to the side of the compensator and p_{C0}^*, zero-sequence instantaneous real power reference of compensator; $p_{C\alpha\beta}^*$, α and β phases instantaneous real

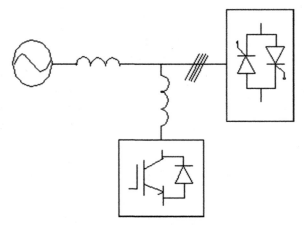

FIGURE 3.10 Four-wire three-phase circuit with an active power filter acting as compensator.

power to reference compensator; $q^*_{C\alpha\beta}$, α and β phases instantaneous imaginary power to reference compensator.

The specific function of conditioning load APLC is determined by the chosen terms (and therefore the compensation strategy chosen) for p^*_{C0}, $p^*_{C\alpha\beta}$, and $q^*_{C\alpha\beta}$ the right side of (3.55), (3.56), and (3.57) and which are introduced in (3.58). In other words, the shunt active conditioner can compensate not only for harmonics but also reactive power and negative-sequence components according to the control objectives. Thus, if the condition that the active filter compensates the instantaneous imaginary power that transfers the load is imposed,

$$q_{C\alpha\beta} = q_{L\alpha\beta} \tag{3.59}$$

and if it is required that the compensator not instantaneous power transfer,

$$p_C = p_{C0} + p_{C\alpha\beta} = 0 \tag{3.60}$$

From (3.59) and (3.60), the currents of compensation in 0, α, β coordinates are followed by the relation (3.61)

$$
\begin{bmatrix} i^*_{c0} \\ i^*_{c\alpha} \\ i^*_{c\beta} \end{bmatrix} = \frac{1}{u_0 u^2_{\alpha\beta}} \begin{bmatrix} u^2_{\alpha\beta} & 0 & 0 \\ 0 & u_0 u_\alpha & -u_0 u_\beta \\ 0 & u_0 u_\beta & u_0 u_\alpha \end{bmatrix} \begin{bmatrix} p_{L0} \\ -p_{L0} \\ q_{L\alpha\beta} \end{bmatrix} \tag{3.61}
$$

Compensating currents (3.61) enable the supply to only the instantaneous real power of α, β circuit transfer ($p_S = p_{L\alpha\beta}$) without zero-sequence current ($i_{S0} = 0$). Other types of compensation are possible [17–21].

EXAMPLE 3.2

To determine the compensation currents to the strategy of constant real power from the supply side under original instantaneous reactive power.

In this example, the compensation objective is to achieve compensation constant value of instantaneous real power in source side and an instantaneous imaginary power null. In any shunt compensation system is verified (3.62),

$$p_S(t) = p_L(t) - p_C(t) \tag{3.62}$$

and the independence of power variables p_0 y $p_{\alpha\beta}$, the relation (3.62) is satisfied also for each of them,

$$p_{S0}(t) = p_{L0}(t) - p_{C0}(t) \tag{3.63}$$

and

$$p_{S\alpha\beta}(t) = p_{L\alpha\beta}(t) - p_{C\alpha\beta}(t) \tag{3.64}$$

Likewise, the instantaneous imaginary power is still (3.65),

$$q_{S\alpha\beta}(t) = q_{L\alpha\beta}(t) - q_{C\alpha\beta}(t) \tag{3.65}$$

For the purpose of constant power compensation is fulfilled after compensation (3.66)

$$p_S(t) = P_L \tag{3.66}$$

That is, the source supplies all the average power (real power) of the load (and the power consumed by the compensator in a real situation); real power transferred by the compensator is zero. Thus, from (3.63) we have the zero-sequence instantaneous power of compensator

$$p_{C0}(t) = p_{L0}(t) - p_{S0}(t) = p_{L0}(t) - P_{L0} = p_{L0\text{ac}}(t) \tag{3.67}$$

where $p_{L0\text{ac}}(t)$ is the oscillatory part of the zero-sequence instantaneous power of the load. Likewise, the instantaneous power of α- and β-axes compensator satisfies the relation (3.68),

$$p_{C\alpha\beta}(t) = p_{L\alpha\beta}(t) - P_{L\alpha\beta} = p_{L\alpha\beta\text{ac}}(t) \tag{3.68}$$

where $p_{L\alpha\beta\text{ac}}(t)$ represents the oscillatory component of the α–β axis instantaneous real power of the load. Finally, the instantaneous imaginary power transferred by the compensator verifies (3.69),

$$q_{C\alpha\beta}(t) = q_{L\alpha\beta}(t) \tag{3.69}$$

Once the pattern of power variables of the compensator have been determined, applying (3.29) allows compensation currents (3.70)

$$i_C = \begin{bmatrix} i_{C0} \\ i_{C\alpha} \\ i_{C\beta} \end{bmatrix} = \frac{p_{L0\text{ac}}(t)}{\mathbf{u}_0 \cdot \mathbf{u}_0} \begin{bmatrix} u_0 \\ 0 \\ 0 \end{bmatrix} + \frac{p_{L\alpha\beta\text{ac}}(t)}{\mathbf{u}_{\alpha\beta} \cdot \mathbf{u}_{\alpha\beta}} \begin{bmatrix} 0 \\ u_a \\ u_\beta \end{bmatrix} + \frac{q_{L\alpha\beta}(t)}{\mathbf{u}_{\alpha\beta} \cdot \mathbf{u}_{\alpha\beta}} \begin{bmatrix} 0 \\ u_{-\beta} \\ u_\alpha \end{bmatrix} \tag{3.70}$$

EXAMPLE 3.3

Determining the compensation currents for strategy of constant instantaneous real power from the source side to allow cancellation of the neutral current under original instantaneous reactive power formulation.

The implementation of compensation strategy given by (3.70) fails to cancel the neutral current. Indeed, the first component of (3.70) yields

$$i_{C0} = \frac{p_{L0\text{ac}}(t)}{u_0^2} u_0 \tag{3.71}$$

then

$$i_{Sn} = \sqrt{3} i_{S0} = \sqrt{3} i_{L0} - \sqrt{3} \frac{p_{L0\text{ac}}(t)}{u_0} \neq 0 \tag{3.72}$$

To ensure the elimination of the neutral current imposes a further restriction, namely

$$i_{C0}(t) = i_{L0}(t) \tag{3.73}$$

This new condition makes it possible to compensate the zero-sequence current in any condition of load voltage and current. However, in this exercise, current compensation will be obtained by the development vector formulation shown in Section 3.2.1. The supply current after compensation is given by (3.74),

$$i_S = \frac{P_L}{u_{\alpha\beta}^2} \boldsymbol{u}_{\alpha\beta} \tag{3.74}$$

and Kirchhoff's law of currents is verified in the compensated system (3.75),

$$
\begin{aligned}
i_C(t) = i_L(t) - i_S(t) &= \frac{p_{L0}}{u_0^2} \boldsymbol{u}_0 + \frac{p_{L\alpha\beta}}{u_{\alpha\beta}^2} \boldsymbol{u}_{\alpha\beta} + \frac{q_{L\alpha\beta}}{u_{\alpha\beta}^2} \boldsymbol{u}_{-\beta\alpha} - \frac{P_{L\alpha\beta} + P_{L0}}{u_{\alpha\beta}^2} \boldsymbol{u}_{\alpha\beta} = \\
&= \frac{p_{L0}}{u_0^2} \boldsymbol{u}_0 + \frac{p_{L\alpha\beta_{ac}} - P_{L0}}{u_{\alpha\beta}^2} \boldsymbol{u}_{\alpha\beta} + \frac{q_{L\alpha\beta}}{u_{\alpha\beta}^2} \boldsymbol{u}_{-\beta\alpha}
\end{aligned}
\tag{3.75}
$$

By the compensation current (3.75) the active filter achieves a constant power in source side with zero transfer of real power from the compensator, and also with zero neutral current. As in all cases of active compensation, the instantaneous imaginary power is compensated.

As has been established, the original formulation gets compensation of nonlinear loads according to (3.58). However, active compensation can be established directly in phase coordinates on the model developed in Subsection 3.3.2. Example 3.4 will help clarify the situation.

EXAMPLE 3.4

Obtain, in phase coordinates, compensation currents injected active power filter for parallel connection of a four-wire three-phase load, consisting of a star configuration formed by branches RL ($R = 20\ \Omega$, $L = 10$ mH) in series with two antiparallel SCRs, (symmetrical thyristor firing with $\alpha = 90°$ for the first SCR), Figure 3.11. The set is supplied with a balanced sinusoidal three-phase voltage system of 400 V, 50 Hz. Two control objectives are developed: (a) Strategy of instantaneous compensation, (b) strategy of constant power compensation.

First expressions of the compensation currents for each compensation strategy will be obtained, then the results of the waveforms of interest will be obtained from simulation models.

1. Instantaneous compensation strategy seeks to compensate the instantaneous imaginary power so that the load on the source side there is only a power variable, the actual instantaneous power before compensation. Also, the neutral current will be required to be cancelled here. Thus, after compensation, the current at the source side will be,

$$i_S = \frac{p(t)}{v^2} v \tag{3.76}$$

using (3.49)–(3.51) and current Kirchhoff's law gives (3.77),

$$
i_C = i_L - i_S = \frac{p_{L0}}{v_0^2}v_0 + \frac{p_{Lv}}{v^2}v + \frac{q_L}{v^2}u_q - \frac{p_{Lv} + p_{L0}}{v^2}v =
$$
$$
= \frac{p_{L0}}{v_0^2}v_0 + \frac{-p_{L0}}{v^2}v + \frac{q_L}{v^2}u_q
$$

(3.77)

Equation (3.77) demonstrates the ease of the method described in Section 3.2.2 to determine the compensation currents in all operating conditions. A simulation model has been implemented in SimPowerSystems. The block diagram of Figure 3.12 presents the generation of the compensation currents (3.77) for each phase.

Figure 3.13 presents the voltage and current of phase 1 in the source side before and after compensation, that is, the current consumed by the load before switching the active compensator ($t < 0.08$ s), and source current once connected the compensator.

Figure 3.14a shows the compensation current of one phase obtained from (3.77). Source currents after instantaneous compensation remain distorted. However, representing the set of currents without zero-sequence components, all possible, with the minimum instantaneous rms value transferred only instantaneous real power; this is the meaning of instantaneous active currents. Figure 3.14b shows the waveform of the neutral current in the source side before and after compensation; is observed as this strategy allows cancellation of zero-sequence currents.

Moreover, the source current allows only instantaneous real power transfer, not instantaneous imaginary power transfer. Figure 3.15 shows the waveforms of power (real and imaginary) before and after compensation; the waveform of the instantaneous real power remains unchanged and that of instantaneous imaginary power drops to zero.

2. The strategy of constant power compensation is intended to compensate the instantaneous imaginary power and oscillatory components of the instantaneous real power. Thus, the instantaneous real power on the source side is average power (active or real) incoming to the load. On this occasion, it is required that the zero-sequence current is zero. From (3.49) to (3.51) and current Kirchhoff's law obtained compensation currents (3.78),

$$
i_C = i_L - i_S = \frac{p_{L0}}{v_0^2}v_0 + \frac{p_{Lv}}{v^2}v + \frac{q_L}{v^2}u_q - \frac{P_{Lv} + P_{L0}}{v^2}v =
$$
$$
= \frac{p_{L0}}{v_0^2}v_0 + \frac{p_{Lvac} - p_{L0}}{v^2}v + \frac{q_L}{v^2}u_q
$$

(3.78)

where p_{Lvac} is the oscillatory component of the instantaneous real power without zero-sequence instantaneous power. A simulation model has been implemented in SimPowerSystems. The block diagram of Figure 3.16 shows the generation of the compensation current of each phase.

The low-pass filter allows the average value of the instantaneous real power to be obtained and from there the oscillatory component thereof is obtained. Figure 3.17a shows the current compensation of phase 1 obtained from (3.78).

Following injection of the compensation current in Figure 3.17a, Figure 3.18 shows the waveforms of voltage and current in phase 1 of the source side before and after compensation.

Source currents after compensation are sinusoidal currents without zero-sequence components; Figure 3.17b shows how the neutral current in the neutral side is zero after compensation. Source currents after compensation, Figure 3.18, are the components of currents of all possible with the smallest rms value; this is the meaning of active currents. Moreover, these currents transfer real power only, they do not transfer instantaneous imaginary power, zero-sequence instantaneous real power, or oscillatory instantaneous real power. Figure 3.19 shows the waveforms of power (real and imaginary) before and after compensation; after compensation, the waveform of the instantaneous real power is constant (real power incoming by the load) and the instantaneous imaginary power drops to zero.

3.3 Modified Instantaneous Reactive Power Formulation

In this section the development of the so-called instantaneous modified reactive power formulation will be addressed. This is a formulation that handles the instantaneous real power as the independent variable, and therefore, does not consider the zero-sequence instantaneous real power separately. On the other hand, the method introduces three variables of instantaneous reactive powers of which only two are independent variables. The formulation was first introduced in 1995 in phase coordinates [22] and later its development was published in $0–\alpha–\beta$ coordinates [24]. Logically, both developments are equivalent.

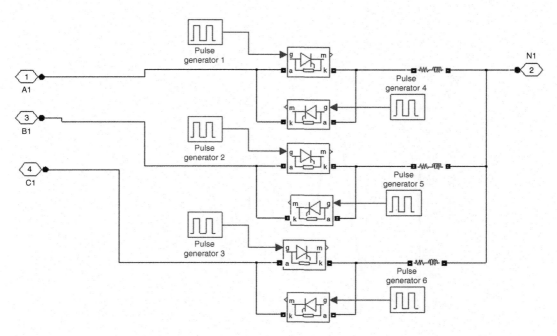

FIGURE 3.11 Nonlinear three-phase load in star formed by branches RL in series with two antiparallel SCRs as SimPowerSystems model.

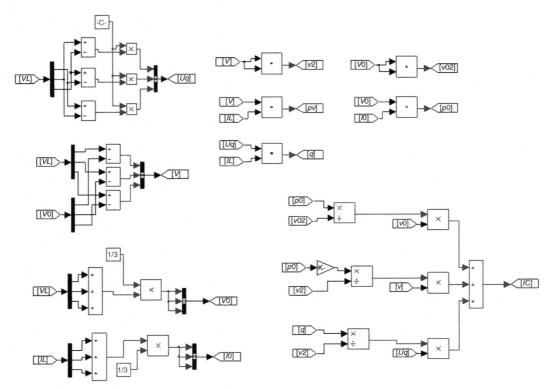

FIGURE 3.12 Block diagram for generating currents of instantaneous compensation without neutral current.

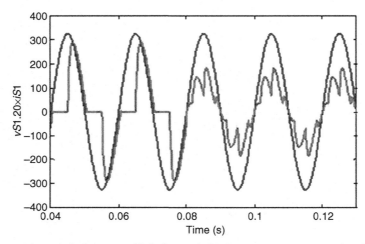

FIGURE 3.13 Voltage and current in the source side before and after instantaneous compensation, (the source current is multiplied by 20).

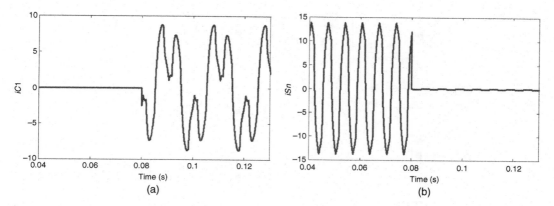

FIGURE 3.14 (a) Current of instantaneous compensation in phase 1 by an active compensator injected at $t = 0.08$ s, (b) Neutral current in the source side before and after instantaneous compensation.

This section will first introduce the analysis in 0–α–β coordinates, and then later in phase coordinates, respecting the scheme followed in Section 3.2. The results of the same will apply to static compensation.

The instantaneous reactive power defined by modified formulation is different from that defined by the original formulation for four-wire three-phase systems which include voltage and current zero-sequence components. Therefore, the modified formulation proposes compensation characteristics different from the original formulation when both are applied to the compensation active power filters. Finally, this section will also contain a discussion of the similarities and differences between the original instantaneous reactive power formulation and modified instantaneous reactive power [25].

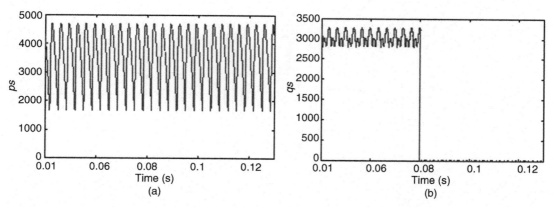

FIGURE 3.15 (a) Instantaneous real power in source side before and after instantaneous compensation, (b) instantaneous imaginary power in source side before and after instantaneous compensation.

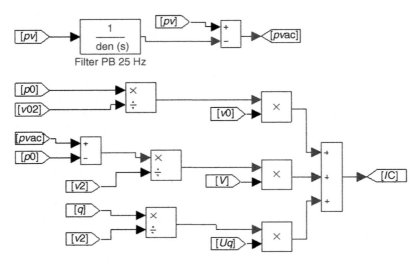

FIGURE 3.16 Block diagram for the compensation currents for the purpose of constant power compensation; oscillatory component of the instantaneous power without zero-sequence power is obtained by a low-pass filter.

3.3.1   0, α, β Coordinates

The modified formulation proposed in 0–α–β coordinates, and therefore with similar language to that used by the original formulation, was presented by Kim–Akagi [24]. This formulation defines an instantaneous real power p and three instantaneous reactive powers q_0, q_α, q_β in the form,

$$
\begin{bmatrix} p \\ q_0 \\ q_\alpha \\ q_\beta \end{bmatrix} = \begin{bmatrix} u_0 & u_\alpha & u_\beta \\ 0 & -u_\beta & u_\alpha \\ u_\beta & 0 & -u_0 \\ -u_\alpha & u_0 & 0 \end{bmatrix} \begin{bmatrix} i_0 \\ i_\alpha \\ i_\beta \end{bmatrix}
\tag{3.79}
$$

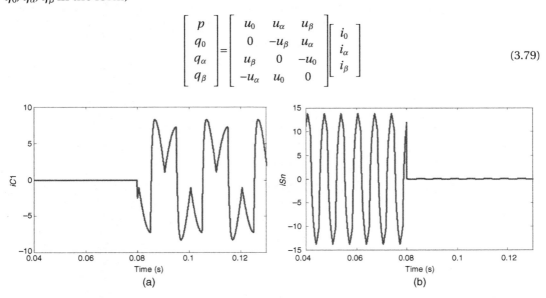

FIGURE 3.17 (a) Current phase 1 of constant power compensation injected by a shunt active compensator, (b) neutral current before and after ($t = 0.08$ s) constant power compensation with elimination of zero-sequence current.

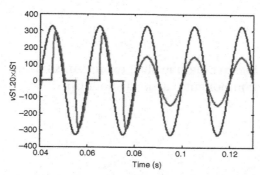

FIGURE 3.18 Voltage and current in phase 1 of the source side before and after constant power compensation (the current is multiplied by 20).

(3.80) follows from equation (3.79)

$$u_0 \, q_0 + u_\alpha \, q_\alpha + u_\beta \, q_\beta = 0 \tag{3.80}$$

The range of the transformation matrix (mapping matrix) in (3.79) is three, that is, the number of independent variables $q_0, q_\alpha,$ and q_β is two. As a result, this is a three-dimensional power space similar to that reported in the original formulation. However, the two transformation matrices defined by (3.6) and (3.79), and the results from its formulation are different as we will see later.

In (3.79), it is always possible to have inverse transformation in the form (3.81) [24],

$$\begin{bmatrix} i_0 \\ i_\alpha \\ i_\beta \end{bmatrix} = \frac{1}{u_{0\alpha\beta}^2} \begin{bmatrix} u_0 & 0 & u_\beta & -u_\alpha \\ u_\alpha & -u_\beta & 0 & u_0 \\ u_\beta & u_\alpha & -u_0 & 0 \end{bmatrix} \begin{bmatrix} p \\ q_0 \\ q_\alpha \\ q_\beta \end{bmatrix} \tag{3.81}$$

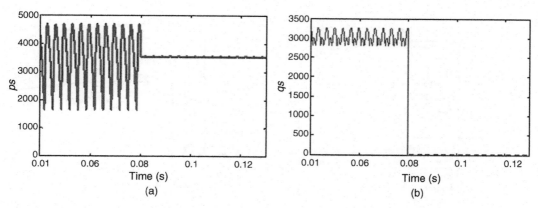

FIGURE 3.19 (a) Instantaneous power on the supply side before and after constant power compensation, (b) instantaneous imaginary power in the supply side before and after compensation.

where

$$u_{0\alpha\beta}^2 = u_0^2 + u_\alpha^2 + u_\beta^2 \tag{3.82}$$

From (3.81) it is possible to obtain explicit expressions for the instantaneous currents of each phase. Thus, the components can be distinguished

$$i_0 = i_{0p} + i_{0q} \tag{3.83}$$

$$i_\alpha = i_{\alpha p} + i_{\alpha q} \tag{3.84}$$

$$i_\beta = i_{\beta p} + i_{\beta q} \tag{3.85}$$

where two new components of zero-sequence current appear

i_{0p}: zero-sequence instantaneous active current
i_{0q}: zero-sequence instantaneous reactive current

Equation (3.83) means that the zero-sequence instantaneous current, i_0, can be divided into two currents, i_{0p}, and i_{0q}; you can also find the following power relationships:

$$p(t) = u_0 i_{0p} + u_\alpha i_{\alpha p} + u_\beta i_{\beta p} = p_{0p} + p_{\alpha p} + p_{\beta p} \tag{3.86}$$

$$0 = u_0 i_{0q} + u_\alpha i_{\alpha q} + u_\beta i_{\beta q} = p_{0q} + p_{\alpha q} + p_{\beta q} \tag{3.87}$$

In these expressions, p_{0p} and p_{0q} are referred to as,

p_{0p} : zero-sequence instantaneous active power
p_{0q} : zero-sequence instantaneous reactive power

Figure 3.20 describes power flow based on the modified formulation. It is clear from (3.83) to (3.85) that the modified formulation is characterized by treating equally the zero-phase circuit, the α- and β-phase circuit. This means the zero-sequence current is divided into two components, one active and one reactive.

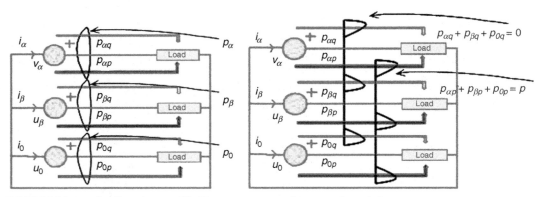

FIGURE 3.20 Power flow based on modified instantaneous reactive power formulation.

Analogous to vector format developed in Section 3.2.1 for the original formulation, a vector format for the modified formulation is possible. In fact, an instantaneous reactive power space vector $\boldsymbol{q}(t)$ is defined in (3.88),

$$q_{0\alpha\beta}(t) = u_{0\alpha\beta} x i_{0\alpha\beta} = \begin{vmatrix} \boldsymbol{0} & \boldsymbol{\alpha} & \boldsymbol{\beta} \\ u_0 & u_\alpha & u_\beta \\ i_0 & i_\alpha & i_\beta \end{vmatrix} = \begin{bmatrix} q_0 \\ q_\alpha \\ q_\beta \end{bmatrix} \qquad (3.88)$$

where $\boldsymbol{0}, \boldsymbol{\alpha}, \boldsymbol{\beta}$ are unit vectors in the direction of the axes $0, \alpha, \beta$. Since the vector identity (3.89)

$$(\boldsymbol{a} \times \boldsymbol{b}) \times \boldsymbol{c} = (\boldsymbol{a} \cdot \boldsymbol{c})\boldsymbol{b} - (\boldsymbol{b} \cdot \boldsymbol{a})\boldsymbol{c} \qquad (3.89)$$

the current components (3.90) and (3.91) are obtained,

$$i_p = \frac{u_{0\alpha\beta} \cdot i_{0\alpha\beta}}{u_{0\alpha\beta} \cdot u_{0\alpha\beta}} u_{0\alpha\beta} \qquad (3.90)$$

and

$$i_q = i_{0\alpha\beta} - i_p = \frac{\left(u_{0\alpha\beta} \cdot u_{0\alpha\beta}\right)}{u_{0\alpha\beta} \cdot u_{0\alpha\beta}} i_{0\alpha\beta} - \frac{u_{0\alpha\beta} \cdot i_{0\alpha\beta}}{u_{0\alpha\beta} \cdot u_{0\alpha\beta}} u_{0\alpha\beta} =$$
$$= \frac{\left(u_{0\alpha\beta} \times i_{0\alpha\beta}\right) \times u_{0\alpha\beta}}{u_{0\alpha\beta} \cdot u_{0\alpha\beta}} = \frac{q_{0\alpha\beta} \times u_{0\alpha\beta}}{u_{0\alpha\beta} \cdot u_{0\alpha\beta}} \qquad (3.91)$$

The first current component (3.90) is the projection of current vector on the voltage vector, that is, the instantaneous active current in $0, \alpha, \beta$ coordinates and the remaining component (3.91) is instantaneous reactive current in $0, \alpha, \beta$ coordinates. The two current components are orthogonal, and verify power relations (3.92),

$$p(t) = u_{0\alpha\beta} \cdot i_p = u_{0\alpha\beta} \cdot i_{0\alpha\beta}; \quad 0 = u_{0\alpha\beta} \cdot i_q \qquad (3.92)$$

From (3.92) it follows that the i_q component does not transmit any instantaneous real power. Both approaches, description of mapping matrix, (3.81), or vector description (3.90)–(3.91), lead to identical results.

3.3.2 Phase Coordinates

Then the modified formulation development in phase coordinates is carried out [20,23]. A vector of phase-neutral voltages $\boldsymbol{u}$ and vector of line currents $\boldsymbol{i}$ expressed in vector form (3.31) is available for a three-phase power system. As has been previously established, the instantaneous real power $p(t)$ is given by the scalar product of two vectors and a new instantaneous vector $\boldsymbol{q}(t)$ is defined by the vector product of voltage and current vectors,

$$p(t) = \boldsymbol{u} \cdot \boldsymbol{i}; \quad \boldsymbol{q}(t) = \boldsymbol{u} \times \boldsymbol{i} \qquad (3.93)$$

This vector is the instantaneous reactive power vector $\boldsymbol{q}$ in phase coordinates, and its modulus q is designated as instantaneous reactive power. The space vector $\boldsymbol{q}$ has three components,

$$\boldsymbol{q}(t) = \begin{vmatrix} \begin{vmatrix} u_2 & u_3 \\ i_2 & i_3 \end{vmatrix} \\ \begin{vmatrix} u_3 & u_1 \\ i_3 & i_1 \end{vmatrix} \\ \begin{vmatrix} u_1 & u_2 \\ i_1 & i_2 \end{vmatrix} \end{vmatrix} = \begin{bmatrix} q_1 \\ q_2 \\ q_3 \end{bmatrix} \qquad (3.94)$$

so that instantaneous reactive power is distributed over the three phases; each phase has a different instantaneous reactive power associated with it. However, the three reactive power components are not independent as verified (3.95),

$$\boldsymbol{u}.\boldsymbol{q}(t) = u_1 q_1 + u_2 q_2 + u_3 q_3 = 0 \qquad (3.95)$$

The modified instantaneous reactive power formulation states that the current space vector is the vector sum of two components:

$$\boldsymbol{i} = \boldsymbol{i_p} + \boldsymbol{i_q} \qquad (3.96)$$

where $\boldsymbol{i_p}$, instantaneous active current vector; $\boldsymbol{i_q}$, instantaneous reactive current vector.

The first is the projection of the current space vector on voltage space vector, that is, component $\boldsymbol{i_p}$

$$\boldsymbol{i_p} = \frac{p(t)}{\boldsymbol{u}.\boldsymbol{u}} \boldsymbol{u} \qquad (3.97)$$

The component (3.97) is the instantaneous current with minimum instantaneous norm that transfer instantaneous real power from source to load.

The second term in (3.96) is the remaining current component, $\boldsymbol{i_q}$

$$\boldsymbol{i_q} = \frac{\boldsymbol{q} \times \boldsymbol{u}}{\boldsymbol{u}.\boldsymbol{u}} \qquad (3.98)$$

and does not transfer any instantaneous real power.

The ratio obtained for the vector $\boldsymbol{i}$ (3.96) follows directly from the application of (3.89) that allows a vector $\boldsymbol{i}$, known scalar product $\boldsymbol{u}.\boldsymbol{i}$ and vector product $\boldsymbol{u} \times \boldsymbol{i}$ by a given vector $\boldsymbol{u}$. The instantaneous active current $\boldsymbol{i_p}$, now called instantaneous power current, was first described by Depenbrock [31]. However, the method shown here is due to Rossetto–Tenti [9], as further popularized by Willems [10].

3.3.3 Applications to Active Compensation

In this subsection, the approach of the modified instantaneous reactive power formulation to compensation by shunt active power filters is applied. The principles initially outlined in Subsection 3.2.3 are applicable here. For simplicity we obtain compensation currents directly for different compensation strategies in phase coordinates according to the approach taken in Subsection 3.3.2. Thus, instantaneous compensation involves two restrictions: the instantaneous real power of compensator is null, and instantaneous reactive powers transferred by the compensator are equal to the instantaneous reactive powers in load side. The first condition implies that the source current after compensation is,

$$i_S = \frac{p_L(t)}{u^2} \boldsymbol{u} \tag{3.99}$$

From (3.99) and Kirchhoff's law of the vector of compensation currents (3.100) is obtained,

$$i_C = i_L - i_S = \frac{p_L(t)}{u^2}\boldsymbol{u} + \frac{q_L(t) \times \boldsymbol{u}}{\boldsymbol{u}.\boldsymbol{u}} - \frac{p_L(t)}{u^2}\boldsymbol{u} =$$
$$= \frac{q_L(t) \times \boldsymbol{u}}{\boldsymbol{u}.\boldsymbol{u}} \tag{3.100}$$

Compensation currents (3.100) injected by the compensator enables only source instantaneous real power supplied through the instantaneous active current (3.99). Other compensation strategies will be examined in the following two examples.

EXAMPLE 3.5

Determine the instantaneous compensation currents to eliminate the neutral current injected by a shunt active power filter to a four-wire three-phase load. Make a proposal for phase coordinates and in the framework of the modified instantaneous reactive power formulation. A nonlinear load consists of a star configuration formed by branches RL in series with two antiparallel SCRs Figure 3.11. The system is supplied by a balanced sinusoidal three-phase voltage. A simulation model for the waveforms of interest is developed.

Compensation currents (3.100) allow source currents (3.99) that meet the instantaneous target compensation, that is, the source transfers only the instantaneous real power of load,

$$p_S(t) = \boldsymbol{u}.i_S = \frac{p_L(t)}{u^2}\boldsymbol{u}.\boldsymbol{u} = p_L(t) \tag{3.101}$$

However, this compensation strategy does not eliminate the current flowing through the neutral of the supply source; indeed,

$$i_{Sn} = i_{S1} + i_{S2} + i_{S3} = \frac{p_L(t)}{u^2}(u_1 + u_2 + u_3) = \frac{p_L(t)}{u^2}\left(\sqrt{3}u_0\right) \neq 0 \tag{3.102}$$

Introducing a vector named instantaneous pseudo-reactive power space vector q_v allows a new approach instantaneous compensation. The instantaneous pseudo-reactive power vector is defined from the voltage vector without zero-sequence components, v (3.43),

$$q_v(t) = v \times i = \begin{Vmatrix} \begin{vmatrix} v_2 & v_3 \\ i_2 & i_3 \end{vmatrix} \\ \begin{vmatrix} v_3 & v_1 \\ i_3 & i_1 \end{vmatrix} \\ \begin{vmatrix} v_1 & v_2 \\ i_1 & i_2 \end{vmatrix} \end{Vmatrix} = \begin{bmatrix} q_{v1} \\ q_{v2} \\ q_{v3} \end{bmatrix} \qquad (3.103)$$

The current $i(t)$ can be decomposed into two orthogonal components of current. A current component i_p, the instantaneous pseudo-active current, carrying actual instantaneous power $p(t)$ without zero-sequence components,

$$i_p = \frac{p(t)}{v^2} v \qquad (3.104)$$

and current component i_q, instantaneous pseudo-reactive current component determined by (3.105),

$$i_q = i - i_p = \frac{(u.v)}{u.v} i - \frac{u.i}{v.v} u =$$
$$= \frac{(v \times i) \times u}{v^2} = \frac{q_v \times u}{v^2} \qquad (3.105)$$

A compensation strategy that compensates instantaneous pseudo-reactive power load gets an instantaneous compensation and eliminates neutral current after compensation. Indeed, the compensation current (3.106),

$$i_C = i_L - i_S = \frac{p_L(t)}{v^2} v + \frac{q_{Lv}(t) \times u}{v^2} - \frac{p_L(t)}{v^2} v =$$
$$= \frac{q_{Lv}(t) \times u}{v^2} \qquad (3.106)$$

determines a source current after compensation without zero-sequence components that transfer only the instantaneous real power of load (3.107),

$$i_S = \frac{p_L(t)}{v^2} v \qquad (3.107)$$

The block diagram of Figure 3.21 implements the compensation currents (3.106). A simulation model for the load compensation of Figure 3.10 has been developed.

Figure 3.22 presents the waveforms corresponding to the instantaneous real power and the instantaneous pseudo-reactive power before and after compensation. Instantaneous pseudo-reactive power is defined by the norm of space vector q_v,

$$|q_v| = \sqrt{q_{v1} + q_{v2} + q_{v3}} \qquad (3.108)$$

Figure 3.22a shows the same instantaneous real power in the supply side after compensation; instantaneous real power does not change, however, the instantaneous pseudo-reactive power is zero after compensation, Figure 3.22b.

Source current and compensation current are shown in Figures 3.13 and 3.14. The results obtained for this compensation strategy from the original and modified formulations are coincident.

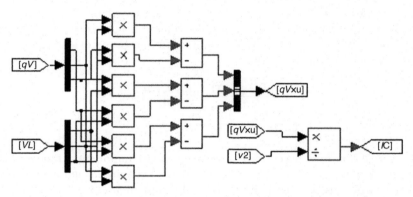

FIGURE 3.21 Block diagram to generate instantaneous pseudo-reactive currents within the modified instantaneous reactive power formulation.

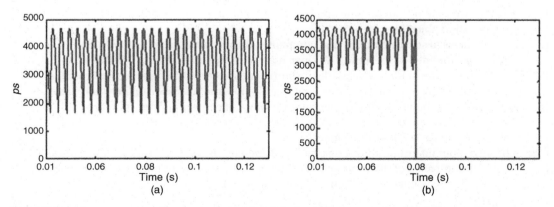

FIGURE 3.22 (a) Instantaneous real power on the supply side before and after compensation for instantaneous compensation under modified instantaneous reactive power formulation, (b) instantaneous pseudo-reactive power in the supply side for instantaneous compensation cancelling neutral current.

EXAMPLE 3.6

Determine compensation currents for a constant power strategy in source side without neutral current. Perform phase coordinates development and under modified instantaneous reactive power formulation. The same load as in Example 3.5 will be used for two types of supply voltages: (a) a three-phase system of balanced sinusoidal voltages, (b) a balanced nonsinusoidal three-phase voltages (a third order harmonic of amplitude 0.08 pu, and fifth order harmonic of amplitude 0.05 pu). A simulation model was developed to obtain the waveforms of interest.

The strategy of constant power is averaged over time is used because it requires determining the average of the instantaneous real power; this concept has already been used in Example 3.4b. Thus, in the same sense in Example 3.2 and according to the new approach adopted in the Example 3.5, the compensator should inject instantaneous pseudo-reactive power and the oscillatory component of the instantaneous real power of load. That is, the compensation current is given by (3.109),

$$i_C = i_L - i_S = \frac{P_L + p_{Lac}(t)}{v^2}\boldsymbol{v} + \frac{\boldsymbol{q}_{Lv}(t) \times \boldsymbol{u}}{v^2} - \frac{P_L}{v^2}\boldsymbol{v} =$$
$$= \frac{p_{Lac}(t)}{v^2}\boldsymbol{v} + \frac{\boldsymbol{q}_{Lv}(t) \times \boldsymbol{u}}{v^2} \tag{3.109}$$

This current will allow a source current no zero-sequence component to transfer a constant instantaneous real power equal to the average value of the load power. Figure 3.23 shows the block diagram that provides the compensation current components (3.109). As in Figure 3.17, the oscillating component of the instantaneous power is determined by a low-pass filter. The results obtained using (3.109) under modified instantaneous reactive power are identical to those obtained by (3.78) under the original instantaneous reactive power. However, various waveforms of interest for two different supply voltages are present.

1. A balanced three-phase sinusoidal voltage identical to that used in Examples 3.4 and 3.5 has been considered. Figure 3.24 shows the waveforms of the instantaneous real power and neutral current in the source side before and after compensation. These results confirm the compensation objectives.

   Waveforms of source currents and compensation currents are coincident with the waveforms obtained in Example 3.4b since it arose compensation target. In this section with balanced sinusoidal voltages, the instantaneous norm (instantaneous rms value) of three-phase voltage is constant and equal to $\sqrt{3}\,V_f$ (V_f is the rms value of the voltage line-to-neutral) and therefore the compensation strategy of constant power coincides with the strategy of unity power factor. Likewise, being the sinusoidal voltages, source currents are sinusoidal and free of harmonic pollution.

2. A balanced but distorted voltage has been considered this time. We used a slightly modified form of the so-called oscillatory wave IEC 61000-4-13 standard for immunity tests distorted supply voltages; voltage applied phase 1 is given by (3.110),

$$v_1(t) = \sqrt{2}\,230\left[\sin(2\pi 50t) + 0.08\sin(3 \times 2\pi 50t) + 0.05\sin(5 \times 2\pi 50t)\right] \tag{3.110}$$

Figure 3.25 shows the waveforms of voltage and current in phase 1 on the source side before and after compensation. This time the source current after compensation is distorted. Its distortion is different from the voltage distortion; this difference is due to two circumstances. First, the source current after compensation is proportional to the supply voltage without the zero-sequence component v, and now the supply voltage, u, includes a harmonic of order 3 with a zero-sequence phase; this does not appear in v. Second, the source current after compensation is inversely proportional to the instantaneous norm of the voltage squared, v^2; now, unlike paragraph (1), this is no longer constant to include the harmonic of order 5. For these conditions the voltage supply, compensation objectives, constant power, unity power factor, and sinusoidal current are different.

Figure 3.26 shows the waveforms of instantaneous real power and current of neutral on the side source before and after compensation. The results verify the compensation objectives of constant power strategy with cancellation of neutral current.

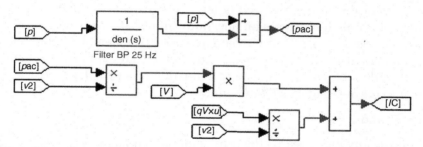

FIGURE 3.23 Block diagram for generating the compensation current to the constant power strategy with cancellation of neutral current within the formulation of the modified instantaneous reactive power.

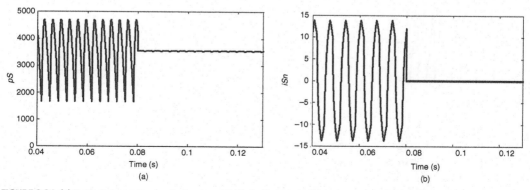

FIGURE 3.24 (a) Instantaneous power in the supply side before and after compensation for constant power compensation with elimination of neutral current within the modified instantaneous reactive power formulation, (b) the neutral current in source side before and after compensation.

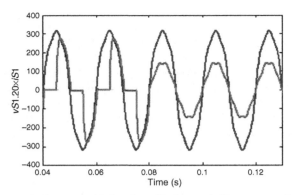

FIGURE 3.25 Voltage and current in the source side for phase 1 before and after compensation of constant power without neutral current to supply nonsinusoidal voltages (the current is multiplied by 20).

3.3.4 Modified Versus Original Instantaneous Reactive Power

The question about the similarities and differences between the modified formulation and the original formulation focuses on how both address the zero-sequence circuit in a four-wire three-phase system [25]. In this situation there is a significant difference in their transformation matrices. Therefore, in what follows there are four possible combinations that can be considered for v_0 and i_0. The coordinate framework 0, α, β places the 0 axis perpendicular to the α–β plane facilitates analysis.

1. Case of $v_0 \neq 0$ and $i_0 \neq 0$.

The original formulation regards the zero-sequence circuit as a separate single-phase circuit of the α-phase circuit and β-phase circuit, just as the method of symmetrical

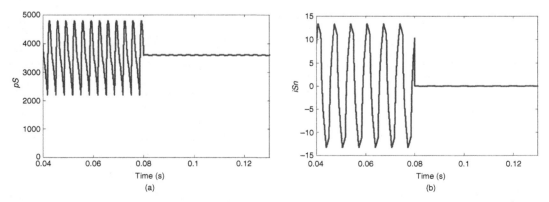

FIGURE 3.26 (a) Instantaneous power in the supply side before and after compensation for constant power strategy with elimination of neutral current for nonsinusoidal voltage, (b) neutral current on source side before and after compensation.

components. A single-phase circuit can define the instantaneous active power as the product of the instantaneous voltage and the current. Furthermore, it is not possible to define uniquely an instantaneous reactive power in a single-phase circuit at a single point in time. The original formulation treats the zero-sequence current, i_0, as an active instantaneous current because the zero-sequence circuit is an instantaneous active power, $p_0 = v_0 i_0$ when $v_0 \neq 0$. Thus no instantaneous reactive current exists in the zero-sequence circuit.

The modified formulation is the zero-sequence circuit, the α- and β-phase circuit, in the same way. This means that the zero-sequence current, i_0, can be divided into zero-sequence instantaneous active current, i_{0p}, and zero-sequence instantaneous reactive current i_{0q}. However, the original formulation and modified formulation assign the same physical meaning to the instantaneous reactive power for each phase.

2. Case of $v_0 = 0$ and $i_0 = 0$.

The original formulation and the modified formulation are identical in a three-wire three-phase system that excludes zero-sequence voltage.

3. Case of $v_0 = 0$ and $i_0 \neq 0$.

If $v_0 = 0$, p_0 is always zero, regardless of the existence of i_0. For the original formulation i_0 is simply an instantaneous current in the zero-sequence circuit when $v_0 = 0$, albeit referred to as instantaneous active current in the zero-sequence circuit when $v_0 \neq 0$. Furthermore, the modified formulation divides i_0 into i_{0p} and i_{0q}. If $v_0 = 0$, it is clear that $i_{0p} = 0$, so $i_0 = i_{0q}$. Therefore, the modified formulation that maintains i_0 is an instantaneous reactive current when $v_0 = 0$. However, experience shows you do not need any energy storage element in an active filter to compensate i_0 when $v_0 = 0$.

4. Case of $v_0 \neq 0$ and $i_0 = 0$.

For this situation the modified formulation produces (3.111),

$$i_0 = i_{0p} + i_{0q}$$
$$i_{0p} = -i_{0q} \tag{3.111}$$

where i_0 is divided into i_{0p} and i_{0q} even a three-wire three-phase system that includes zero-sequence voltage. However, it is impossible to control i_{0p} and i_{0q} independently, since there is no zero-sequence current in a three-wire system.

The analysis along with examples of four-wire three-phase systems compensation, in which a zero-sequence voltage can exist, through both formulations (the formulation of original instantaneous reactive power and the modified instantaneous reactive power formulation) leads to the following conclusions [25]: when $v_0 = 0$, the control strategy based on the modified formulation is identical to that based on the original formulation; therefore both have the same characteristics compensation. When $v_0 \neq 0$, both formulations lead to different control strategies, which are different characteristics of compensation. In particular, the modified formulation in its initial version cannot fully compensate for the zero-sequence current, $i_{C0} \neq i_{L0}$.

3.4 Other Approaches: Synchronous Frames

Previous sections have presented two formulations of the instantaneous reactive power developed coordinate axes considered a "fixed", that is, nonrotating. Specifically, we used the 0, α, β, coordinates or the 1, 2, 3 phase coordinates. In this section other formulations of the instantaneous reactive power in rotating coordinate systems, either in solidarity with the vector of voltage, or at no. This requires performing the appropriate coordinate transformation [26–29].

Mathematical transformations to perform coordinate changes are often used in the field of electric power systems. In general, when a coordinate transformation is achieved, all electrical variables refer to a common frame of reference, and usually a decoupling of the variables is achieved. A singular case is Park's transformation for the study of electrical machines and power systems. This is a transformation from a three-phase system to a two-phase system rotating at angular velocity ω (synchronous speed). Typically, a third variable known as the zero-sequence component is added to determine the inverse transformation matrix, Figure 3.27. The matrix determines from Park's transformation 1, 2, 3 phase variables to the 0, d, q variables is given by (3.112),

$$[T] = \sqrt{\frac{2}{3}} \begin{bmatrix} \dfrac{1}{\sqrt{2}} & \dfrac{1}{\sqrt{2}} & \dfrac{1}{\sqrt{2}} \\ \cos\theta_1(t) & \cos\theta_2(t) & \cos\theta_3(t) \\ -\sin\theta_1(t) & -\sin\theta_2(t) & -\sin\theta_3(t) \end{bmatrix} \tag{3.112}$$

where

$$\theta_k(t) = \theta\left(t - \frac{2\pi}{3}(k-1)\right), \ k = 1,2,3 \tag{3.113}$$

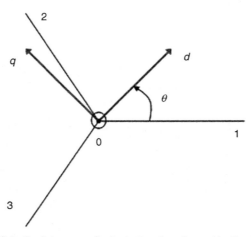

FIGURE 3.27 Coordinate system 0dq; 0 axis is perpendicular to the plane formed by the dq axes and having direction perpendicular out of the plane.

and $\theta(t)$ is an angular arbitrary function. For example, Park processing for analysis of the synchronous machine is a choice of

$$\theta_k(t) = \omega t - \frac{2\pi}{3}(k-1), \quad k = 1,2,3 \tag{3.114}$$

where ω is the synchronous speed of the machine. If, on the other hand the following is chosen

$$\theta_k(t) = -\frac{2\pi}{3}(k-1), \quad k = 1,2,3 \tag{3.115}$$

coordinate transformation to 0, α, β, fixed axes are obtained (3.1)–(3.2), that is, the Concordia–Clarke transformation. The choice of the coefficient $\sqrt{(2/3)}$ ensures that the transformation is a power transformation invariant.

3.4.1 *dq* Formulation

The application of Park's transformation (3.112) to the set of a system of three-phase currents of a power system is what has come to be called the *dq* formulation of the instantaneous reactive power [26]. This formulation is taken as the angular velocity of rotation of the 0, *d*, *q* axes the fundamental harmonic pulsation value of the supply voltage. Thus the transformation (3.112) emerges as a double transformation of axes. First, a transformation from a 1, 2, 3 phase coordinates system, to a fixed coordinate system, 0, α, β. Then, a transformation of 0, α, β coordinate system to a 0, *d*, *q* rotating coordinate system at constant angular velocity corresponding to the frequency of the fundamental harmonic of the voltage, ω.

$$
\begin{bmatrix} i_0 \\ i_d \\ i_q \end{bmatrix} =
\begin{bmatrix} 1 & 0 & 0 \\ 0 & \cos\omega t & \sin\omega t \\ 0 & -\sin\omega t & \cos\omega t \end{bmatrix}
\sqrt{\frac{2}{3}}
\begin{bmatrix} \frac{1}{\sqrt{2}} & \frac{1}{\sqrt{2}} & \frac{1}{\sqrt{2}} \\ 1 & -\frac{1}{2} & -\frac{1}{2} \\ 0 & \frac{\sqrt{3}}{2} & -\frac{\sqrt{3}}{2} \end{bmatrix}
\begin{bmatrix} i_1 \\ i_2 \\ i_3 \end{bmatrix} =
$$

$$
= \sqrt{\frac{2}{3}}
\begin{bmatrix} \frac{1}{\sqrt{2}} & \frac{1}{\sqrt{2}} & \frac{1}{\sqrt{2}} \\ \cos(\omega t) & \cos\left(\omega t - \frac{2\pi}{3}\right) & \cos\left(\omega t + \frac{2\pi}{3}\right) \\ -\sin(\omega t) & -\sin\left(\omega t - \frac{2\pi}{3}\right) & -\sin\left(\omega t + \frac{2\pi}{3}\right) \end{bmatrix}
\begin{bmatrix} i_1 \\ i_2 \\ i_3 \end{bmatrix}
\tag{3.116}
$$

The coordinate system of 0, *d*, *q* axes rotates at an angular velocity of $\omega = 2\pi/T$ (*T* is the period of the voltage waveform) for fixed axes 1,2,3 or 0, α, β. The angular position of the axis "*d*" from the axis "α" is determined by the angle $\theta(t) = \omega t$ that changes with time. If the voltages are balanced and sinusoidal, the voltage vector rotates at the same speed as the axes *dq*. If voltages are unbalanced and nonsinusoidal, the speed of rotation of the voltage vector is not constant, it becomes influenced by the imbalance and distortion of the voltage. In any case, the *dq* formulation chosen for rotational speed of *dq* plane taking as pivot axis 0, the frequency of the fundamental harmonic of the supply voltage.

EXAMPLE 3.7

A three-phase voltage system whose line-to-neutral voltages include a negative sequence component at the fundamental frequency, and harmonic order *h* of negative sequence is considered.

$$u_1(t) = \sqrt{2}\,V^+\cos(\omega t + \psi_1^+) + \sqrt{2}\,V^-\cos(\omega t + \psi_1^-) + \sqrt{2}\,V_h^-\cos(h\omega t + \psi_h^-)$$
$$u_2(t) = \sqrt{2}\,V^+\cos(\omega t + \psi_1^+ - 120) + \sqrt{2}\,V^-\cos(\omega t + \psi_1^- + 120) + \sqrt{2}\,V_h^-\cos(h\omega t + \psi_h^- + 120) \quad (3.117)$$
$$u_3(t) = \sqrt{2}\,V^+\cos(\omega t + \psi_1^+ + 120) + \sqrt{2}\,V^-\cos(\omega t + \psi_1^- - 120) + \sqrt{2}\,V_h^-\cos(h\omega t + \psi_h^- - 120)$$

Determine the 0, *d*, *q* coordinates of voltages service indicated.

The transformation matrix (3.116) applies to three-phase voltages service (3.117) for the components 0, *d*, *q* of the voltage.

$$\begin{bmatrix} u_0 \\ u_d \\ u_q \end{bmatrix} = T \begin{bmatrix} u_1 \\ u_2 \\ u_3 \end{bmatrix} = \sqrt{3}V^+ \begin{bmatrix} 0 \\ \cos\psi_1^+ \\ \sin\psi_1^+ \end{bmatrix} + \sqrt{3}V^- \begin{bmatrix} 0 \\ \cos(2\omega t + \psi_1^-) \\ -\sin(2\omega t + \psi_1^-) \end{bmatrix} + \sqrt{3}V_5^- \begin{bmatrix} 0 \\ \cos((h+1)\omega t + \psi_5^-) \\ -\sin((h+1)\omega t + \psi_5^-) \end{bmatrix}$$

$$(3.118)$$

That is, the fundamental harmonic of positive-sequence phase becomes a constant term, and the negative sequence of the fundamental harmonic and higher harmonics of any sequence phases become oscillatory terms with a frequency of at least twice the fundamental frequency.

The conclusions drawn from Example 3.7 can be applied to the active compensation of a nonlinear unbalanced load. Thus, if we consider any load, applying (3.116) allows the waveforms of load current $i_d(t)$ and $i_q(t)$ to be derived. The mentioned $i_d(t)$ and $i_q(t)$ signals are time variable, and their average values are I_d and I_q, respectively. So, each one can be divided into two terms, a constant value and a variable component.

$$i_d(t) = I_d + \tilde{i}_d(t) \tag{3.119}$$

$$i_q(t) = I_q + \tilde{i}_q(t) \tag{3.120}$$

The constant I_d and I_q components are, respectively, the fundamental harmonic positive sequence of $i_d(t)$ and $i_q(t)$. The other components are the fundamental harmonic negative sequence and the rest of $i_d(t)$ and $i_q(t)$ harmonics. The average values of these last components are null [27].

This current decomposition is used to generate the compensation current reference in shunt active power filter applications. So, the source current will include the constant components I_d and I_q, and the shunt active conditioner will inject the variable components. In practical implementation, it is possible to separate both parts with a low-pass filter. In 1–2–3 coordinates, the source currents will be (3.121):

$$
\begin{bmatrix} i_{S1} \\ i_{S2} \\ i_{S3} \end{bmatrix} = \sqrt{\frac{2}{3}} \begin{bmatrix} \dfrac{1}{\sqrt{2}} & \cos\theta & -\sin\theta \\ \dfrac{1}{\sqrt{2}} & \cos(\theta-120) & -\sin(\theta-120) \\ \dfrac{1}{\sqrt{2}} & \cos(\theta+120) & -\sin(\theta+120) \end{bmatrix} \begin{bmatrix} 0 \\ I_d \\ I_q \end{bmatrix} \tag{3.121}
$$

and the compensation currents will be:

$$
\begin{bmatrix} i_{C1} \\ i_{C2} \\ i_{C3} \end{bmatrix} = \begin{bmatrix} i_{L1} \\ i_{L2} \\ i_{L3} \end{bmatrix} - \begin{bmatrix} i_{S1} \\ i_{S2} \\ i_{S3} \end{bmatrix} = \sqrt{\frac{2}{3}} \begin{bmatrix} \dfrac{1}{\sqrt{2}} & \cos\theta & -\sin\theta \\ \dfrac{1}{\sqrt{2}} & \cos(\theta-120) & -\sin(\theta-120) \\ \dfrac{1}{\sqrt{2}} & \cos(\theta+120) & -\sin(\theta+120) \end{bmatrix} \begin{bmatrix} i_0(t) \\ \tilde{i}_d(t) \\ \tilde{i}_q(t) \end{bmatrix} \tag{3.122}
$$

In conclusion, this theory allows sinusoidal and balanced currents to be obtained, but independent of a symmetrical or distorted voltage supply. Moreover, the active power supplied by the compensator only be null if the voltages are balanced and sinusoidal [58].

3.4.2 i_d–i_q Method

A variant of the formulation dq is called the i_d–i_q method [27]. This is based on the following consideration: the speed of rotation of the system $0dq$ about the axis 0 is the pulsation of the fundamental harmonic of the supply voltage; thus this synchronous rotating system rotates at the same speed as the voltage vector only in the case of balanced sinusoidal positive-sequence voltages. In general, the rotation speed of the system $0dq$ and voltage vector will be different. The method i_d–i_q places the axis d in the direction of the projection of the voltage vector on the $\alpha\beta$-plane. This ensures that the dq axes rotates in solidarity with the projection voltage vector, $u_{\alpha\beta}$, Figure 3.28.

The process is summarized as follows. The coordinate system 1–2–3 is transformed into the coordinate system 0, α, β by the Clarke–Concordia transformation (3.1)–(3.2), and then a transformation is performed, (3.123), to a coordinate system rotating at one speed $\omega(t)$ such that the angular position of the d axis relative to the α axis is given by θ,

$$
\begin{bmatrix} u_0 \\ u_d \\ u_q \end{bmatrix} = \begin{bmatrix} 1 & 0 & 0 \\ 0 & \cos\theta & \sin\theta \\ 0 & -\sin\theta & \cos\theta \end{bmatrix} \begin{bmatrix} u_0 \\ u_\alpha \\ u_\beta \end{bmatrix} \tag{3.123}
$$

Where now θ represents the angle between the projection of the voltage vector in the plane α, β to the α axis, that is,

$$
\theta = tg^{-1}\frac{u_\beta}{u_\alpha} \tag{3.124}
$$

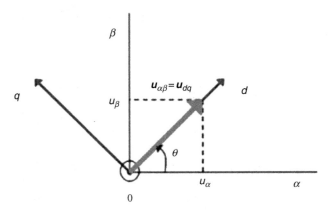

FIGURE 3.28 System of coordinates for method i_d–i_q; d axis is matched with $u_{\alpha\beta}$ vector.

thus:

$$\cos\theta = u_\alpha / u_{\alpha\beta} \tag{3.125}$$

$$\sin\theta = u_\beta / u_{\alpha\beta} \tag{3.126}$$

So, i_d–i_q method may be expressed as follows, (3.127):

$$\begin{bmatrix} i_0 \\ i_d \\ i_q \end{bmatrix} = \frac{1}{u_{\alpha\beta}} \begin{bmatrix} u_{\alpha\beta} & 0 & 0 \\ 0 & u_\alpha & -u_\beta \\ 0 & -u_\beta & u_\alpha \end{bmatrix} \begin{bmatrix} i_0 \\ i_\alpha \\ i_\beta \end{bmatrix} \tag{3.127}$$

With this transformation, the direct voltage component is

$$u_d = |u_{\alpha\beta}| = u_{\alpha\beta} = \sqrt{u_\alpha^2 + u_\beta^2} \tag{3.128}$$

and the q voltage component is always null, $u_q = 0$. Current components verify the next propositions:

The positive sequence of the first harmonic becomes a dc quantity. All the other higher order current harmonics including the negative-sequence fundamental component become variable along the time quantities and they receive a frequency displacement in the spectrum. So, they comprise the oscillatory current components.

From the viewpoint of the active compensation, a shunt active filter must compensate variables terms of current over time, both the dq current components and the zero-sequence current. The compensation current is in phase coordinates therefore,

$$\begin{bmatrix} i_{C1} \\ i_{C2} \\ i_{C3} \end{bmatrix} = \sqrt{\frac{2}{3}} \begin{bmatrix} \frac{1}{\sqrt{2}} & 1 & 0 \\ \frac{1}{\sqrt{2}} & -\frac{1}{2} & \frac{\sqrt{3}}{2} \\ \frac{1}{\sqrt{2}} & -\frac{1}{2} & -\frac{\sqrt{3}}{2} \end{bmatrix} \frac{1}{u_{\alpha\beta}} \begin{bmatrix} u_{\alpha\beta} & 0 & 0 \\ 0 & u_\alpha & -u_\beta \\ 0 & -u_\beta & u_\alpha \end{bmatrix} \begin{bmatrix} i_{L0} \\ \tilde{i}_{Ld} \\ \tilde{i}_{Lq} \end{bmatrix} \tag{3.129}$$

The compensation currents (3.129) can eliminate neutral currents; however, although the source currents present low distortion after compensation, they never achieve a complete elimination of distortion from a practical point of view, except in the case of balanced and sinusoidal voltages. Furthermore, one-unity displacement factor will not be achieved. This latter circumstance can be overcome by a slight modification to the compensation strategy (3.129). Indeed, under the method i_d–i_q, the instantaneous real power is

$$p(t) = \boldsymbol{u}_{0dq} \cdot \boldsymbol{i}_{0dq} = u_0\, i_0 + u_d\, i_d + u_q\, i_q = u_0\, i_0 + u_d\, i_d \qquad (3.130)$$

The load current component i_q does not participate in the transfer of instantaneous real power, making their total compensation possible. The compensation currents for a nonlinear load in the method i_d–i_q are given by (3.131),

$$\begin{bmatrix} i_{C0} \\ i_{C\alpha} \\ i_{C\beta} \end{bmatrix} = \frac{1}{u_{\alpha\beta}} \begin{bmatrix} u_{\alpha\beta} & 0 & 0 \\ 0 & u_\alpha & -u_\beta \\ 0 & -u_\beta & u_\alpha \end{bmatrix} \begin{bmatrix} i_{L0} \\ \tilde{i}_{Ld} \\ i_{Lq} \end{bmatrix} \qquad (3.131)$$

The cancellation of i_{Lq} allows us to obtain a unity displacement factor, as the average reactive power is eliminated. The advantage of i_d–i_q method over the dq method is that compensating currents are determined directly from supply voltages, therefore, independent of frequency, face difficulties arising from the use of PLLs in the control circuit of the compensator.

3.4.3 *p–q–r* Formulation

As has been established in a system with unbalanced voltages and distortion, the rotational speed of the voltage vector is not constant but is influenced by harmonic components and negative sequences thereof. In such cases, a formulation of the instantaneous reactive power in a rotating coordinate system attached to the voltage vector is presented, that is the *p–q–r* system. The resulting transformation of coordinates defines three independent power variables more simply (specifically an instantaneous real power and two instantaneous reactive power) that support its independent compensation [28,29].

In the *p–q–r* formulation a new coordinate system in which the p axis coincides with the direction of the voltage vector, the q-axis is perpendicular to p and is in the α–β plane is defined, and the r axis is perpendicular to the previous two. In this system, the current vector has three components and only one voltage vector. Thus, only a component of current vector, the component i_p is collinear with the voltage; the aim of compensation consists of compensate remaining components, components i_q and i_r, which do not carry real power. Two compensation strategies, a call reference power control strategy, which also compensates the part of i_p current that does not transport active power, and another strategy called reference current control compensation are distinguished in which it compensates the alternating part of the i_p current. In both cases, supply current carrying the active power, and collinear with the voltage will be achieved. Thus, if they were nonsinusoidal voltages, supply current obtained would not be sinusoidal.

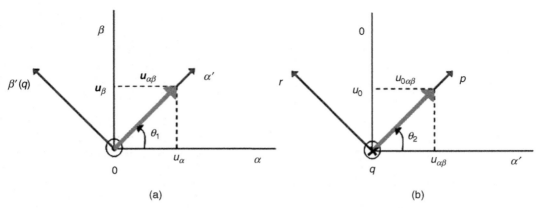

FIGURE 3.29 Construction *pqr* coordinates system. (a) Rotation of $\alpha\beta$ plane taking as axis of rotation zero axis, (b) rotation of $0\alpha'$ plane taking as the axis of rotation $\beta'\equiv q$-axis. The q-axis enters perpendicular to the *pr* plane.

In what follows, the new framework will be achieved by two successive rotational transformations. A new system of coordinates $0\alpha'\beta'$ will be obtained from the rotation of the $\alpha\beta$ plane through the axis 0 angle $\theta_1 = tg^{-1}(u_\beta/u_\alpha)$ allowing the α axis to be aligned with the projection of the vector of voltage in the $\alpha\beta$ plane (transformation of i_d–i_q method), Figure 3.29a shows that the transformation for the current is given by (3.132),

$$
\begin{bmatrix} i_0 \\ i_{\alpha'} \\ i_{\beta'} \end{bmatrix} = \begin{bmatrix} 1 & 0 & 0 \\ 0 & \cos\theta_1 & \sin\theta_1 \\ 0 & -\sin\theta_1 & \cos\theta_1 \end{bmatrix} \begin{bmatrix} i_0 \\ i_\alpha \\ i_\beta \end{bmatrix} = \frac{1}{u_{\alpha\beta}} \begin{bmatrix} u_{\alpha\beta} & 0 & 0 \\ 0 & u_\alpha & -u_\beta \\ 0 & -u_\beta & u_\alpha \end{bmatrix} \begin{bmatrix} i_0 \\ i_\alpha \\ i_\beta \end{bmatrix}
\tag{3.132}
$$

The *pqr* reference frame is formed by a rotation of the plane $0-\alpha'$ angle $\theta_2 = tg^{-1}(u_0/u_{\alpha\beta})$ against rotation β' axis such that the α' axis coincides with the direction vector voltage $\boldsymbol{u}_{0\alpha\beta}$, new p axis Figure 3.29b.

$$
\begin{bmatrix} i_p \\ i_q \\ i_r \end{bmatrix} = \begin{bmatrix} \sin\theta_2 & \cos\theta_2 & 0 \\ \cos\theta_2 & -\sin\theta_2 & 0 \\ 0 & 0_1 & 1 \end{bmatrix} \begin{bmatrix} i_0 \\ i_{\alpha'} \\ i_{\beta'} \end{bmatrix} = \frac{1}{u_{\alpha\beta}} \begin{bmatrix} \dfrac{u_0}{u_{0\alpha\beta}} & \dfrac{u_{\alpha\beta}}{u_{0\alpha\beta}} & 0 \\ \dfrac{u_{\alpha\beta}}{u_{0\alpha\beta}} & -\dfrac{u_0}{u_{0\alpha\beta}} & 0 \\ 0 & 0 & 1 \end{bmatrix} \begin{bmatrix} i_0 \\ i_{\alpha'} \\ i_{\beta'} \end{bmatrix}
\tag{3.133}
$$

The β' axis will be the new q-axis and the *zero* axis is the new r axis.

Finally, currents and voltages are translated from $0\alpha\beta$ coordinates to *pqr* coordinates by means of the next matrix [7, 8]:

$$
\begin{bmatrix} i_p \\ i_q \\ i_r \end{bmatrix} = \frac{1}{u_{0\alpha\beta}} \begin{bmatrix} u_0 & u_\alpha & u_\beta \\ 0 & -\dfrac{u_{0\alpha\beta}u_\beta}{v_{\alpha\beta}} & \dfrac{u_{0\alpha\beta}u_\alpha}{v_{\alpha\beta}} \\ u_{\alpha\beta} & -\dfrac{u_0 u_\alpha}{u_{\alpha\beta}} & -\dfrac{u_0 u_\beta}{u_{\alpha\beta}} \end{bmatrix} \begin{bmatrix} i_0 \\ i_\alpha \\ i_\beta \end{bmatrix}
\tag{3.134}
$$

$$\begin{bmatrix} u_p \\ u_q \\ u_r \end{bmatrix} = \begin{bmatrix} u_{0\alpha\beta} \\ 0 \\ 0 \end{bmatrix} \tag{3.135}$$

Equation (3.135) presents the voltage transformation. Now the voltage vector has only one component that is not null

$$u_p = u_{0\alpha\beta} = \sqrt{u_0^2 + u_\alpha^2 + u_\beta^2} \tag{3.136}$$

Once defined, the instantaneous real power $p(t)$ (dot product of voltage/current space vectors) and the instantaneous reactive power vector $\boldsymbol{q}(t)$ (cross product of voltage/current space vectors), three power variables are disposed: instantaneous real power $p(t)$ and two instantaneous reactive powers $q_r(t)$ and $q_q(t)$ (Figure 3.30).

Current components i_p, i_q, i_r respect to those three power variables are calculated as follows:

$$\begin{bmatrix} i_p \\ i_q \\ i_r \end{bmatrix} = \frac{1}{u_p} \begin{bmatrix} 1 & 0 & 0 \\ 0 & 1 & 0 \\ 0 & 0 & -1 \end{bmatrix} \begin{bmatrix} p \\ q_r \\ q_q \end{bmatrix} \tag{3.137}$$

The p–q–r formulation presents two control strategies: Reference Current Control and Reference Power Control [28]. Results are very similar in both cases. In the Reference Power Control, the compensation current equation is as follows:

$$\begin{bmatrix} i_{C0} \\ i_{C\alpha} \\ i_{C\beta} \end{bmatrix} = \frac{1}{u_{0\alpha\beta}} \begin{bmatrix} u_0 & 0 & u_{\alpha\beta} \\ u_\alpha & -\dfrac{u_{0\alpha\beta} u_\beta}{u_{\alpha\beta}} & -\dfrac{u_0 u_\alpha}{u_{\alpha\beta}} \\ u_\beta & \dfrac{u_{0\alpha\beta} u_\alpha}{u_{\alpha\beta}} & -\dfrac{u_0 u_\beta}{u_{\alpha\beta}} \end{bmatrix} \begin{bmatrix} \dfrac{\tilde{p}_L}{u_p} \\ \dfrac{q_{Lr}}{u_p} \\ \dfrac{-q_{Lq}}{u_p} + \dfrac{u_0}{u_{\alpha\beta}} \mathrm{val}\left(\dfrac{p_L}{u_p}\right) \end{bmatrix} \tag{3.138}$$

where $\mathrm{val}(p_L / u_p)$ means the numerical value of the ratio of the instantaneous real power load and voltage.

As can be seen in (3.138), the variable instantaneous real power, the whole component q of the instantaneous reactive power and the necessary part of the instantaneous reactive power component r to eliminate the neutral current are compensated.

The equation of Reference Current Control is as follows:

$$\begin{bmatrix} i_{cp} \\ i_{cq} \\ i_{cr} \end{bmatrix} = \begin{bmatrix} \tilde{i}_{Lp} \\ i_{Lq} \\ i_{Lr} - \dfrac{u_0}{u_{\alpha\beta}} \mathrm{val}(\tilde{i}_{Lp}) \end{bmatrix} \tag{3.139}$$

where $\mathrm{val}(\tilde{i}_{Lp})$ means the numeric value of the load current component p variable part.

Here, $i_{cq} = i_{Lq}$ to compensate its constant part as its variable part and to get a source current in phase to the voltage supplied.

3.5 Dual Instantaneous Reactive Power

In this section the theory of dual formulations of the instantaneous reactive power are developed. They are widespread in the literature, but have traditionally been applied to active power filters parallel connection: the original formulation and modified formulation. Hereinafter it is shown that these developments allow the use of such formulations in designing strategies compensation nonlinear loads of type Harmonic Voltage Source (HVS) using series active power filters.

So far, all formulations have been presented and applied to the case of active power filters in parallel connection. These formulations are not directly applicable to the case of series active filters. Now a dual formulation of the two formulations indicated will be developed and implemented to the compensation of the HVS type that constitutes the most appropriate load for the use of compensation equipment of series connection [4, 30].

3.5.1 Dual Original Instantaneous Reactive Power Formulation

The formulation of the instantaneous reactive power is applied since its inception in controlling compensation equipment in parallel connection. Thus, the components of the compensation currents are obtained from the voltages at $0\alpha\beta$ coordinates, the instantaneous real power, and the instantaneous imaginary power. However, when it comes to equipment of series compensation a problem statement is necessary from the point of dual view. Thus, the voltages are those to be determined from the $0,\alpha,\beta$ components of current, and power variables of load [4].

It was initially considered a three-wire three-phase system. Instantaneous real power within the $\alpha\beta$ coordinates is calculated by the expression (3.140),

$$p_{\alpha\beta} = u_\alpha \, i_\alpha + u_\beta \, i_\beta \tag{3.140}$$

This power can be written in vector form by the scalar product (3.141),

$$p_{\alpha\beta} = i_{\alpha\beta}^T \, u_{\alpha\beta} \tag{3.141}$$

where $i^T{}_{\alpha\beta}$ is the transposed vector of currents in α–β coordinates,

$$i_{\alpha\beta} = \left[\begin{array}{cc} i_\alpha & i_\beta \end{array} \right]^T \tag{3.142}$$

In like manner $u_{\alpha\beta}$ is the vector of voltages in the same coordinates

$$u_{\alpha\beta} = \left[\begin{array}{cc} u_\alpha & u_\beta \end{array} \right]^T \tag{3.143}$$

The instantaneous imaginary power is defined by the expression

$$q_{\alpha\beta} = u_\alpha \, i_\beta - u_\beta \, i_\alpha \tag{3.144}$$

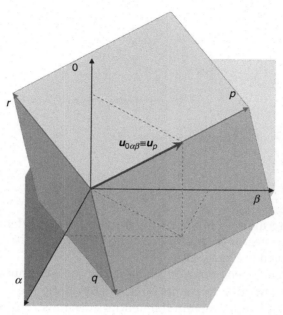

FIGURE 3.30 *Pqr* axes arrangement where the voltage vector is in the direction of the axis *p*.

Similar to the instantaneous real power, the instantaneous imaginary power can be expressed by the dot product (3.145),

$$q_{\alpha\beta} = i_{\alpha\beta\perp}^{T}\, u_{\alpha\beta} \tag{3.145}$$

where $i^{T}_{\alpha\beta\perp}$ is the normal or perpendicular vector defined by (3.146),

$$i_{\alpha\beta\perp} = \begin{bmatrix} i_{\beta} & -i_{\alpha} \end{bmatrix}^{T} \tag{3.146}$$

Thus, the instantaneous real power and instantaneous imaginary power can be expressed in matrix form by (3.147),

$$\begin{bmatrix} p_{\alpha\beta} \\ q_{\alpha\beta} \end{bmatrix} = \begin{bmatrix} i_{\alpha\beta}^{T} \\ i_{\alpha\beta\perp}^{T} \end{bmatrix} u_{\alpha\beta} \tag{3.147}$$

In view of the expression (3.147), it is possible to obtain the voltage in $\alpha\beta$ coordinates based on the instantaneous power as

$$u_{\alpha\beta} = \frac{1}{i_{\alpha}^{2} + i_{\beta}^{2}} \begin{bmatrix} i_{\alpha\beta}^{T} \\ i_{\alpha\beta\perp}^{T} \end{bmatrix} \begin{bmatrix} p_{\alpha\beta} \\ q_{\alpha\beta} \end{bmatrix} \tag{3.148}$$

or else

$$u_{\alpha\beta} = \frac{p}{i_{\alpha\beta}^{2}}\, i_{\alpha\beta} + \frac{q}{i_{\alpha\beta}^{2}}\, i_{\alpha\beta\perp} \tag{3.149}$$

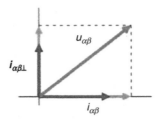

FIGURE 3.31 Split of voltage vector along the axes defined by $i_{\alpha\beta}$ e $i_{\alpha\beta\perp}$.

where

$$i_{\alpha\beta}^2 = i_\alpha^2 + i_\beta^2 \tag{3.150}$$

The analysis in $\alpha\beta$ terms of (3.149) allows two coordinate axes defined by the vectors $i_{\alpha\beta}$ e $i_{\alpha\beta\perp}$. Thus, the voltage vector is decomposed into its orthogonal axes defined by the two current vectors, as shown in Figure 3.31.

This formulation developed part of a voltage vector and a current vector to effect transformation of Clarke–Concordia to put its components along the α and β axes. This development is valid for systems wherein the neutral current is zero because the system is four-wire or three-wire balanced. Analogously to Section 3.2, to take into account the zero-sequence component, it is necessary to add a third coordinate (0 axis) perpendicular to the plane formed by the $\alpha\beta$ coordinates. In the new $0\alpha\beta$ reference system of interest, following vectors are defined: the vector of zero-sequence current i_0, which is the only non-zero component of the zero-sequence current, that is,

$$i_0 = \begin{bmatrix} i_0 & 0 & 0 \end{bmatrix}^T \tag{3.151}$$

Furthermore, in this situation $i_{\alpha\beta}$ and $i_{\alpha\beta\perp}$ include a zero-sequence component null, which are both defined in the form

$$i_{\alpha\beta} = \begin{bmatrix} 0 & i_\alpha & i_\beta \end{bmatrix}^T \tag{3.152}$$

and

$$i_{\alpha\beta\perp} = \begin{bmatrix} 0 & i_\beta & -i_\alpha \end{bmatrix}^T \tag{3.153}$$

Instantaneous real power is decomposed into two terms

$$p(t) = p_{\alpha\beta} + p_0 \tag{3.154}$$

where $p_{\alpha\beta}$ is the instantaneous real power defined by (3.140) and p_0 is the zero-sequence instantaneous power defined in the usual way,

$$p_0 = u_0\, i_0 \tag{3.155}$$

In matrix form the three power variables can be expressed by (3.156),

$$
\begin{bmatrix} p_0 \\ p_{\alpha\beta} \\ q_{\alpha\beta} \end{bmatrix} = \begin{bmatrix} i_0^T \\ i_{\alpha\beta}^T \\ i_{\alpha\beta\perp}^T \end{bmatrix} u_{0\alpha\beta} \tag{3.156}
$$

In (3.156) the matrix including the components of currents is given by

$$
\begin{bmatrix} i_0 & 0 & 0 \\ 0 & i_\alpha & i_\beta \\ 0 & i_\beta & -i_\alpha \end{bmatrix} \tag{3.157}
$$

It is a square matrix whose inverse is

$$
\frac{1}{i_0\left(i_\alpha^2+i_\beta^2\right)} \begin{bmatrix} \left(i_\alpha^2+i_\beta^2\right) & 0 & 0 \\ 0 & i_0\, i_\alpha & i_0\, i_\beta \\ 0 & i_0\, i_\beta & -i_0\, i_\alpha \end{bmatrix} \tag{3.158}
$$

Thus, in this situation you can obtain the vector of voltages $u_{0\alpha\beta}$ in the form given by (3.159),

$$
u_{0\alpha\beta} = \begin{bmatrix} u_0 \\ u_\alpha \\ u_\beta \end{bmatrix} = \begin{bmatrix} i_0^T/i_0^2 \\ i_{\alpha\beta}^T/i_{\alpha\beta}^2 \\ i_{\alpha\beta\perp}^T/i_{\alpha\beta}^2 \end{bmatrix} \begin{bmatrix} p_0 \\ p_{\alpha\beta} \\ q_{\alpha\beta} \end{bmatrix} \tag{3.159}
$$

or alternatively expressed

$$
u_{0\alpha\beta} = \frac{p_0}{i_0^2} i_0 + \frac{p_{\alpha\beta}}{i_{\alpha\beta}^2} i_{\alpha\beta} + \frac{q_{\alpha\beta}}{i_{\alpha\beta}^2} i_{\alpha\beta\perp} \tag{3.160}
$$

The vectors i_0, $i_{\alpha\beta}$, e $i_{\alpha\beta\perp}$ define three orthogonal coordinate axes. The expression (3.160) determines the components of the $u_{0\alpha\beta}$ voltage vector from the projections on these axes.

The different dual formulations allow instantaneous reactive power compensation to set different strategies. The decomposition of the power variables gets these terms of power with one or other compensation objectives. The dual original instantaneous reactive power formulation is applied to the series constant power compensation.

The objective of the constant power strategy is that the source supplies a constant instantaneous power. The instantaneous powers $p_0(t)$, $p_{\alpha\beta}(t)$, y $q_{\alpha\beta}(t)$ defined in (3.156), can be divided by a constant term and a variable term over time. The first is obtained as the integral average of the instantaneous value; the second term will have an average value of zero (on this we discussed in Subsection 3.2.3). Thus, the instantaneous power $p_{\alpha\beta}(t)$ can be expressed in the form

$$
p_{\alpha\beta}(t) = P_{\alpha\beta} + \tilde{p}_{\alpha\beta}(t) \tag{3.161}
$$

where $P_{\alpha\beta}$ is the average value and $\tilde{p}_{\alpha\beta}(t)$ is the variable term.

The zero-sequence instantaneous power may also decompose in a similar way, that is,

$$p_0(t) = P_0 + \tilde{p}_0(t) \tag{3.162}$$

where, P_0 is the constant term and $\tilde{p}_0(t)$ is the variable term.

Similarly, the instantaneous imaginary power is decomposed into a constant term $Q_{\alpha\beta}$ and other variable $\tilde{q}_{\alpha\beta}(t)$, that is,

$$q_{\alpha\beta}(t) = Q_{\alpha\beta} + \tilde{q}_{\alpha\beta}(t) \tag{3.163}$$

Thus, for an arbitrary load their powers can be expressed in the form (3.164),

$$\begin{aligned}
p_{L0}(t) &= P_{L0} + \tilde{p}_{L0}(t) \\
p_{L\alpha\beta}(t) &= P_{L\alpha\beta} + \tilde{p}_{L\alpha\beta}(t) \\
q_{L\alpha\beta}(t) &= Q_{L\alpha\beta} + \tilde{q}_{L\alpha\beta}(t)
\end{aligned} \tag{3.164}$$

The compensation objective is to set the system so as to supply a constant instantaneous real power and $q_{\alpha\beta}(t) = 0$. Therefore, references to voltage and current shown in Figure 3.32, the source power will be

$$p_S(t) = -P_{L\alpha\beta} - P_{L0} \tag{3.165}$$

So that the series compensation equipment must supply the rest of the powers of (3.164), that is,

$$\begin{aligned}
p_{C0}(t) &= -\tilde{p}_{L0}(t) \\
p_{C\alpha\beta}(t) &= -\tilde{p}_{L\alpha\beta}(t) \\
q_{CL\alpha\beta}(t) &= -Q_{L\alpha\beta} - \tilde{q}_{L\alpha\beta}(t) = -q_{L\alpha\beta}(t)
\end{aligned} \tag{3.166}$$

In matrix form, (3.166) is

$$\begin{bmatrix} p_{C0}(t) \\ p_{C\alpha\beta}(t) \\ q_{C\alpha\beta}(t) \end{bmatrix} = \begin{bmatrix} -\tilde{p}_{L0}(t) \\ -\tilde{p}_{L\alpha\beta}(t) \\ -q_{L\alpha\beta}(t) \end{bmatrix} \tag{3.167}$$

In this, expression (3.159) is applied, allowing determination of the vector of voltages, $u_{C0\alpha\beta}$ the compensator must generate in $0\alpha\beta$ coordinates, that is,

$$u_{C0\alpha\beta} = \begin{bmatrix} u_{C0} \\ u_{C\alpha} \\ u_{C\beta} \end{bmatrix} = \begin{bmatrix} i_0^T / i_0^2 \\ i_{\alpha\beta}^T / i_{\alpha\beta}^2 \\ i_{\alpha\beta\perp}^T / i_{\alpha\beta}^2 \end{bmatrix} \begin{bmatrix} -\tilde{p}_{L0}(t) \\ -\tilde{p}_{L\alpha\beta}(t) \\ -q_{L\alpha\beta}(t) \end{bmatrix} \tag{3.168}$$

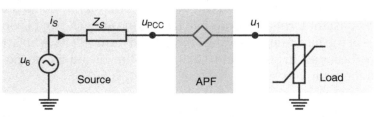

FIGURE 3.32 Single-phase equivalent circuit for series compensation where voltage and current references are indicated.

From (3.30) vector components of voltages in phase coordinates are obtained. It is applied to become (3.169),

$$
\boldsymbol{u}_C = \begin{bmatrix} u_{C1} \\ u_{C2} \\ u_{C3} \end{bmatrix} = \sqrt{\frac{2}{3}} \begin{bmatrix} \dfrac{1}{\sqrt{2}} & 1 & 0 \\ \dfrac{1}{\sqrt{2}} & -\dfrac{1}{2} & \dfrac{\sqrt{3}}{2} \\ \dfrac{1}{\sqrt{2}} & -\dfrac{1}{2} & -\dfrac{\sqrt{3}}{2} \end{bmatrix} \begin{bmatrix} u_{C0} \\ u_{C\alpha} \\ u_{C\beta} \end{bmatrix}
\tag{3.169}
$$

3.5.2 Dual Modified Instantaneous Reactive Power Formulation

The modified formulation begins with the 123 phase coordinate transformation to $0\alpha\beta$ coordinates given by (3.1)–(3.2). Thus voltage vector ($\boldsymbol{u}$) and current ($\boldsymbol{i}$), and power variables defined in (3.79) and (3.88) are obtained. However, this section will have for convenience instantaneous reactive power in the form given in (3.170),

$$
\begin{bmatrix} q_0 \\ q_\alpha \\ q_\beta \end{bmatrix} = \begin{bmatrix} 0 & i_\beta & -i_\alpha \\ -i_\beta & 0 & i_0 \\ i_\alpha & -i_0 & 0 \end{bmatrix} \begin{bmatrix} u_0 \\ u_\alpha \\ u_\beta \end{bmatrix}
\tag{3.170}
$$

To obtain the components of the voltage as a function of the four variables of power (instantaneous real power and three instantaneous reactive power) a dual reasoning set out in Section 3.3 shall be followed.

The square matrix (3.170) is of rank 2, whereby a row is a linear combination of the other two. Thus, by multiplying q_0 for i_0, q_α for i_α, and q_β for i_β is obtained (3.171),

$$
\begin{bmatrix} i_0 q_0 \\ i_\alpha q_\alpha \\ i_\beta q_\beta \end{bmatrix} = \begin{bmatrix} 0 & i_0 i_\beta & -i_0 i_\alpha \\ -i_\alpha i_\beta & 0 & i_0 i_\alpha \\ i_\alpha i_\beta & -i_0 i_\beta & 0 \end{bmatrix} \begin{bmatrix} v_0 \\ v_\alpha \\ v_\beta \end{bmatrix}
\tag{3.171}
$$

From (3.171) on verification

$$
0 = i_0 q_0 + i_\alpha q_\alpha + i_\beta q_\beta
\tag{3.172}
$$

The different powers introduced here can be grouped in matrix form, this is

$$
\begin{bmatrix} p \\ q_0 \\ q_\alpha \\ q_\beta \end{bmatrix} = \begin{bmatrix} i_0 & i_\alpha & i_\beta \\ 0 & i_\beta & -i_\alpha \\ -i_\beta & 0 & i_0 \\ i_\alpha & -i_0 & 0 \end{bmatrix} \begin{bmatrix} u_0 \\ u_\alpha \\ u_\beta \end{bmatrix}
\tag{3.173}
$$

From (3.173) it can always find the inverse transformation of dual form as proposed in Section 3.3.1; so you get to the expression (3.174),

$$
\begin{bmatrix} u_0 \\ u_\alpha \\ u_\beta \end{bmatrix} = \frac{1}{i_{0\alpha\beta}^2} \begin{bmatrix} i_0 & 0 & -i_\beta & i_\alpha \\ i_\alpha & i_\beta & 0 & -i_0 \\ i_\beta & -i_\alpha & i_0 & 0 \end{bmatrix} \begin{bmatrix} p \\ q_0 \\ q_\alpha \\ q_\beta \end{bmatrix}
\tag{3.174}
$$

where $i_{0\alpha\beta}^2 = i_0^2 + i_\alpha^2 + i_\beta^2$. The relation (3.174) gives the vector voltages depending on the variables of power, with the only restriction that is verified $i_{0\alpha\beta}^2 \neq 0$.

Then the dual modified formulation for nonlinear loads compensation by series active filter is applied; namely the constant power strategy will be discussed.

According to the formulation of modified instantaneous reactive power, for any load it is possible to define an instantaneous power (p_L) and a vector of instantaneous reactive power ($\boldsymbol{q_L}$), whose components are q_{L0}, $q_{L\alpha}$, $q_{L\beta}$. As in the original formulation, instantaneous active power can be divided into a constant term (P_L) given by the average value of p_L and a variable term $\tilde{p}_L$, that is

$$
p_L = P_L + \tilde{p}_L
\tag{3.175}
$$

Compensation objectives to constant power are established:

- The source instantaneous reactive power vector must be zero, that is $\boldsymbol{q_S} = 0$. Therefore for this goal to be met, the instantaneous reactive power vector ($\boldsymbol{q_C}$) the compensator should be

$$
\boldsymbol{q_C} = -\boldsymbol{q_L}
\tag{3.176}
$$

that is, the compensator must transfer the instantaneous reactive power of the load.

- The source must transfer a constant instantaneous active power (p_S). Considering that the average power transferred by the compensator is zero, the following applies

$$
p_S = -P_L
\tag{3.177}
$$

then attending to this criterion, the compensation equipment must provide an instantaneous power (p_C) given by (3.177),

$$
p_C = -\tilde{p}_L
\tag{3.178}
$$

whose average value is zero.

Accordingly, the compensator transfer power space vector (3.179),

$$
\begin{bmatrix} p_C \\ q_{C_0} \\ q_{C_\alpha} \\ q_{C_\beta} \end{bmatrix} = \begin{bmatrix} -\tilde{p}_L \\ -q_{L_0} \\ -q_{L_\alpha} \\ -q_{L_\beta} \end{bmatrix}
\tag{3.179}
$$

After obtaining the components of current vector in 0–α–β coordinates, the vector of voltages that must generate the compensation equipment in the same coordinates is given by (3.180),

$$
\begin{bmatrix} u_{C0} \\ u_{C\alpha} \\ u_{C\beta} \end{bmatrix} = \frac{1}{i_{0\alpha\beta}^2} \begin{bmatrix} i_0 & 0 & -i_\beta & i_\alpha \\ i_\alpha & i_\beta & 0 & -i_0 \\ i_\beta & -i_\alpha & i_0 & 0 \end{bmatrix} \begin{bmatrix} -\tilde{p}_L \\ -q_{L_0} \\ -q_{L_\alpha} \\ -q_{L_\beta} \end{bmatrix}
\tag{3.180}
$$

The compensation voltages in phase coordinates are obtained from the application of the inverse coordinates transformation (3.180).

This strategy is now applied to a practical case.

EXAMPLE 3.8

Figure 3.33 shows a nonlinear load compensated by a series active filter, which is considered ideal for the purposes of this exercise. The active filter applies the strategy of constant power control based on the formulation of the dual instantaneous reactive power. This strategy is applied to a load of HVS type, that is, an uncontrolled three-phase rectifier with a 2200 μF capacitor in parallel with a 50/3 Ω resistor in the dc side. This is a three-wire nonlinear balanced load. As in previous examples, the system is simulated in Matlab–Simulink environment using Simulink SymPowerSystem toolbox.

The block diagram for calculating the compensation reference voltage is shown in Figure 3.34. The voltage vector components in the load side and current of source in coordinates are obtained 0$\alpha\beta$. The matrix of currents is constructed and multiplied by the vector of voltages $\boldsymbol{u}_{0\alpha\beta}$. The result of this product provides the vector of instantaneous reactive power and instantaneous real power. The latter is decomposed into a variable term and constant term through a low-pass filter (LPF). The vector of compensation powers is formed and multiplied by the matrix currents defined in (3.174), which compensates the voltages vector in 0$\alpha\beta$ coordinates are obtained as defined in (3.180). The transformation to phase coordinates enables the components of the compensation voltage to be obtained.

When a series active filter is connected (set at $t = 0.08$ s) to a load with this compensation strategy, the waveforms of the voltage at the point of common coupling, PCC, before and after compensation shown in Figure 3.35a are obtained. Figure 3.35b shows the voltage of phase 1 in the PCC and the load current before and after compensation. The waveforms presented correspond to the sums of compensation voltage reference and load voltage considering an ideal active filter; in Chapter 5 the results of the operation with series active power filter are analyzed.

Finally, Figure 3.36 shows the waveforms of power variables in the PCC. Figure 3.36a shows the waveform of the instantaneous real power before and after compensation; the instantaneous power is constant after compensation. Figure 3.36b shows the instantaneous reactive power (3.108) before and after compensation; instantaneous reactive power associated with each coordinate axis is zero after compensation. Figure 3.36 verifies the compensation objectives originally specified.

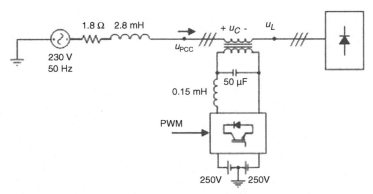

FIGURE 3.33 Single-phase equivalent circuit for a load of HVS type compensated by a series active power filter for Example 3.8.

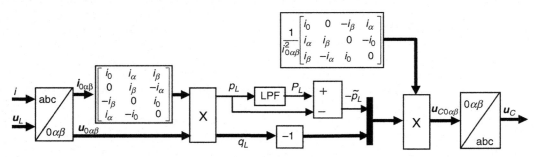

FIGURE 3.34 Block diagram for calculating the voltage reference constant power compensation for a series active filter when the formulation of the dual modified instantaneous reactive power is applied.

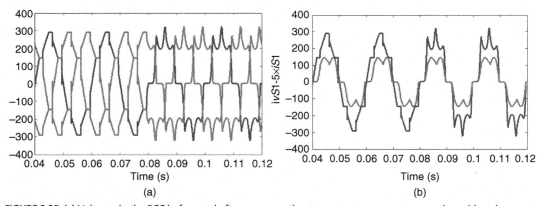

FIGURE 3.35 (a) Voltages in the PCC before and after compensation to constant power compensation with series active filter, (b) voltage of phase 1 in the PCC and current of phase 1 (multiplied by 5) for a HVS load before and after compensation.

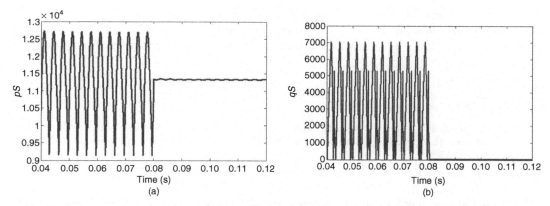

FIGURE 3.36 (a) Instantaneous real power in the PCC before and after constant power compensation with series active filter, (b) instantaneous reactive power at the PCC before and after constant power compensation.

3.6 Summary

This chapter has described the theory of instantaneous reactive power for three/four-wire three-phase systems in the most general terms of asymmetry and distortion. The theory of instantaneous reactive power today corresponds to a set of formulations with the same common base, that is, the partition of current in components of only instantaneous real power transfer, and one or more components that do not transfer instantaneous real power. Thus began stating the basis for the formulation of Akagi–Nabae [1,2], whose original formulation was designated as the first in the world to introduce the concept of instantaneous reactive power in three-phase systems. He followed this with an analysis of the formulation's greatest impact, which was subsequently published as Refs [6–30]. In particular, a comprehensive analysis of the so-called modified instantaneous reactive power formulation, and formulations that require a rotating framework of reference, dq formulation, i_d–i_q method, and pqr formulation was made. From the presentation made in this chapter, one can deduce that the formulation of modified instantaneous reactive power or using rotating reference frames from the transformation of Park, essentially have the same pattern as the original formulation: a mathematical transformation that enables finding three independent power variables from which it is possible to explain the flow of supply source and load. Over the years there have been other formulations that have had less impact and can be found in references [31–33]. It has also paid special attention to the developments of the original and modified formulations in phase coordinates allowing a description of the theory and its subsequent application in load compensation more simply.

The formulation of the instantaneous reactive power arose from finding compensating control strategy-based power converters (active power filters), so this chapter has presented the basis for obtaining the currents of reference in designing a shunt APLC with and without energy storage elements. In that order of things, the theory of instantaneous reactive power can also be applied in the design of series active compensators control.

One section was dedicated to the dual formulation of the original instantaneous reactive power and the modified formulation and its application in load compensation of HVS type using series active filter. Several application examples have been included throughout the chapter.

References

[1] Akagi H, Kanazawa Y, Nabae A. Generalized theory of the instantaneous reactive power in three-phase circuits. In: Proceedings of the IEEJ International Power Electronics Conference (IPEC-Tokio); 1983. p. 1375–1386.

[2] Akagi H, Kanazawa Y, Nabae A. Instantaneous reactive power compensators comprising switching devices without energy storage components. IEEE Trans Ind Appl 1984;IA-20(3):625–30.

[3] Akagi H, Nabae A, Atoh S. Control strategy of active power filters using multiple voltage source PWM converters. IEEE Trans Ind Appl 1986;IA-22(3):460–5.

[4] Aredes M, Watanabe EH. New control algorithms for series and shunt three-phase four-wire active power filters. IEEE Trans Power Deliv 1995;10(3):1649–55.

[5] Cavallini A, Montanari GC. Compensation strategies for shunt active filter control. IEEE Trans Power Electr 1994;9(6):587–93.

[6] Furuhashi T, Okuma S, Uchikawa Y. A study on the theory of instantaneous reactive power. IEEE Trans Ind Electr 1990;37(1):86–90.

[7] Takahashi I. Analysis of instantaneous current and power using space switching functions. In: Conference Records of IEEE/PESC; April 1988. p. 42–49.

[8] Ferrero A, Superti-Furga G. A new approach to the definition of power components in three-phase systems under nonsinusoidal conditions. IEEE Trans Instrum Meas 1991;40(3):568–77.

[9] Rossetto L, Tenti P. Using AC-fed PWM converters as instantaneous reactive power compensators. IEEE Trans Power Electr 1992;7(1):224–9.

[10] Willems JL. A new interpretation of the Akagi–Nabae power components for nonsinusoidal three-phase situations. IEEE Trans Instrum Meas 1992;41(4):523–27.

[11] Nabae A, Tanaka T. A new definition of instantaneous active-reactive current and power based on instantaneous space vectors on polar coordinates in three-phase circuits. IEEE Trans Power Deliv 1996;11:1238–43.

[12] Salmerón P, Montaño JC. Instantaneous power components in polyphase systems under nonsinusoidal conditions. IEE Proc Sci Meas Technol 1996;143:151–5.

[13] Montaño JC, Salmerón P. Instantaneous and full compensation in three-phase systems. IEEE Trans Power Deliv 1998;13:1342–7.

[14] Montaño JC, Salmerón P. Identification of instantaneous current components in three-phase systems. IEE Proceedings of Science, Measurement and Technologies; October 1999. p. 227–233.

[15] Cristaldi L, Ferrero A. Mathematical foundations of the instantaneous power concepts: an algebraic approach. Eur Trans Elec Power 1996;6(5):305–9.

[16] Nabae A, Cao L, Tanaka T. A universal theory of instantaneous active/reactive current and power including zero sequence component. Proceedings of the seventh ICHQP; October 1996. p. 90–95.

[17] Salmerón P, Herrera RS, Vázquez JR. Mapping matrices against vectorial frame in the instantaneous reactive power compensation. IET Elec Power Appl 2007;1(5):727–36.

[18] Herrera RS, Salmerón P, Kim H. Instantaneous reactive power theory applied to active power filter compensation: different approaches, assessment, and experimental results. IEEE Trans Ind Electr 2008;55(1):184–96.

[19] Herrera RS, Salmerón P. Instantaneous reactive power theory: a reference in the nonlinear loads compensation. IEEE Trans Ind Electr 2009;56:2015–22.

[20] Salmerón P, Herrera RS. Instantaneous reactive power theory – a general approach to polyphase systems. Elec Power Syst Res 2009;79:1263–70.

[21] Herrera RS, Salmerón P. A present point of view about the instantaneous reactive power theory. IET Elec Power Appl 2009;2(4):25–31.

[22] Nabae A, Nakano H, Togasawa S. An instantaneous distortion current compensator without any coordinate transformation. Proceedings of IEEJ, International Power Electronics Conference (IPEC-Tokyo); 1995. p. 1651–1655.

[23] Peng FZ, Lai JS. Harmonic and reactive power compensation based on the generalized instantaneous reactive power theory for three-phase four-wire systems. IEEE Trans Power Electr 1998;13:1174–81.

[24] Kim H, Akagi H. The instantaneous power theory based on mapping matrices in three-phase four-wire systems, Proceedings of IEEE/IEEJ Power Conversion Conf. (Nagaoka); 1997. p. 361–66.

[25] Akagi H, Ogasawara S, Kim H. The theory of instantaneous in three-phase four-wire systems: a comprehensive approach, Presented at IEEE IAS Annual Meeting; 1999. p. 431–39.

[26] Bhattacharya S, Divan DM, Banerjee B. Synchronous frame harmonic isolator using active series filter. European Power Electronic, Firenze; 1991. p. 3030–3035.

[27] Soares V, Verdelho P. An instantaneous active and reactive current component method for active filters. IEEE Trans Power Electr 2000;15(4):660–9.

[28] Kim H, Blaabberg F, Bak-Jensen B, Choi J. Instantaneous power compensation in three-phase systems by using p–q–r theory. IEEE Trans Power Electr 2002;17(5):701–10.

[29] Kim H, Blaabjerg F, Bak-Jensen B. Spectral analysis of instantaneous powers in single-phase and three-phase systems UIT use of p–q–r theory. IEEE Trans Power Electr 2002;17(5):711–20.

[30] Kim YS, Kim JS, Ko SH. Three-phase three-wire series active power filter, which compensates for harmonics and reactive power. IEEE Proc Elec Power Appl 2004;151(3):276–82.

[31] Depenbrock M. The FBD method, a generally applicable tool for analyzing power relations. IEEE Trans Power Syst 1993;8(2):381–87.

[32] Gengyin L, Zhiye Ch, Xinwel W, Yihan Y. An advanced active power filter based on generalized instantaneous reactive power theory in d–q–0 coordinates. ICEE'96 Proceedings of the International Conference on Electrical Engineering. vol. 1, Beijing, China; 1996.

[33] Komatsu Y, Kawabata T. Control of three phase active power filter using the extension *pq* theory PEMC'98, Proc Praga Rep Checa; 1998; 5: p. 66–71.

4

Shunt Active Power Filters

CHAPTER OUTLINE

Electrical energy is now considered as a product and, as such, it is subject to a number of rules for evaluating quality. As it has been established in Chapter 1, the electric power quality (EPQ) is associated with the whole set of changes or supply disturbances of electrical energy that can result in malfunction or damage to the equipment that depends on it. Specifically, in electronic loads, unlike conventional loads, the power conversion control is made by successive cuts of the voltage and current waves. Thus, distorted waveforms appear to impair the quality of electric power. The resulting "electrical pollution" may be produced by a high power nonlinear load or by the accumulation of many low power loads; it spreads throughout the electrical distribution system in the form of voltage and current harmonics. Several solutions have been proposed in engineering practice to eliminate current harmonics.

Progress in the power capacity and switching speed of electronic devices, such as IGBTs and GTOs, and the availability of new DSP techniques has made possible the use of active power filters (APFs). The APF concept is to use a power electronic converter to produce harmonics that cancel the load current harmonics, and in general, all the undesired components of the load current. Easy to assemble, an APF can be installed at any point in a low voltage ac system in order to compensate for one or more loads, thereby preventing

P. Salmerón Revuelta, S.P. Litrán, and J.P. Thomas: Active Power Line Conditioners. http://dx.doi.org/10.1016/B978-0-12-803216-9.00004-3

current harmonic flow through the system. The development and subsequent implementation of APFs has made it possible to compensate reactive power and unbalanced currents, in addition to current harmonics, which results in a more generic term to designate them: active power line conditioners (APLC). This chapter focuses on the study of APFs or APLCs of parallel or shunt connection.

4.1 Introduction

The current need for the compensation of nonlinear loads has motivated the emergence of a large number of works on static compensation based on power electronic converters in the last 40 years. More specifically, from the 1980s, it has been developed especially for use in the so-called APFs for compensation of nonlinear three-phase loads, and today, it has a usual application as a compensation system to avoid many of the problems arising from the lack of quality of electrical supply [1–4].

An APF of parallel connection is a power electronic circuit (power inverter) that is connected in parallel with the load, acting as a controlled current source to inject a current in opposite phase to eliminate the harmonics, the current imbalance, and the reactive component of the load current. To date, there have been many proposals for control strategies, and especially for the method used for obtaining the reference waveforms. Thus, two strategies have been considered mainly for parallel active filters: sinusoidal compensation and unit power factor compensation.

This chapter provides a tour through the operating principles, the power circuit structure, and control elements of the APFs as correction equipment for the lack of electric power quality. Thus, it starts with the operation principle of APFs and how they enable an increase in the load power factor and mitigation of source current harmonics. A complete description of the pulse width modulation (PWM) power inverters as a central element of the power circuit of the APFs is provided. Its operation as part of the APF is highlighted. Another essential element in the operation of the APF is the control strategy; instant compensation strategies, unit power factor compensation, and balanced and sinusoidal current compensation are described in this chapter. Different simulation examples help to complement this material. Finally, two important practical issues have been included. Some practical criteria for choosing the passive components of a final design and implementation of an experimental prototype to validate the contents exposed.

4.2 Fundamentals of Shunt Active Power Filters, APFs

The performance of an APF is based on three basic design criteria:

1. The characteristics of the power converter: topology of the circuit, electronic device type, etc.
2. The method used to obtain the reference current.
3. The method of PWM control used.

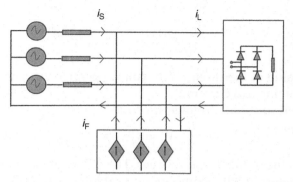

FIGURE 4.1 Schematic of a system with an ideal compensator (APF) modeled with current sources.

The power circuit of parallel connection APFs used here is the current-controlled voltage source inverter, CC-VSI. An APF in which it has been decided to use a CC-VSI performs two major tasks: to generate the reference signal and obtain the appropriate switching pattern, in order to follow the reference as closely as possible. However, the problem of the generation of the reference (criterion 2) is not completely uncoupled from the tracking method (criterion 3); which is where the problem lies. Sections 4.3 and 4.4 will describe in detail each of the above criteria and several examples will be provided to illustrate its operation. First, the basic principles governing the compensation parallel active filters will be introduced in this section.

A parallel (shunt) APF acts as a controlled current source that, when connected in parallel with a nonlinear unbalanced load or time-variant, is capable of injecting (out of phase) the unwanted current components that are present in the load current. In principle this action enables the supply source current to take any waveform of interest to us, for example, a current free of unbalance and distortion.

Figure 4.1 shows the basic principle of an ideal shunt APF, where the APF is modeled using three ideal controlled current sources.

In Figure 4.1 a nonlinear unbalanced load supplied from a power source of balanced sinusoidal voltage is shown. The load current, i_L, will present certain harmonics, unbalance between phases, and a displacement power factor lower than unity. That is, the load currents in each phase include a harmonic component and the fundamental component presents a given phase difference with respect to the voltage. Subsequently a zero-sequence component may circulate through the neutral wire.

The shunt APF in its role of compensator supplies a compensation current i_F, which must satisfy the current Kirchhoff law, CKL, in the Point of Common Coupling (PCC).

$$i_S = i_L - i_F \tag{4.1}$$

At this point we must say that we aim for a source current that is in phase with the voltage, that is, the set APF + load behave with unity power factor at the source. If we demand that the compensator does not provide active power, the source current must transfer all the active power consumed by the load, P_L. Thus, the source current will be,

$$i_{S1} = \frac{P_L}{V^2}u_1; \quad i_{S2} = \frac{P_L}{V^2}u_2; \quad i_{S3} = \frac{P_L}{V^2}u_3 \tag{4.2}$$

where,

$$V^2 = \frac{1}{T}\int_0^T (v_1^2 + v_2^2 + v_3^2)\,\mathrm{d}t \tag{4.3}$$

is the squared rms value of the three-phase voltage. Source currents (4.2) are the active current component of the load current.

If (4.3) is a balanced sinusoidal voltage, it verifies

$$V^2 = 3V_1^2 \tag{4.4}$$

where V_1 is the rms value of the phase voltages. The difference between the load currents and active currents determine the components that are not carrying useful power to the load and therefore those currents that are compensable,

$$i_{F1} = i_{L1} - i_{S1}; \quad i_{F2} = i_{L2} - i_{S2}; \quad i_{F3} = i_{L3} - i_{S3} \tag{4.5}$$

Example 4.1 will help to clarify the situation.

EXAMPLE 4.1

A set of balanced three-phase voltages given by the following waveforms are considered,

$$v_1 = \sqrt{2}\,230\cos(\omega t); \quad v_2 = \sqrt{2}\,230\cos\left(\omega t - \frac{2\pi}{3}\right); \quad v_3 = \sqrt{2}\,230\cos\left(\omega t + \frac{2\pi}{3}\right) \tag{4.6}$$

with $\omega = 100\pi$. They are applied to a load consisting of three single-phase rectifiers with a RL branch in the dc side and connected in a star with parameters, $R_a = 10\ \Omega$, $R_b = 8\ \Omega$, $R_c = 12\ \Omega$, $L = 100$ mH. Figure 4.2 shows the configuration of the load in Matlab SimPowerSystems model. The load imbalance is achieved with the different resistance values of the RL branch on the dc side of each rectifier.

The voltages v_1, v_2, and v_3 are applied to the PCC from an ideal power supply, that is, we assume a stiff supply and therefore the load currents will not disturb the supply voltages. The compensator is connected in parallel with the load (shunt APF), eight cycles after the start of the simulation to show the operation and dynamics of the system.

In Figure 4.3a the source currents in the three phases are shown. They present distorted and unbalanced waveforms before compensation. By contrast, the current waveforms are sinusoidal and balanced after compensation. Figure 4.3b shows the voltage of phase 1 with the current of the same phase; current is in phase with the voltage and therefore the displacement factor is unity. The set load + compensator has unity power factor.

Figure 4.3c shows the neutral current, and it is observed that after the activation of the compensator, the neutral current in the side of the power source is cancelled. That is, a four-wire topology as that used for the compensator of Figure 4.1 allows the elimination of zero-sequence currents flowing through the neutral. Finally, Figure 4.3d shows the instantaneous power view from the PCC. It is observed that after compensation, the instantaneous power is constant and equal to the active power on the load.

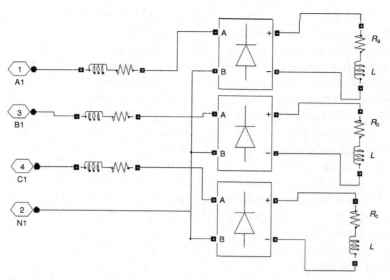

FIGURE 4.2 Nonlinear, unbalanced three-phase load for Example 4.1.

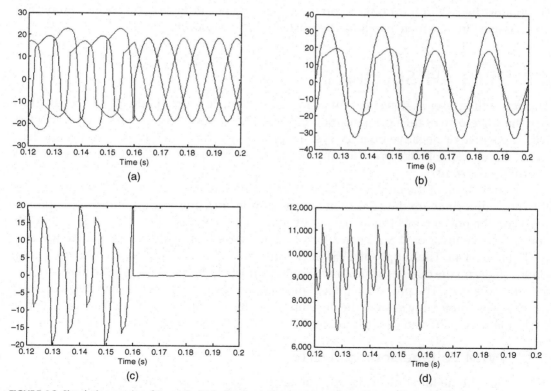

FIGURE 4.3 Simulation results of Example 4.1 before and after compensation at t = 0.16 s. (a) Source currents. (b) Voltage and current of phase 1. (c) Neutral source current. (d) Instantaneous source power.

In Example 4.1, the compensation currents were obtained through three ideal current sources, but in practice these currents are generated from a power electronic converter. This, in addition to the injection of high-frequency harmonics due to the switching mode operation and the increase of the system losses, will present a dynamic that will result in a delay in the injection of current compensation; compensation currents do not apply instantaneously as in Figure 4.3. Thus, the currents (4.5) are reference currents for controlling the real APF, which injects compensation currents.

Regardless of the particular example that has been used in Figure 4.1, there are various possible strategies to solve the compensation problem [5]. These are of particular interest: sinusoidal compensation, unit power factor compensation (considered in this section), and the approach of constant power. Using the first strategy, the line current is sinusoidal and in phase with the balanced positive-sequence voltage at the fundamental frequency. The second approach makes the waveform of the line current equal in shape and phase with the voltage waveform. Finally, the constant power compensation can be applied in two ways. Compensation under the approach of maintaining the same instantaneous power is the cancellation of the instantaneous imaginary power. And the compensation of constant average power in the line preserving the mean value of the instantaneous power. In all cases, the active filter, for harmonics, decouples the load from the power supply: this eliminates the risk of harmonic resonances, and the power factor correction equipment and passive filters, which are within the distribution system, are protected.

4.3 Shunt APF Structure

The active filter uses as power circuit an inverter with a dc source in the dc side, to synthesize the waveform of the compensation current. The dc source is alternatively connected with a positive or negative polarity. A static power inverter or dc-ac converter is an electronic circuit capable of converting the energy in the form of direct current to the form of alternating current.

This operation of the circuit has allowed its use in many industrial applications; among the most important are the control of electric motors and uninterruptible power supplies. Based on the power converter, two types of active filters can be chosen: an active filter with voltage source inverter (VSI), Figure 4.4b, and an active filter with current source inverter (CSI), Figure 4.4a. The supply of a VSI filter consists of a capacitor, which maintains a constant voltage on the dc side, while an active filter with an inductor CSI has high inertia for the current variations. The power circuit and devices used by both types of converters are similar. VSIs were preferably used in the development of active filters from the 70s. Thus, higher performance is achieved with a lower initial cost of APFs with VSI versus APFs with CSI.

4.3.1 PWM Voltage Source Inverter

The basic scheme for converting a dc voltage into a single-phase ac voltage, without presupposing a particular waveform, is presented in Figure 4.5. This is a single-phase full

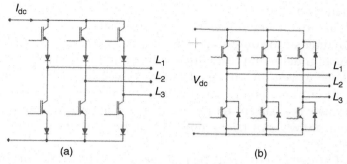

FIGURE 4.4 (a) Configuration of a current source inverter, CSI. (b) Configuration of a voltage source inverter, VSI.

wave inverter or a full bridge inverter, that is, a circuit that includes the four power elements described later in the chapter, and for a preliminary analysis will be assumed as ideal switches.

Sequential connection and disconnection of each pair of opposite elements of the bridge provides a rectangular ac voltage at the output. Figure 4.6 shows various options for the waveforms of the output voltage. The signal frequency can be varied by varying the switching sequence of the switches. Furthermore, the amplitude may be regulated either by changing the voltage level of the dc side, or by modulating the pulse width. The latter method is more practical, and therefore usually adopted, it can be performed by constant PWM or sinusoidal pulse width within a period (PWM), Figure 4.6. In the latter case, a triangular

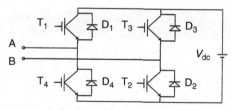

FIGURE 4.5 Single-phase inverter (dc-ac converter).

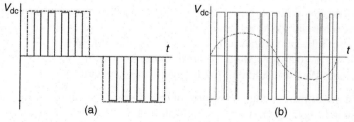

FIGURE 4.6 (a) Modulation of several equal width pulses per period. (b) Modulation width proportional to a sine wave alternating positive and negative pulses.

wave (carrier signal) is compared directly with a sinusoidal modulation wave (or modulating control signal) in order to determine the switching instants [6].

The voltage obtained at the output of the circuit of Figure 4.5 is ac but not sinusoidal. Voltage consists of a fundamental component and a number of harmonics. If they are attenuated, the voltage waveform will get closer and closer to the sine wave. By contrast, the waveform of the current depends on the load connected to its output (ac side). If the load is purely resistive, the current reproduces the voltage waveform. If instead the load is inductive, the current waveform is smoothed and the current harmonic content will be lower than the voltage.

Each branch of bridge inverter may be realized with BJTs, MOSFETs, or IGBTs. In Figure 4.5, the converter is designed with IGBTs electronic devices, this is a hybrid component between the bipolar junction transistor and the power FET. Its behavior is similar to the BJT but with much lower excitation current, being for this purpose as a FET.

An inductive load connected to the inverter output of Figure 4.5, requires the inclusion of an antiparallel diode with each transistor; power transistors used to include this diode in the encapsulation device. In fact, the current in an inductance is delayed with respect to the voltage waveform. A change in the conduction state of two legs of the bridge requires the presence of two separate diodes in the other two branches; thus the current can continue to flow through the load in the same direction. In this time interval the load returns energy to the dc side source. Therefore these diodes are called energy recovery diodes or freewheeling diodes. If the load was purely resistive it will be dispensed with such diodes.

EXAMPLE 4.2

A single-phase bridge inverter will be considered with a constant voltage at 100 V in the dc side and a RL branch in the ac side whose parameters are $R = 10\ \Omega$ and $L = 100$ mH. The control signal is a sinusoidal waveform of amplitude $V_m = 10$ V and frequency $f_m = 50$ Hz. A modulation rate was considered $m_a = V_m / V_c = 0.9$ (V_c amplitude of the carrier signal), and a frequency modulation rate $m_f = f_c / f_m = 25$. Through a simulation model, the waveforms of interest are obtained.

Figure 4.7 shows the simulation model in the SimPowerSystems environment. Figure 4.8 shows the model of the PWM control circuit.

Figure 4.9 shows two cycles of the waveforms of the voltage and the output current; voltage appears as a typical bipolar PWM waveform, that is, the output takes positive and negative values, and does not take zero values as in the case of a unipolar PWM.

Figure 4.10 allows us to appreciate the pattern of the frequency spectrum of the output voltage. Indeed, in a PWM sinusoidal fit to highlight three key features:

1. Harmonics in the output voltage waveform of the inverter appear as sidebands centered around the switching frequency and its multiples, that is, around the harmonics orders m_f,

$2m_f$, $3m_f$, etc. This pattern holds true for all linear range, $m_a \leq 1$. The frequency spectrum is distributed in the form,

$$h = j(m_f) \pm k \qquad (4.7)$$

where h is the harmonic order and j, k are integer numbers. For odd values of j, harmonics exist only for even values of k ($j = 1$; $h = m_f$, $m_f \pm 2$, $m_f \pm 4$, ...). For even values of j harmonics exist only for odd values of k ($j = 2$; $h = 2m_f \pm 1$, $2m_f \pm 3$, $2m_f \pm 5$....). The fundamental frequency corresponds to $h = 1$. Figure 4.10 identifies the pattern of frequencies suitable for $m_f = 25$ (with $j = 1, 2, 3, ...$).

2. The fundamental frequency component of the output voltage varies linearly with m_a for the linear zone, $m_a \leq 1$; therefore, the range of m_a between 0 and 1 is called linear range. Specifically, the fundamental frequency component of the output voltage is

$$v_{ac}|_1 = m_a V_d \sin \omega t \qquad (4.8)$$

where $\omega = 2\pi f_m$.

3. If the frequency modulation index m_f is an odd integer, for a time origin appropriately chosen, the output voltage has odd and half wave symmetry. Therefore it includes only odd harmonics, and sine coefficients only appear in the Fourier analysis of the waveform. Figure 4.9 also shows the waveform of the load current. An inductive load filters the higher order harmonics of the current, and the current in the RL branch is smooth and approaches a sinusoidal wave.

When the amplitude modulation index m_a rises to a value above 3.24, then abandoning the linear and over-modulation region; thus, now we choose $m_a = 4$. Figure 4.11 shows the waveform of the output voltage, which degenerates to a square wave. In the method of square wave switching the inverter switch conducts for half cycle of the output frequency. In addition, the voltage reaches its highest value of peak, $(4/\pi)V_{dc}$. By contrast, the waveform of the load current worsens, Figure 4.11.

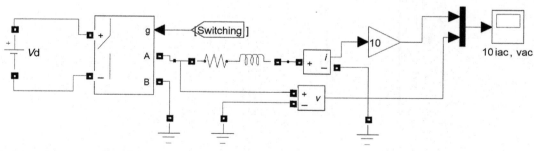

FIGURE 4.7 Simulation model of a single-phase bridge inverter.

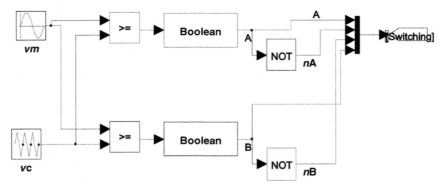

FIGURE 4.8 Control circuit model of bipolar PWM.

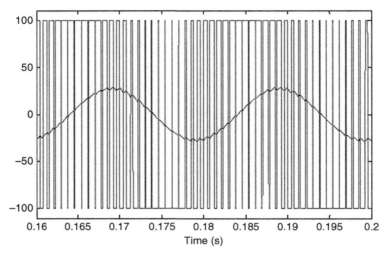

FIGURE 4.9 Waveform of the output voltage of a PWM bridge inverter and output current in an inductive branch.

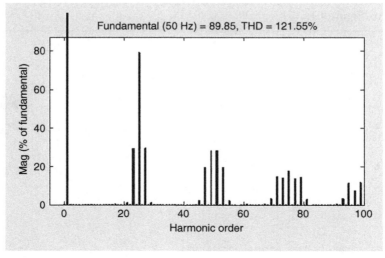

FIGURE 4.10 Harmonic spectrum of the output voltage of a bipolar PWM bridge inverter.

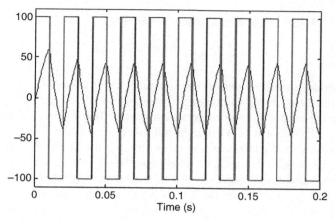

FIGURE 4.11 Waveforms of voltage and current in a bridge inverter by square wave switching mode.

At this point it is noted that, from a practical point of view, the switching signals and trigger signals (gate drive signals) are complementary but must include a small deadband that is intentionally introduced to prevent the two switches of each active leg of inverter from being ON, simultaneously.

Another example will now be developed using a bridge inverter to compensate a nonlinear inductive load.

EXAMPLE 4.3

An active compensator consists of a bridge inverter to compensate a nonlinear load that is formed by two antiparallel SCRs in series with a *RL* load whose parameters are $R = 20\ \Omega$ and $L = 10$ mH; the firing angle of the SCRs is $\pi/2$ rad. A sinusoidal source of 230 V and 50 Hz was applied. Get the waveforms of the system before and after compensation.

The load is nonlinear and has inductive character, therefore, the current includes a fundamental component and a harmonic series; with a displacement factor lower than unity. Thus,

$$i_{L} = i_{La} + i_{Lr} + i_{Lh} \tag{4.9}$$

where i_{La} and i_{Lr} are the active and reactive fundamental components of the load current and i_{Lh} is the harmonic current drawn by the load. The goal of the compensator is to inject a current i_{F} to cancel or at least mitigate the harmonic and reactive parts of the load current. Because the source must supply the real power required by the load, the source current has no harmonics and the displacement factor is unity,

$$i_{S}^{*} = \sqrt{2}\,\frac{P_{L}}{V^{2}}\sin \omega t \tag{4.10}$$

with $\omega = 2\pi 50$ rad/s. The compensator reference current is

$$i_F^* = i_L - i_S^* = i_L - \sqrt{2}\,\frac{P_L}{V^2}\sin \omega t \tag{4.11}$$

A single-phase bridge inverter is connected in parallel with the load. The gate drive signals of the power switches have been obtained by a hysteresis band (HB) control; that is, the difference between the output and the reference currents ($e(t)$) is the input of a comparator with HB. The inverter is then switched using the following logic: if $e(t) \geq +h$ then devices 3 and 4 are on; else if $e(t) \leq -h$ then devices 1 and 2 are on, (h is the width of the band). Since the load current cannot change very quickly, it is likely that the compensation current regularly exceeds the HB. This explains the ripple waveforms that are present when the compensator is working. The compensator parameters are $R_f = 0$ $L_f = 8$ mH, and $V_{dc} = 500$ V, Figure 4.12a shows the waveforms of the current source and the voltage at the terminals of the load before and after compensation; the active compensator is connected at 0.1 s.

The source current becomes sinusoidal and in phase with the voltage. Figure 4.12b presents the instantaneous power of the compensator, which has a zero mean value. The compensator provides the harmonic and reactive power required by the load but no real power.

The previous example shows how an inverter circuit can be used to reproduce compensation current and therefore operate as an active compensator, that is, an APF. However, two hypotheses have been assumed. The compensator is supplied by a dc source, which in practical cases is replaced by a dc capacitor. Also system losses (R_f in the example) must be considered. Therefore, a control loop to avoid dc capacitor discharge and to replenish the inverter losses from the supply must be considered.

A three-phase inverter can be constructed from three single-phase inverters connected to the same dc source, Figure 4.26. In this case, the control of each inverter must be such that the voltage waveforms in the outputs of each are arranged with a phase difference of 120° between every two of them. A three-phase inverter can also be designed as a bridge configuration, Figure 4.13, which is the most common configuration.

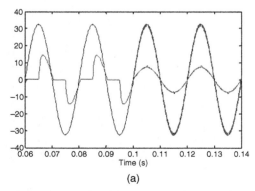

(a)

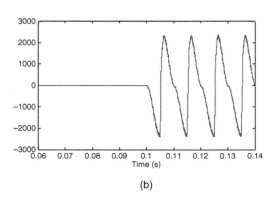

(b)

FIGURE 4.12 Results of Example 4.3 before and after compensation. (a) Voltage (scaled by 1/10) and source current. (b) Instantaneous power of the compensator.

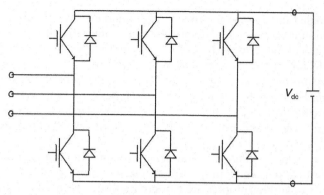

FIGURE 4.13 Three-phase inverter (dc-ac converter) according to a bridge configuration.

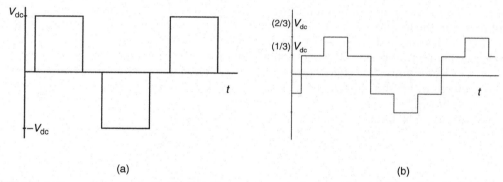

(a)   (b)

FIGURE 4.14 (a) Line-to-line output voltage of a full-bridge inverter for switching in square wave mode. (b) Phase to neutral voltage in the same situation (a).

Figure 4.14 shows a single-phase voltage and line-to-line voltage for a full-bridge three-phase inverter switched in square wave/six-step mode. With these waveforms we can appreciate how the use of three-phase systems significantly reduces the harmonic content of the output waveforms. Also, the use of the technique of PWM further reduces distortion of the output voltage.

EXAMPLE 4.4

Figure 4.15 shows the simulation model of a three-phase inverter, which is to generate three sinusoidal currents of 50 Hz frequency from a dc signal. The bridge inverter has a dc voltage source with a value of 100 V. Three RL branches arranged in star with parameters $R = 0.5\ \Omega$, $L = 1$ mH are connected in the ac side.

In the carrier-based sinusoidal PWM method, SPWM, three symmetrical sinusoidal waveforms are used as three-phase control signals or modulating signals for each phase, and compared with a single triangular waveform of higher frequency. On each side of the inverter, the upper switch conducts when the modulating signal is larger than the carrier triangular wave. The lower switch of the same leg has a complementary operation. Figure 4.16 presents the simulation model for the PWM inverter control of Figure 4.15.

Here a peak value for the modulating signals of $V_m = 10$ V, a frequency of 50 Hz, an amplitude modulation index $m_a = 0.9$, and frequency modulation index $m_f = 9$ will be considered. Figure 4.17a shows a line-to-line voltage, and Figure 4.17b the single voltage and current in the corresponding inductive load; the inductance of the RL branch filters out the high harmonics of the current.

Harmonics of the output voltage of the ac side of the inverter, for example, v_{AN} in Figure 4.15, are identical to those found in the single-phase inverter output voltage. That is, for odd values of m_f, odd harmonics are centered around m_f and its multiples only. However, where the harmonics are specifically considered in m_f and odd multiples, the phase difference, for example, between v_{AN} and v_{BN} is $m_f(2\pi/3)$ rad. This phase difference is equal to zero (2π) if m_f is odd and multiple of 3. So, in this case m_f harmonics are suppressed in the line-to-line voltage v_{AB}. In general, if m_f is chosen as an odd multiple of 3, m_f harmonics and odd multiples of m_f in the linear region $(m_a \leq 1)$ are suppressed. Thus, some prevalent harmonics in the single-phase inverter are eliminated in the line-to-line voltage of a three-phase inverter. Figure 4.18 shows the frequency spectrum of the line-to-line voltage where the pattern of the characteristic harmonics of a SPWM three-phase inverter is shown.

In Figure 4.18 odd harmonics that are multiples of $m_f = 9$ do not appear.

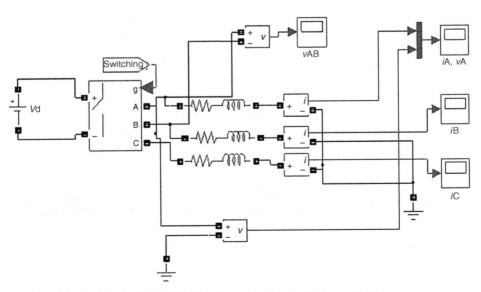

FIGURE 4.15 Simulation model of a SPWM full-bridge three-phase inverter for Example 4.4.

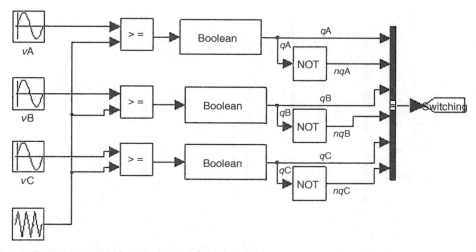

FIGURE 4.16 Simulation model of SPWM control of Example 4.4.

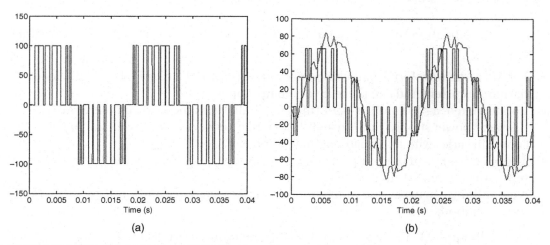

FIGURE 4.17 (a) Line-to-line voltage v_{ab}, (b) Single-phase voltage v_{an} and current i_a.

4.3.2 Current Control

The function of the inverter current controller is to force the output to follow the reference current as closely as possible. The currents measured at the active filter output are compared with the reference currents, and the error signals are operated by the current controller to generate the firing pulses that switch the state of the inverter power devices so as to reduce the current error. When a CC-VSI is used as APF, some considerations

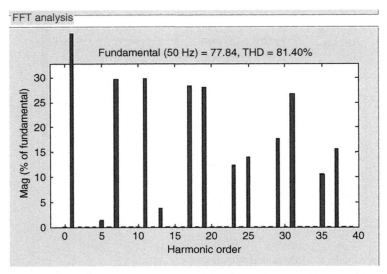

FIGURE 4.18 Frequency spectrum of the line-to-line voltage of three-phase inverter of Example 4.4.

about the modulation method should be taken into account [7–11]. And this is mainly due to:

- For APFs, the waveforms of the reference currents are not sinusoidal; includes load harmonic current and can present very sharp slopes.
- Connection of a shunt APF is performed in parallel with the load and therefore with the same power supply; this is usually the public network that is essentially a voltage source in series with an impedance of low value.
- The effectiveness of the compensation depends on the capacity of the compensator to follow the reference signal as fast as it can.
- A compromise between the generation of a perfect reference and a fast-track method must be found.

The technique of PWM is generally used for generating the trigger pulses of power electronic devices. There are different strategies to generate a PWM signal [8–11], however for APFs this is usually limited to:

- Periodical sampling (PS).
- Comparison with triangular carrier (TC) wave.
- Hysteresis Band (HB).

In the method of periodic sampling, the switching signal is obtained as a result of comparing the inverter output signal with the reference signal to be generated (Figure 4.19). If the output signal is greater than the reference level, the switches are switched to decrease the value of the output signal of the inverter. However, if the output signal is less than the reference, the switches commute to increase the output signal value. The switching

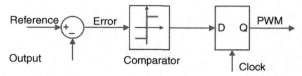

FIGURE 4.19 PWM control scheme for periodic sampling.

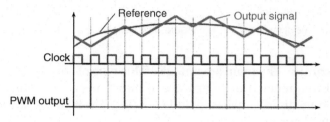

FIGURE 4.20 Generating a PWM signal with periodic sampling on rising edge.

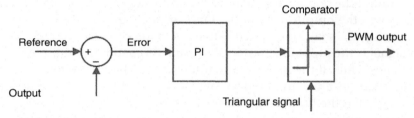

FIGURE 4.21 PWM control scheme triangular wave comparison.

is performed only when a clock signal changes, which lets you set the switching frequency depending on the power electronic devices of the inverter. Figure 4.20 shows how this method generates the PWM signal; here it is considered that the output is switched at the rising edge of the clock signal.

In the scheme for triangular wave modulation, Figure 4.21, the PWM switching sequence is determined by comparing the error signal with a TC signal as shown in Figure 4.22. It often includes a proportional-integral block (PI) to process the error signal. Adjusting the parameters k_p and k_i we can modify the steady-state error and the transient

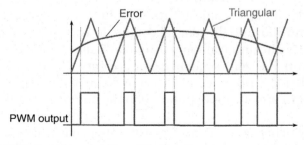

FIGURE 4.22 Generating a PWM signal by comparison with triangular.

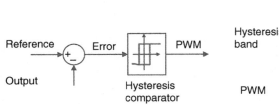

FIGURE 4.23 Control scheme of PWM HB.

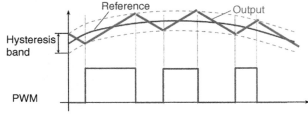

FIGURE 4.24 Generating a PWM signal by HB.

response of the control. The main advantage of this type of controller is that, since this system uses a fixed-frequency triangular wave, it maintains a constant switching frequency in the inverter. Some disadvantages include possible errors in the amplitude and phase of the resultant output current.

In Figure 4.23 the control scheme by HB is shown. This method involves the imposition of a band around the reference signal. If the output signal is outside the band, forcing the device to switch the output back to the limits of the band. The process is performed as follows: the output signal is compared with the reference to follow (Figure 4.24); the resulting error signal is the input of a hysteresis comparator, and the output is the firing pulses of the switches. Thus, if the measured output is larger than the reference signal plus half the HB, the switches are commuted so that the signal of the inverter provides a negative output voltage, and the output signal decreases. Moreover, if the actual signal falls below the reference by the same amount, the devices switch to provide a positive signal at the output. As a result, the output signal remains within the boundaries of the HB around the reference signal.

HB control has the advantage of simplicity and a dynamic response that allows fast controllability of the output signal. Its main drawback is that this method is a variable switching frequency for power electronic devices.

4.3.3 Shunt Three-Phase Four-Wire Active Filter

Different alternatives have been proposed for when three-phase four-wire systems have to be compensated. One of them is to include a fourth branch to the inverter as shown in Figure 4.25a. Therefore, this configuration requires two additional power devices, which is its main drawback. Another solution proposed in the literature [3,4,12] is to divide the dc side supply into two; and the neutral conductor is connected at this midpoint, as shown in Figure. 4.25b. The disadvantage of this configuration appears when the current injected to the network has zero-sequence components. The dc side typically includes two capacitors so that the presence of the zero-sequence component of the current causes one capacitor to provide more power than the other, resulting in an imbalance of their voltages.

Some situations with heavy unbalanced consumption, have been known to use topologies with three or four single-phase inverters, Figure 4.26, which share the same dc source.

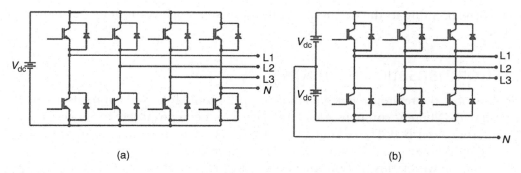

FIGURE 4.25 VSI inverter topologies for four-wire systems: (a) four branches; (b) split supply to the dc side.

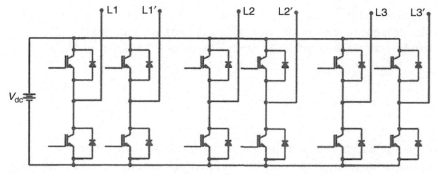

FIGURE 4.26 VSI inverter topologies with three single-phase inverters.

The disadvantage of this configuration is the high number of power electronic devices required. A more detailed analysis of different topologies can be found in [2].

This chapter will consider the VSI configuration of Figure 4.25b for the construction of an APF for three-phase four-wire systems. This structure will be controlled by a PWM current control, preferably HB, which will make the APF behave as a controlled current source. However, the presence of dc components and/or zero-phase sequence in the reference signal requires further analysis of the three legs VSI. In effect, if the current references include zero-sequence components, the sum of line currents will return via the neutral conductor. In the split capacitor inverter topology, this forces each phase current to flow through the upper capacitor, or through the lower capacitor, to return by the neutral conductor. The currents can flow in both directions through the switches and capacitors. When the current of the active filter is positive, the upper capacitor voltage increases and lower capacitor voltage decreases but with a different ratio of variation, since the positive and negative values of the di_f/dt are different and depend on the instantaneous values of the phase voltages on the ac side. The opposite occurs when $i_f < 0$. The voltage variation on the dc side is also dependent on the shape of the reference current and the hysteresis bandwidth. Therefore, the total dc voltage and the voltage difference in both capacitors oscillate not only with the switching frequency but also with those corresponding to the frequency of the zero-sequence components generated by the VSI.

4.4 Compensation Strategies

In this section, the reference currents that the APF should follow to compensate the load are obtained. Different compensation strategies have been proposed; in Section 4.2 three of them were described. The formulations of the instantaneous reactive power theory introduced in Chapter 3 enable the compensation currents for each of the defined strategies to be found. However, this section will address the problem from another point of view to complement the above [13,14].

In general, a three-phase four-wire system will be treated as a four-conductor system. Chapter 2 discussed this issue extensively; here we recall the most relevant topics. The line current is the current in the conductors of the system, and the vector of line currents i_{L1}, i_{L2}, i_{L3}, i_{L4} can be represented, with the help of the vector $1 = [1\ 1\ 1\ 1]^T$, as:

$$\mathbf{i} = \begin{bmatrix} i_{L1} & i_{L2} & i_{L3} & i_{L4} \end{bmatrix}^T; \quad \mathbf{1.i} = \begin{bmatrix} 1 & 1 & 1 & 1 \end{bmatrix} \begin{bmatrix} i_{L1} \\ i_{L2} \\ i_{L3} \\ i_{L4} \end{bmatrix} = 0 \tag{4.12}$$

Because of the Kirchhoff current law, some of these components are a linear combination of the others, since the conductor currents sum to zero. Mathematically, only (4-1) components are sufficient to describe the current vector. However, in three-phase four-wire circuits, using all four components often leads to a more straightforward and convenient formulation of the equations describing the system behavior. In the same way, with regard to the voltage, it is convenient and interesting to characterize the line voltages at a set of terminals by the voltages from the conductors to a virtual start-point, Figure 4.27. This virtual star-point is chosen such that the voltages sum to zero, that is, a set of equal impedances with star connection, where the common point is the reference of voltages. A more suitable notation change will be used in this section for clarity. Thus, the conductor or line voltages u_{L1}, u_{L2}, u_{L3}, u_{L4} can be represented as a vector,

$$\mathbf{u} = \begin{bmatrix} u_{L1} & u_{L2} & u_{L3} & u_{L4} \end{bmatrix}^T; \quad \mathbf{1.u} = 0 \tag{4.13}$$

Let $\|\mathbf{u}(t)\| = u_\Sigma(t)$, $\|\mathbf{i}(t)\| = i_\Sigma(t)$, be the norm of the instantaneous vector of line voltages and line currents in a four-conductor system,

$$u_\Sigma(t) = \sqrt{\mathbf{u.u}} = \sqrt{\sum_{j=1}^{4} u_{Lj}^2(t)}; \quad i_\Sigma(t) = \sqrt{\mathbf{i.i}} = \sqrt{\sum_{j=1}^{4} i_{Lj}^2(t)} \tag{4.14}$$

It is usual to give them the name "collective" instantaneous norm and term the associated subscript "Σ" in order to distinguish the four-conductor values. Thus, the collective

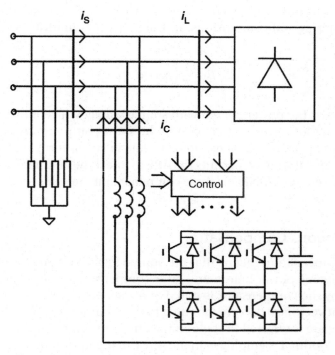

FIGURE 4.27 Four-wire power system schematic that includes compensation with a Shunt APF.

instantaneous power at a cross section of a four-conductor system is the dot product of the collective voltages vector and collective currents vector at that section,

$$p_\Sigma(t) = \mathbf{u}.\mathbf{i} \qquad (4.15)$$

The average value concept for three-phase systems, in contrast to that of the instantaneous value, has been based on time-averaged quantities developed in both the frequency-domain and the time-domain. Under stationary and periodic conditions, the average value of the instantaneous active power and the rms value of voltage and current vectors, to the fundamental period T, are usually defined. In fact, under periodic conditions, for an observation interval T that coincides with the period or an integer multiple of the period, the mean square value or rms value over the time interval T is given by:

$$U_\Sigma = \sqrt{\overline{\|u(t)\|^2}} = \sqrt{\frac{1}{T}\int_{t-T}^{t}\left(\sum_{j=1}^{4}u_{Lj}^2(\tau)\right)d\tau} = \sqrt{\sum_{j=1}^{4}U_{Lj}^2} \qquad (4.16)$$

for the voltage vector, (4.2), and

$$I_\Sigma = \sqrt{\overline{\|i(t)\|^2}} = \sqrt{\frac{1}{T}\int_{t-T}^{t}\left(\sum_{j=1}^{4}i_{Lj}^2(\tau)\right)d\tau} = \sqrt{\sum_{j=1}^{4}I_{Lj}^2} \qquad (4.17)$$

for the current vector, (4.1).

The instantaneous power (4.4) can be averaged over an observation interval T equal to the period under periodic conditions. In this case, the quantity is the active or real power,

$$P_\Sigma = \frac{1}{T} \int_T p_\Sigma(\tau) \, \mathrm{d}\tau \qquad (4.18)$$

Figure 4.27, shows the power system with source voltages, a generic three-phase load, the virtual star-point for voltage reference and the static compensator (a shunt APF). The voltage source, and load and compensation currents in vector form are also shown.

A convenient control strategy permits the APF to inject the compensating currents that improve the source energy efficiency; three compensation methods are possible, as will be described in this section.

4.4.1 Instantaneous Compensation

Instantaneous compensation means that the source instantaneous power is the same as the load instantaneous power before and after compensation. Therefore, instantaneous power delivered by the compensator is null. For a given source voltage vector, $u(t)$, there are infinity current vectors, $i(t)$, corresponding to the incoming instantaneous power $p_\Sigma(t)$, at time t. Here, we are interested in the optimal current vector that delivers the instantaneous power $p_\Sigma(t)$; optimal is employed in the sense of the minimum instantaneous norm. This yields an alternative method called time-instantaneous compensation (TIC) [13]. When the load is time-variant and the transient state is prevalent, this type of compensation mode is especially appropriate.

According to this approach, for a given source voltage $u(t)$, the objective consists of obtaining the current vector $i_p(t)$ with minimum instantaneous norm, which delivers the incoming instantaneous power to the load. The optimization problem is solved by means of the Lagrange multiplier technique.

The procedure is outlined as follows,

- Objective function: minimum of squared instantaneous norm

$$i.i \qquad (4.19)$$

- Equality constraints:

  Instantaneous power

$$p_\Sigma(t) = u.i \qquad (4.20)$$

  Kirchhoff's current law

$$1.i = 0; \quad 1 = \begin{bmatrix} 1 & 1 & 1 & 1 \end{bmatrix}^T \qquad (4.21)$$

The solution of the Lagrange's method gives,

$$i_p = \frac{p_\Sigma(t)}{\|u(t)\|^2} u(t)$$

(4.22)

The current vector defined by (4.22) is called the vector of the instantaneous power currents, which is considered the technically correct terminology. In general, line current waveforms of $i_p(t)$ are different from the corresponding line voltages waveforms of $u(t)$. According to the first approach, maximum time-instantaneous compensation is obtained using an APF that generates the compensating current vector,

$$i_c = i - i_p$$

(4.23)

thus, after compensation the source current i_s coincides with the instantaneous power current $i_p(t)$

$$i_s \equiv i_p = \frac{p_\Sigma}{u_\Sigma^2} u$$

(4.24)

The instantaneous power generated by the compensator is zero, that is, according to (4.23),

$$p_{c\Sigma} = u.i_c = u.i - \frac{p_\Sigma}{u_\Sigma^2} u.u = 0$$

(4.25)

In this compensation process, the instantaneous power delivered within the compensator is zero, $p_{c\Sigma}(t) = 0$. It is considered as a compensation method, which does not use energy storage elements.

Although the instantaneous norm of the supply current is minimized without affecting the instantaneous power, when the TIC is applied, the zero-sequence current component still exists, and it is proportional to the zero-sequence voltage. In fact, the neutral supply current, after compensation, is

$$i_{s4} = \frac{p_\Sigma}{u_\Sigma^2} u_{L4} \neq 0$$

(4.26)

which implies that a no null neutral conductor current can still exist after compensation. In [13] an approach that achieves a neutral zero current is proposed.

4.4.2 Unit Power Factor

On the other hand, in unity power factor, compensation is used to ensure that the source current has the same distortion and symmetry conditions as the source voltage. It means that the source current is collinear to the supply voltage. In this situation the source supplies the load active power, but the instantaneous power is not constant after compensation. This time, after compensation, the supply current will carry the real power consumed by the load (and converter losses). Therefore, in contrast to the method of time-instantaneous compensation, the unit power factor approach is called time-average compensation (TAC) [13].

The principal objective of TAC is the optimization of the average power transfer from source to load, that is, the source current must deliver only average power to the compensator and load.

According to this approach, the objective is to obtain the current vector $i_a(t)$ with minimum rms value that transfers to the load the incoming average power for a given source voltage. The problem is solved by means of the Lagrange multiplier technique.

- Objective function: minimum of current rms value

$$\frac{1}{T}\int_T \boldsymbol{i}.\boldsymbol{i}\,\mathrm{d}t \tag{4.27}$$

- Equality constraints:
  Real power

$$\frac{1}{T}\int_T \boldsymbol{u}.\boldsymbol{i}\,\mathrm{d}t = P_\Sigma \tag{4.28}$$

Kirchhoff's current law

$$1.\boldsymbol{i}=0; \quad 1=\begin{bmatrix} 1 & 1 & 1 & 1 \end{bmatrix}^T \tag{4.29}$$

The solution of the Lagrange's method gives

$$\boldsymbol{i}_a = G\cdot\boldsymbol{u} \tag{4.30}$$

which defines those conductor currents that would be absorbed by an equivalent symmetric resistive load having the same average power consumption as the actual load at voltage $\boldsymbol{u}$. So, the load presents an equivalent conductance defined by:

$$G=\frac{P_\Sigma}{U_\Sigma^2} \tag{4.31}$$

The compensating current

$$\boldsymbol{i}_c = \boldsymbol{i}-\boldsymbol{i}_a \tag{4.32}$$

constitutes the current vector supplied by the APF, and the source current after compensation is

$$\boldsymbol{i}_S = G\cdot\boldsymbol{u} \tag{4.33}$$

Moreover,

$$p_{c\Sigma}(t)=\boldsymbol{u}.\boldsymbol{i}_c = p_\Sigma(t)-\frac{P_\Sigma}{U_\Sigma^2}u_\Sigma^2 \tag{4.34}$$

is the instantaneous active power flowing through the compensator, according to (4.31)–(4.33), which has an average value equal to zero,

$$\frac{1}{T}\int_T p_{c\Sigma}(t)\,\mathrm{d}t = 0 \tag{4.35}$$

However, although no power dissipation exists in the compensator, the source-side neutral current is not perfectly controlled to zero, thus according (4.31),

$$i_{S4} = G\,u_{L4} \neq 0 \tag{4.36}$$

In Ref. [13] an approach that achieves zero neutral current for time-average compensation is proposed.

4.4.3 Balanced and Sinusoidal Source Current

When the source voltage is sinusoidal and balanced, unity power factor compensation results in a balanced and sinusoidal current after compensation, furthermore the instantaneous power supplied by the source is constant. However, when supply voltage is unbalanced and/or nonsinusoidal, the situation will not be as described previously. To overcome this, a further compensation objective is possible. Thus, since the electrical utilities produce, in general, electrical power with sinusoidal and positive-sequence phase voltages, this has been established as reference condition in the supply. A resistive load, balanced and linear is considered as the ideal reference load. A reference voltage applied to an ideal reference load will generate sinusoidal and balanced currents, in phase with the voltages. Section 4.4.2 established the control strategy derived from unity power factor compensation. There, if the voltage supply is balanced and sinusoidal, the source current is also balanced and sinusoidal. Nevertheless, if the voltage supply is balanced and nonsinusoidal, the source current is balanced and nonsinusoidal. The voltage at the fundamental frequency is balanced and therefore the voltage of the neutral conductor to the reference is zero. In this case, to obtain a source current balanced and sinusoidal, it is necessary to modify the control strategy as follows [14]:

$$i_s = G\,u_1 \tag{4.37}$$

where u_1 represents the fundamental wave of the three-phase voltage. Assuming that the average real power exchanged by the compensator is null:

$$G = \frac{P_\Sigma}{U_{\Sigma 1}^2} \tag{4.38}$$

where

$$U_{\Sigma 1}^2 = \frac{1}{T}\int_T \left(u_{11}^2 + u_{21}^2 + u_{31}^2\right)\mathrm{d}t \tag{4.39}$$

represents the square of the fundamental voltage rms value.

In this case, the compensation current is

$$i_C = i - G u_1 \tag{4.40}$$

On the other hand, the achievement of balanced and sinusoidal current when voltage supply is unbalanced and sinusoidal requires the next modification in the control strategy:

$$i_s = G u^+ \tag{4.41}$$

where u^+ represents the positive sequence of the three-phase supply voltage vector. And, to make null the average real power exchanged by the compensator:

$$G = \frac{P_\Sigma}{U_\Sigma^{+2}} \tag{4.42}$$

where

$$U_\Sigma^{+2} = \frac{1}{T} \int_T (u_1^{+2} + u_2^{+2} + u_3^{+2}) \, \mathrm{d}t \tag{4.43}$$

represents the square of the symmetrical positive-sequence voltage rms value.

In the case of unbalanced and nonsinusoidal voltage supply, the control strategy necessary to obtain balanced and sinusoidal source current is:

$$i_C = i - G u_1^+ \tag{4.44}$$

where

$$G = \frac{P_\Sigma}{U_{\Sigma 1}^{+2}} \tag{4.45}$$

and

$$U_{\Sigma 1}^{+2} = \frac{1}{T} \int_T \left(u_{11}^{+2} + u_{12}^{+2} + u_{13}^{+2} \right) \mathrm{d}t \tag{4.46}$$

represents the square of the rms value of positive-sequence fundamental voltage.

Figure 4.28, shows a diagram of the procedure used to get the positive-sequence fundamental component corresponding to the voltage vector. The block named $T/3$ represents a delay of the third of the fundamental period T. The block named $-T/3$ is the same but with advancement instead of a delay. Finally, the blocks LPF are low-pass filters, and the blocks of constant values 1/3 and 2. The other phases are obtained by phase shifts in delay of 120 and 240 degrees, respectively.

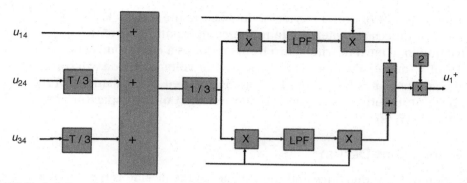

FIGURE 4.28 Block diagram for determining the symmetrical positive component of the fundamental voltage.

4.5 Practical Design Considerations

As has already been described in previous sections, the compensation process will be more effective, the more accurately the power converter is capable of reproducing the current reference [15–18]. If we refer to the method of unit power factor compensation,

$$i_{C\,ref} = i_L - Gv_S \tag{4.47}$$

where G represents the equivalent conductance, v_s, the supply voltage and the HB method is selected for the inverter switching control, with a width ΔI around the current reference i_{Cref}, the current generated by the APF, i_C, fluctuate around i_{Cref}. The APF is connected to the PCC in parallel with the load through an inductive link L_F. The value of this inductance and the value of the dc-side voltage of the converter determine the current ratio of compensator, that is,

$$\frac{d\,i_C}{d\,t} = \frac{\pm V_C - v_S}{L_F} \tag{4.48}$$

where V_C is the voltage at dc side, and where the inductance resistance, R_F is null.

A suitable tracking of the current reference requires the ratio of current compensator to be higher than the ratio of the current reference,

$$\left|\frac{d\,i_C}{d\,t}\right| > \left|\frac{d\,i_{Cref}}{d\,t}\right| = \left|\frac{d\,i_L}{d\,t} - G\frac{d\,v_s}{d\,t}\right| \tag{4.49}$$

The value of the derivative of the current compensator will be greater than the derivative of the reference signal at the ramp up. That relationship will be contrary to the down ramp. That is so, according to the path of the current compensation ripple where we are. The optimal tracking reference current requires a high value di_C/dt, which implies a small L_F. However, decreasing the value of inductance causes higher ripple of the output current as it easily reaches the limits of the HB, and therefore, the current at the supply side exhibits a high value of THD. The same effect occurs with respect to the value of the voltage on the dc side of the converter. An increase in the voltage V_C increases di_C/dt, but otherwise it produces a higher ripple in the compensation current. Thus, the optimal compensation load current with sharp fronts requires a compromise in the selection of the values of V_C and L_F. Furthermore, the capacitor voltage dc side will suffer discharges due to the compensator losses and fluctuations caused by oscillations of the instantaneous power at its terminals, governing the selection of the capacitor and the voltage control circuit [17,18].

4.5.1 Component Design

The selection of the inductance and the voltage on the dc side is based on the following criteria [16]:

1. Limiting the high-frequency components of the currents injected by the APF; therefore, the amplitude of the current at the maximum switching frequency of the converter is limited to 5% of the rated load.

   For a harmonic of order h (switching frequency), with the usual hypothesis, we will have,

$$\frac{I_{Ch}}{I_1} = \frac{V_{Ch}}{1+(f_h/f_1)\omega L_F} < 0.05 \tag{4.50}$$

   where I_{Ch} is the rms value of the current of the converter at the frequency f_h, V_{ch} is the rms value of the output voltage of the converter at the frequency f_h, I_1 is the rms value of the load current at the fundamental frequency f_1.

2. The di_C/dt of the current injected by the APF must be greater than the maximum value of di_L/dt of the load,

$$\sqrt{2}V_S + L_F\left(\frac{di_L}{dt}\right)_{máx} < \frac{V_{dc}}{2} \tag{4.51}$$

Further discussion requires the choice of the capacitor of the dc side. Assuming that the power supplies the power consumed by the load and the converter, the real power transferred by the APF is zero, and hence the average current I_{dc} on the capacitor is zero; the average capacitor voltage is kept constant at steady state. The power balance is established then,

$$p_c(t) = p_L(t) - p_S(t) = P_L + \tilde{p}_L(t) - P_L - \tilde{p}_S(t) = \tilde{p}_C(t) \tag{4.52}$$

The average values in one cycle of the instantaneous power of supply and load, P_S and P_L, do not affect the average voltage of the capacitor, but the oscillatory component of the two can result in a fluctuation of the capacitor voltage,

$$v_{dc}(t) = V_{dc} + \tilde{v}_{dc}(t) \tag{4.53}$$

where $\tilde{v}_{dc}(t)$ represents the inevitable fluctuation of the capacitor voltage. The design of the capacitor should be chosen to limit the voltage ripple to a prespecified value, typically less than 2%. The voltage variation on the dc side is,

$$\Delta V_{dc} = v_{cmax} - v_{cmin} \simeq \frac{\Delta \int \tilde{p}_C(t) dt}{CV_{dc}} = \frac{\Delta \int \sum_{\forall j,h} v_{Sjh} i_{Cjh} \, dt}{CV_{dc}} \tag{4.54}$$

for three phases $j = 1, 2, 3$ and prevalent harmonics h. Given a capacitor voltage ripple r,

$$r = \frac{\Delta V_{dc}}{V_{dc}} \tag{4.55}$$

capacitor value is determined by,

$$C = \frac{\Delta \int \sum_{\forall j,h} v_{Sjh} i_{Cjh} \, dt}{rV_{dc}^2} \tag{4.56}$$

The value of the capacitor depends on the load and should be chosen for the worst case; a simulation study will help in the design.

As has been stated in the beginning of this section, the dc link voltage of the converter cannot remain constant due to the power losses of the APF [17,18]. This problem can be solved by a voltage control circuit that allows control of the amplitude of the supply current and subsequently ensures that the grid provides that additional power. It is common practice to include a PI control loop for the error between the measured voltage on the dc side and a value of dc voltage reference. Its transfer function can be represented by

$$H(s) = K_p + \frac{K_{PI}}{s} \tag{4.57}$$

where K_p is the proportional gain and determines the dynamic response of the dc capacitor voltage, and K_{PI} is the integral gain that determines the settling time.

The energy stored in the capacitor at a time t may be denoted as,

$$E_C = \frac{1}{2} C v_{dc}^2(t) \tag{4.58}$$

If V_{dcRef} is the reference voltage to the dc-side voltage, the energy discharged by the capacitor during a period of the fundamental frequency is

$$\Delta E_{dc}(t) = \frac{1}{2} C V_{dc}^{*2} - \frac{1}{2} C v_{dc}^2(t) = \frac{C}{2}(V_{dc}^* + v_{dc}(t))(V_{dc}^* - v_{dc}(t)) \tag{4.59}$$

If one considers a small variation of the voltage variation in the capacitor $\Delta v_{dc}(t)$ during a cycle, then

$$\Delta E_{dc}(t) \simeq C V_{dc}^*(V_{dc}^* - v_{dc}(t)) \tag{4.60}$$

This variation in energy in the capacitor must be compensated by a current component at the fundamental frequency and in phase with the voltage, that is, an active component that is added to the current source after compensation. The variation of energy that must be restored in the capacitor is given by,

$$\Delta E_{dc} = \int_0^T \sum_{\forall j=1,2,3} (\sqrt{2}V_{Sj} \sin \omega t \sqrt{2}I_{Caj} \sin \omega t) \mathrm{d}t \qquad (4.61)$$

where I_{Caj} refers to the rms value of the additional component in each phase to compensate the APF losses. Equation (4.61) yields,

$$I_{Ca} = \frac{1}{TV_S} \Delta E_{dc} \qquad (4.62)$$

From (4.60) and (4.62), the control used to maintain the charge on the capacitor is PI-type with proportional integral constant

$$K_{PI} = \frac{CV_{dc}}{TV_S} \qquad (4.63)$$

Finally we have to say that the current harmonics injected by the APF include high-frequency harmonics. These harmonic currents, when injected into the system, excessively distort the voltage at PCC. Now the terminal voltage is not sinusoidal and the generation of the reference current based on this voltage will not operate properly. Consequently, the erroneous compensating currents injected increase the system distortion. To evacuate these components at the switching frequency from the voltage of the PCC, a path for the flow of high-frequency current has to be supplied. Therefore a capacitor is located in parallel at the output of PCC.

For circuit analysis consider Figure 4.29. The impedance offered by the capacitor ($1/C\omega$) for the low-frequency components of the inverter output voltage is much higher than that of the inductance ($L\omega$). Therefore these voltage components will be present in the capacitor. However, for high-order harmonics, the impedance of the inductance is much greater than that of the capacitor, so that high-frequency components of the output voltage will appear on the terminals of the inductance and practically null in the capacitor. As a result there will be an output voltage at the capacitor terminals without the

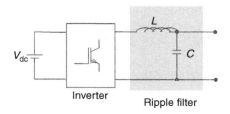

FIGURE 4.29 Output filter for eliminating high-frequency components.

high-frequency components. On the other hand, to obtain a good dynamic response, one has to choose the values of L and C carefully. Thus, if the values of *L* and *C* are too small the ripple of the commutations may not be sufficiently reduced, however if *L* and *C* have high values, they do not only increase the physical size of the equipment but the system response will be slower and will result in a poorer tracking capability. Therefore, the choice of the parameters *L, C* require a compromise between the response speed and the ripple of the output signal.

The criteria proposed here allows you to set margins for these elements with a fairly wide range in many situations. Computer simulations are helpful for this task.

4.5.2 Simulation Models

In Section 4.6, a laboratory prototype of a power system including an APF in parallel with the load will be developed. In the previous section a detailed simulation of the compensated system is performed. The modeling and simulation from previous studies will help in the task of final design. An unbalanced nonlinear three-phase load is considered. A single-phase bridge rectifier with capacitive load on the dc side and a single-phase bridge rectifier with inductive load on the dc side has been considered in phases 1 and 2, respectively. A linear resistor is located in phase 3. Figure 4.30 shows the simulation model.

An APF in parallel with the load is connected to the PCC. A three-phase four-wire APF using a conventional three-leg converter, split-capacitor inverter topology, in order to have the fourth conductor was used, Figure 4.27. Two control strategies were considered: unity power factor and sinusoidal balanced current. Three building blocks are distinguished in

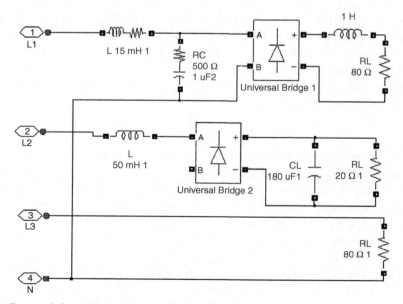

FIGURE 4.30 Nonlinear unbalanced three-phase load for Example 4.5.

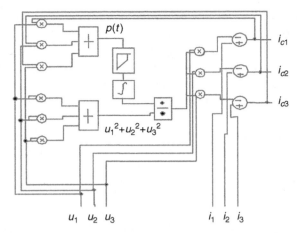

FIGURE 4.31 Block diagram of the compensation current estimator of the active filter of Example 4.5.

the control system: the block to generate the sinusoidal direct sequence-phase voltages (only relevant for the strategy of sinusoidal balanced current), Figure 4.28; the block to generate the reference of the compensating currents, Figure 4.31, and the block to set the trigger signals of each power electronic device, Figure 4.19. Simulation models for the control system blocks and the power circuit models can be found in Appendix I.

A specific description is required for the reference currents generator used in Example 4.5, Figure 4.31, which is the same as that used in the experimental prototype of Section 4.6 [19]. Figure 4.31 shows the block diagram of the system that determines the reference currents in the control, (4.23) or (4.32), depending on the type of compensation: unity power factor or sinusoidal balanced current. In the first case the voltage inputs are the voltages measured at the PCC, and in the second case they will be the outputs of sinusoidal direct sequence-phase voltages generator, Figure 4.28. The determination of the real power of the load, P, requires a low-pass filter whose input is the instantaneous power $p(t)$. This is expressed as,

$$p(t) = P + p_{ac}(t) \tag{4.64}$$

where $p_{ac}(t)$ is an oscillating component around the mean value P and with a frequency of at least twice the frequency of the voltage. By suitable choice of the cut-off frequency for the low-pass filter, the average power P is obtained at its output. This part of the circuit produces unavoidable phase shifts in its operation, and to overcome these problems, Figure 4.31 includes a feedback loop with an integrator block. In effect, the response of the low-pass filter is the input of an integrator; that in steady state, after obtaining current compensation (4.23) or (4.32), the low-pass filter reaches the sum $u_1 i_{c1} + u_2 i_{c2} + u_3 i_{c3}$, whose average value is zero. As a result, the integrator output remains unchanged in the previously estimated value of P. This is the situation until the operation of the load varies. An example illustrates the performance of the system for the two compensation strategies.

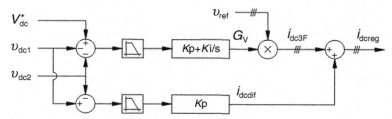

FIGURE 4.32 Block diagram for the regulation of the dc side voltages of Example 4.5.

Another important control block, from the practical point of view, is the regulation of the dc side voltages. Figure 4.32 shows the block diagram for this subsystem. As discussed in Section 4.3.3, a set of currents, i_{dc3F}, in phase with the supply voltages is necessary to provide for the internal losses of the converter. The difference between the total dc-side voltage and its reference V^*_{dc} is passed through a low-pass filter, to compensate only the average value and allow the compensation of the oscillating instantaneous power of the load compensation. The proportional gain K_P can be adjusted to provide an adequate dynamic behavior; and an integral gain can be included to achieve a fine regulation of the dc voltage in steady state. On the other side, the difference in the voltages of the two capacitors must be regulated to avoid imbalance between them in the presence of zero-sequence currents phase with unbalanced loads. The dc component of these zero-sequence currents can produce an accumulative effect in the imbalance of these voltages, even with small errors, for example, in the average values in the switching control of the output current. With a feedback of this voltage difference and a low-pass filter to obtain its dc component, a proportional control can be made to introduce a dc current, i_{dcdif}, in the current references to compensate these offsets and balance the voltages of the capacitors.

EXAMPLE 4.5

A sinusoidal unbalanced three-phase voltage supplies to the unbalanced nonlinear three-phase load of Figure 4.30. The rms values of the phase voltages are 214, 198, and 225 V, respectively, and the parameters of the source impedance are $R= 1 \Omega$, $L= 1$ mH. The system is compensated by a shunt APF according to the scheme of Figure 4.27, by two compensation methods: (a) applying a unit power factor compensation, (b) applying a sinusoidal balanced current compensation.

A simulation platform consisting of the blocks described in this section has been designed in the SimPowerSystems environment. The generator of the reference currents includes two essential subsystems. One for estimation of the compensation current due to the load, Figure 4.31, and another for the regulation of the capacitors of the dc side of the converter, Figure 4.32. Detailed schemes of the different subsystems are provided in Appendix I.

1. Figure 4.33 shows the waveforms of the three-phase voltages at PCC for unit power factor compensation before and after compensation. A shunt compensation (current

compensation) does not compensate for the voltages of the PCC; voltages remain unbalanced. It is true that the presence of the shunt APF provides a slight improvement in the voltages of PCC. In effect, correcting the imbalance of currents reduces the imbalance due to the voltage drops in the source impedances, although the high-frequency harmonics introduced by the converter also affect the PCC voltage.

The shunt APF is connected at time $t = 0.1$ s; Figure 4.34 shows the source currents. Before compensation the source currents match the load currents; after compensation source currents are sinusoidal and have the same imbalance as the supply voltages. However, the power factor of the APF + load set has a value close to unity.

Figure 4.35a shows the waveforms of the voltage and current of phase 1. After compensation the phase voltages and currents are collinear. However, as discussed in Section 4.2, the compensated neutral current is not zero; Figure 4.35b shows the neutral current for this compensation strategy.

2. Now, the compensation strategy of sinusoidal balanced current is applied. The subsystem called generator of sinusoidal direct sequence-phase voltages is included in the control system of the APF. Figure 4.36 shows the waveforms of the source current. After compensation ($t = 0.1$ s) the currents are balanced and sinusoidal. The limitations of the converter along with the passive elements to track the reference current, determine the operation of the APF.

Figure 4.37a shows the waveforms of voltage and current of phase 1. This time the currents are not collinear with the voltages although their phase differences are very small. The mitigation of the neutral current with this compensation strategy is higher, as shown in Figure 4.37b.

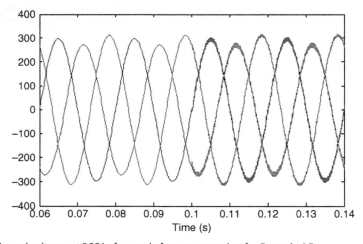

FIGURE 4.33 Unbalanced voltages at PCC before and after compensation for Example 4.5.

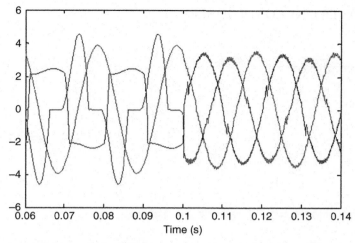

FIGURE 4.34 Source currents before and after with compensation strategy of unity power factor.

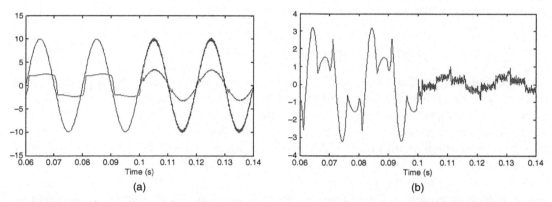

FIGURE 4.35 Waveforms of interest for unit power factor compensation before and after compensation. (a) Voltage (scale 1/30) and current of phase 1. (b) Neutral current.

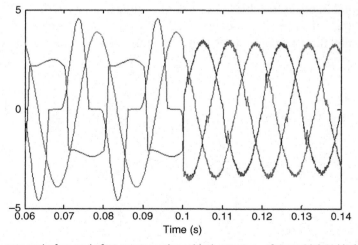

FIGURE 4.36 Source currents before and after compensation with the strategy of sinusoidal and balanced current.

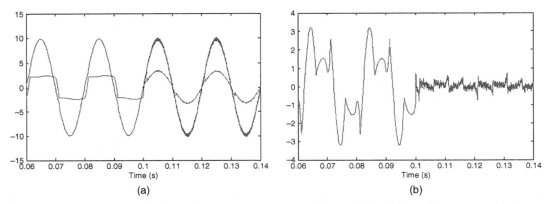

FIGURE 4.37 Waveforms of interest with the compensation strategy of sinusoidal and balanced source current before and after compensation. (a) Voltage (1/30 scale) and current of phase 1. (b) Neutral current.

4.6 Experimental Prototype of APF

An experimental platform has been set to perform a validation of the proposals and to verify the behavior of the control strategies. A parallel compensation has been applied to an unbalanced and nonlinear load supplied by an unbalanced set of voltages. Figure 4.38 shows the power circuit per phase of the shunt APF with their corresponding matching transformers and the passive elements for the filtering of the high-frequency components.

A shunt three-phase IGBT converter (Semikron SKM50GB123D) has been selected for practical implementation. The converter comprises a common dc link composed by two electrolytic capacitors, C_{dc+} and C_{dc-} each of 2200 μF, 400 V [4,5]. The middle point of the dc link is connected with the neutral wire of the three-phase line. The turn ratio of the matching transformer T_P is 1:2 in order to achieve reasonable voltage values in the dc link with respect to the output voltage of the shunt converter. The filtering inductance L_P is 50 mH, and the values of the parallel filtering capacitor C_P and resistance R_P are 1 μF and 22 Ω, respectively.

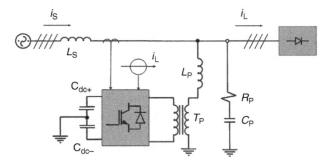

FIGURE 4.38 Scheme of compensation experimental prototype with shunt APF.

The switching control of the active converters and the calculation of the compensation references are made with a data acquisition and control system (dSPACE-DS1103) [20]. The sampling time of the main processor was set to 50 µs in the case of the shunt compensation. Detailed information is provided in Appendix II. The target references for the compensation are the fundamental positive-sequence components of the load voltages v_L in order to achieve the lowest unbalance and harmonic components in the compensated system. In this case, when the parallel compensation is performed, the reference for the compensating current i_C is:

$$i_C^* = i_L - i_S^*$$
(4.65)

where i_S^* is a set of balanced currents in phase with the fundamental direct-sequence component of the load voltage that transports the average active power of the load and the losses of the conditioner (4.41). This way, the unbalanced and harmonic components of the load current are compensated.

The compensating current of the shunt converters is built with a periodic sampling control, PS, at 20 kHz, in an external circuit, using the deviation of the compensating current i_C respect to its reference i_C^*. On the other side, an independent measurement system [9] to calculate different values of voltage, current, and power is implemented in another data acquisition card (dSPACE-DS1005), with a signal conditioning system formed of six voltage sensors (LEM-LV25-P) and eight current sensors (LEM LA35-NP). These signals are taken simultaneously to avoid phase differences between the measurements, in order to enhance the accuracy of the calculated results. The configuration of the virtual instrument was made following the EN 61000-4-7 and EN 61000-4-30 recommendations [21,22], using a window equal to ten cycles of the fundamental component and a sampling frequency of 6400 Hz to avoid problems of aliasing and leakage errors.

Finally, the load is composed by three different single-phase loads. In phase 1, it is a single-phase diode rectifier with a high capacitor and a resistive load in the dc side, and a smoothing reactor in the ac side. For phase 2, it is a single-phase diode rectifier with a series RL load in the dc side. And a resistive load branch is in phase 3. This load is fed through a variable autotransformer with an unbalanced set of voltages (214,198,225 V), consisting of an rms nominal value of 230 V with a small distortion similar to the existing in the supply network of the laboratory.

Figure 4.39 shows the voltage and current waveforms of the load fed with this power supply, as well as the corresponding magnitudes in the supply side during compensation. The supply currents i_{Sabcn} are practically balanced, sinusoidal, and in phase with the fundamental positive-sequence component of the supply voltage, despite its distortion and unbalance.

Figure 4.40 shows the harmonic spectra of the waveforms of voltage and current of one phase after compensation. The shunt compensation mitigates the current harmonics but does not cancel the voltage harmonics. The spectra are another way to visualize the results shown by the waveforms, Figure 4.39.

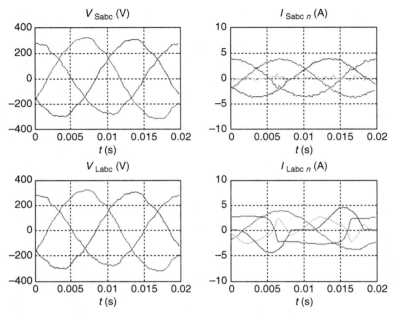

FIGURE 4.39 Experimental results for the voltage and current waveforms at the supply and load sides, for the compensated system with the sinusoidal and balanced current strategy.

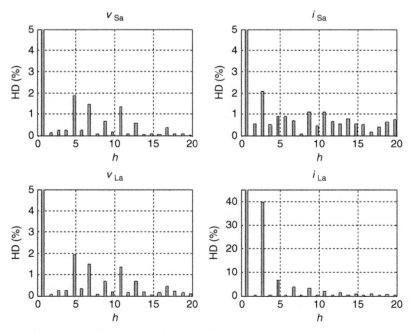

FIGURE 4.40 Harmonic spectra for the voltage and current of phase 1, at the supply and load sides.

Table 4.1 presents the calculations obtained with the virtual instrument, corresponding to parameters and factors in the three phases before compensation in the PCC [23]: total harmonic distortion of voltage, VTHD; current, ITHD; the real power, P; the apparent power, S; power factor of displacement, PF1; and power factor, PF. Table 4.2 shows the same indices after compensation. In Chapter 2 you can find details on these magnitudes and indices.

Comparison of Tables 4.1 and 4.2 shows a significant reduction in harmonic current distortion, ITHD; indicated through phases 1 and 2, which is where nonlinear loads are located. The active power consumption is balanced in all three phases. The displacement factors and power factors of each phase are close to one. Table 4.3 presents the results of overall unbalance indices.

In Table 4.3, VUF is the voltage unbalance fundamental index and IUF is the current unbalance fundamental index defined as,

Table 4.1   Measured per Phase Parameters Before Compensation

	Phases		
	1	**2**	**3**
VTHD	3.168	3.343	2.969
ITHD	40.520	33.090	2.880
P (W)	417.9	427.9	607.0
S (VA)	536.6	475.0	607.0
PF1	0.8399	0.9508	1.0000
PF	0.7788	0.9008	1.0000

Table 4.2   Measured per Phase Parameters After Compensation

	Phases		
	1	**2**	**3**
VTHD	3.029	3.360	2.948
ITHD	4.792	6.322	3.880
P (W)	557.7	516.6	587.9
S (VA)	559.7	518.6	588.8
PF1	0.9981	0.9987	0.9995
PF	0.9964	0.9961	0.9985

Table 4.3   Unbalance Parameters Before and After Compensation

(%)	Before	After
VUF	4.677	4.752
IUF	37.950	2.077
LU	38.280	5.187

$$\text{VUF} = \frac{V_{u1}}{V_1^+}, \quad \text{IUF} = \frac{I_{u1}}{I_1^+} \tag{4.66}$$

Similarly, Table 4.3 lists the load unbalance index, LU, as stated in the Std. 1459 [23].

$$\text{LU} = \frac{S_{U1}}{S_1^+} \tag{4.67}$$

The results confirm the strong compensation of load imbalance after compensation.

4.7 Summary

This chapter has described the operating principles, structure, control circuits, compensation strategies, and design rules of the so-called APFs in parallel connection as corrector equipment of the load current. First, the operational principle of a shunt APF was presented, as a controlled current source for compensation of unbalanced nonlinear three-phase loads. The proposed APF as compensation system was effective in harmonic elimination, displacement factor correction, cancellation of unbalances and neutral current mitigation. Second, an APF circuit structure was described. Thus, the fundamentals and operation modes of PWM-type power inverters, showing how they construct a voltage wave with different tranches of voltage from a constant value at the dc side, are established. The ability of an inverter to synthesize different waveforms made possible its use as an active compensator. Next, by means of different schemes, PWM techniques common in the APFs design were presented. Third, the compensation strategies of instant compensation, unit power factor, and compensation of balanced sinusoidal current were developed through an optimizing process of the instantaneous norm or rms value of the source current. From these principles, as final part, it was possible to approach the design of a control circuit for an APF to compensate for nonlinear three-phase four-wire loads. First, the design rules for energy storage elements and capacitor voltage regulation at the dc side of the inverter were established. Second, a complete power system compensated through a shunt APF was developed. To accomplish this task, a simulation platform was developed, which enabled the testing of the different compensation strategies. Strategies of unity power factor and sinusoidal and balanced current were put into practice. The simulation platform was the basis for an experimental implementation that enabled visualization of the waveforms of interest and measurements of the most relevant results of the system operation.

In this way, the chapter has shown how the APF is one of the active power conditioner configurations for load compensation and how it contributes to improving the quality of the electric wave.

References

[1] Akagi H. Active harmonic filters. Proc. IEEE 2005;93(12):2128–41.

[2] Peng FZ. Harmonic sources and filtering approaches. IEEE Ind Appl Mag 2001;7:31–7.

[3] Akagi H, Watanabe EH, Aredes M. Instantaneous power theory and applications to power conditioning. Wiley-IEEE Press; New York, USA, 2007.

[4] Ghosh A, Ledwich G. Power quality enhancement using custom power devices. New York: Kluwer Academic Publishers; 2002.

[5] Cavallini A, Montanari GC. Compensation strategies for shunt active-filter control. IEEE Trans Power Electr 1994;9(6):587–93.

[6] Rashid MH. Power electronics handbook. London: Academic Press; 2001.

[7] Perales MA, Carrasco JM, Sánchez JA, Terrón L, García L. Predictive middle point modulation: a new modulation method for parallel active filters. EPE Graz.

[8] Dixon JW, Tepper S, Moran LA. Practical evaluation of different modulation techniques for current-controlled voltage source inverters. IEE Proc Electric Power Appl 1996;143(4):301–6.

[9] Holmes DG, Lipo TA. Pulse width modulation for power converters. New York: IEEE Press; 2003.

[10] Buso S, Malesani L, Mattavelli P. Comparison of current control techniques for active filter applications. IEEE Trans Ind Electr 1998;45(5):722–9.

[11] Malesani L, Tenti P. A novel hysteresis control method for current controlled voltage-source PWM inverters with constant modulation frequency. IEEE Trans Ind Appl 1990;26(1):88–92.

[12] Aredes M, Häfner J, Heumann K. Three-phase four-wire shunt active filter control strategies. IEEE Trans Power Electr 1997;12(2):311–18.

[13] Salmerón P, Montaño JC, Vazquez JR, Prieto J, Perez A. Compensation in nonsinusoidal, unbalanced three-phase four-wire systems with active power-line conditioner. IEEE Trans Power Deliv 2004;19(4):1968–2197.

[14] Herrera RS, Salmerón P, Kim H. Instantaneous reactive power theory applied to active power filter compensation: different approaches, assessment, and experimental results. IEEE Trans Ind Electr 2008;55(1):184–96.

[15] Superti Furga G, Tironi E, Ubezio G. Shunt active filter for four wire low-voltage systems: theoretical operating limits and measures for performance improvement. Eur Trans Elec Power 1997;7(1):41–48.

[16] Thomas T, Haddad K, Joos G, Jaafari A. Design and performance of active power filters. IEEE Ind Appl Mag 1998;4(5):38–46.

[17] Jou HL, Wu JC, Chu HY. New single-phase active power filter. IEEE Proc Elec Power Appl 1994;141(3):129–34.

[18] Wu JC, Jou HL. Simplified control method for the single-phase active power filter. IEEE Proc Electric Power Appl 1996;143(3):219–24.

[19] Tepper JS, Dixon JW, Venegas D, Morán L. A simple frequency-independent method for calculating the reactive and harmonic current in a nonlinear load. IEEE Trans Ind Electr 1996;43(6):647–54.

[20] Jacobs J, Detjen D, Karipidis CU, De Doncker RW. Rapid prototyping tools for power electronic systems: demonstration with shunt active power filters. IEEE Trans Power Electr 2004;19(2):500–7.

[21] IEC 61000-4-7, Testing and measurement techniques. General guide on harmonics and interharmonics measurements and instrumentation, for power supply systems and equipment connected thereto, IEC.

[22] European Standard, UNE-EN 61000-4-30. Técnicas de ensayo y de medida. Métodos de medida de la calidad de suministro, AENOR 2009.

[23] IEEE Power and Energy Society, IEEE Standards definitions for the measurement of electric power quantities under sinusoidal, nonsinusoidal, balanced or unbalanced conditions, 2010.

5

Series Active Power Filters

This chapter describes the most common control strategies applicable to active filters of series connection, SAF. From the equivalent single-phase circuit, the behavior of the active filter in the steady state is analyzed. This allows theoretical analysis of three different control strategies:

- Control by detection of the source current.
- Control by detection of the load voltage.
- Hybrid control, which combines source current and load voltage detection.

These strategies are analyzed for two types of nonlinear loads of dual behavior against distortion: load that generates current harmonics and load that generates voltage harmonics. The analysis is performed from the point of view of the distortion produced by the load and the distortion due to other nonlinearities than those produced by the load.

In addition, the state model for a series active filter (SAF) configuration has been obtained. The behavior of the three proposed strategies is analyzed by using system state equations, which allows design rules to be established for each of them.

The theoretical analysis is contrasted with the results obtained using different simulation examples. A simulation platform based on MATLAB–Simulink has been developed for this purpose. The results obtained have allowed us to establish the most appropriate strategy and the active filter topology for each type of nonlinear load from the point of view of harmonic elimination.

P. Salmerón Revuelta, S.P. Litrán, and J.P. Thomas: Active Power Line Conditioners. http://dx.doi.org/10.1016/B978-0-12-803216-9.00005-5

Finally, to verify the proposed analysis, an experimental prototype is presented. Here, the design criteria established throughout the chapter are applied.

5.1 Introduction

Harmonic distortion is an issue that has been present since the beginning of electrical engineering [1,2], although it has become more relevant in the last decades with the proliferation of power electronic loads that exhibit general nonlinear behavior [3]. In the last few years, the use of active power filters (APF) has been developed as a method to compensate for the harmonic distortion of an electrical power system. An APF is a static compensation system based on a pulse width modulation (PWM) electronic converter, which can be connected in parallel or in series with the load [4].

APFs can provide harmonic, reactive power, and/or neutral current compensation in electrical networks [5]. This is possible nowadays, mainly due to the evolution of the power electronic devices technology, the development of different configurations and topologies, and the different control strategies proposed [6]. On the other hand, APFs have been also used to compensate for voltage harmonics, to regulate the voltage supply, to eliminate flicker, and to enhance the voltage balance in three-phase systems [7]. This wide range of targets is achieved individually or in combination with other compensation devices, depending on the compensation strategy, the configuration, and the requirements of the problem that needs solving.

The most extended APF in the scientific literature, and the most used in low-voltage installations is the parallel APF [8–10]. This APF produces a harmonic current with the same amplitude and opposite phase to the current harmonics of the load. This type of configuration has been demonstrated to be very effective with loads such as SCR-controlled rectifiers with high inductances at the dc side, and cycloconverters or regulators made with branches of two antiparallel SCRs.

These types of loads can be considered as nonlinear loads that generate harmonics like the current sources, named harmonic current source (HCS). However, other loads such as diode rectifiers with high capacitors at the dc side are considered as nonlinear loads that generate harmonics in the voltage source, named harmonic voltage source (HVS). It has been demonstrated that the compensation of HVS-type loads with shunt APFs do not completely cancel the load current harmonics, often producing a very incomplete compensation [11].

The utilization of a series of active filters has been proposed [12] to compensate for these HVS loads. It has been demonstrated that this is the most suitable configuration for compensating for this kind of load, although they are less common in low-voltage installations.

Besides the "pure" active filters, shunt APF, and series APF, other topologies have been tested, such as a combination of active filters and passive filters. The passive filters are LC branches tuned to the most significant harmonics of the load. These configurations that include associations of APF and passive filters are generically termed hybrid filters [13,14].

In this chapter, the use of SAFs for improving power quality is studied. Thus, in Section 5.2 of the chapter, the basic concepts of the APF and control strategies to be applied are introduced. This section analyzes each of them in a circuit model in steady state. In Section 5.3 the state model of a system with SAF is explored. Here, a dynamic behavior analysis of the system for each strategy is performed and design criteria are proposed depending on the objectives of the design. Finally, in Section 5.4 an experimental prototype of a series of APF for harmonic compensation of HVS-type load is developed. The results allow us to compare the effectiveness of each of the strategies and verify the performance of the compensation equipment in two different voltage situations: where the supply voltage is sinusoidal and where it is nonsinusoidal.

5.2 Series Active Power Filters

In the SAF topology, the APF is connected in series with the load, as shown in Figure 5.1. The connection to the system is through a coupling transformer. An LC filter is connected at the inverter output (L_{FR} inductance and C_{FR} capacity). Its function is to eliminate high-frequency components that occur due to the switching of power electronic devices. The inverter is voltage source type, using IGBTs as switching devices. At the dc side of the inverter, it is possible to connect one dc source for a three-wire system, or two dc sources, as shown in Figure 5.1, to access the fourth wire needed for a three-phase four-wire system.

To determine the reference signal three main strategies have been proposed in the literature, to date [12,15,16]:

- Control strategy by detection of the source current. This consists of generating a voltage proportional to source current harmonics at the APF output.
- Control strategy by detection of the load voltage, with the idea that the APF should generate the voltage with the same harmonic content of the voltage across the load, but with opposite phase.

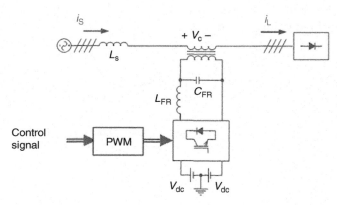

FIGURE 5.1 SAF topology and connection.

- Hybrid strategy, where the APF generates a voltage that combines the previous two strategies.

In what follows, each strategy and its operational principle will be described and analyzed.

5.2.1 Source Current Detection Control

With this compensation strategy, the SAF generates a voltage proportional to the harmonics of the source current [16], this is

$$v_{Ch} = k \, i_{Sh} \tag{5.1}$$

Figure 5.2 shows the single-phase equivalent circuit for an h harmonic different from fundamental harmonic. In the same, Z_{Sh} represents the source impedance at the frequency of the h order harmonic and V_{Lh} is the source voltage value that models the load that produces voltage harmonics (HVS). For the h harmonic, the expression for the source current is given by

$$I_{Sh} = \frac{V_{Sh}}{(Z_{Sh} + k)} - \frac{V_{Lh}}{(Z_{Sh} + k)} \tag{5.2}$$

Expression (5.2) shows how the current source depends on two voltage terms: voltage harmonic at the supply side and the voltage harmonic at the load side. A value of k such that $k \gg Z_{Sh}$ reduces the value of the harmonics of the source current this is

$$I_{Sh} \approx 0; \quad k \gg Z_{Sh} \tag{5.3}$$

On the other hand, the harmonic voltage at the point of common connection (PCC) can be obtained by the equation

$$V_{PCCh} = \frac{k}{(Z_{Sh} + k)} V_{Sh} + \frac{Z_{Sh}}{(Z_{Sh} + k)} V_{Lh} \tag{5.4}$$

This equation includes two terms, one related to the supply voltage harmonics and other related to the load voltage harmonics.

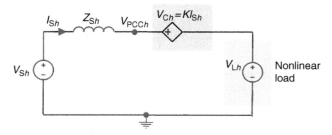

FIGURE 5.2 Single-phase circuit of a system with SAF and HVS load, control strategy: $V_{Ch} = k i_{Sh}$.

When $k \gg Z_{Sh}$ it is possible to reduce the value of the term that depends on the load side voltage. This is not the case for the first term of (5.4), which always includes the harmonics due to the supply voltage. Such a situation would lead to an ideal value, $k = \infty$, but from the viewpoint of control would be an impossible goal to reach. Therefore, the optimal value of k is difficult to obtain due to its dependence on the source impedance, which is usually a variable parameter and not easy to determine. However, the primary purpose of the active filter is to reduce the harmonic content of the voltage at the PCC due to the nonlinearity of the load. Therefore, this control can achieve this goal so that we can say that the SAF "isolates" the PCC of voltage harmonics produced by the load.

EXAMPLE 5.1

The strategy $v_{Ch} = ki_{Sh}$ control will be applied to a HVS-type load. Figure 5.3 shows the circuit scheme to be simulated. The load consists of an uncontrolled three-phase rectifier with a capacitor of 2200 μF and a resistor of 50/3 Ω connected in parallel at the dc side, which is a typical HVS load. The source is sinusoidal, with an rms value of the phase voltage of 100 V and a frequency of 50 Hz. A resistance of 1.8 Ω and an inductance of 2.8 mH in series have been included to model the equivalent impedance from the PCC. The active filter is connected through three transformers of ratio 1:1 with an LC ripple filter of 13.5 mH inductance and 50 μF capacity. The inverter is a three-phase IGBT Bridge. At the dc side a constant source of 100 V has been connected. This example has been simulated in MATLAB–Simulink using device models of the SymPowerSystem library.

Figure 5.4 shows the waveforms of the voltage at the common connection point, v_{PCC}, and source current, i_S, when the active filter is not connected. The waveform THDs are: 20.95% for the current and 13.59% for the voltage. The most significant harmonics for the voltage and the current are the 5th, 7th, 9th, and 11th.

Figure 5.5 shows the waveforms when the SAF is connected and the control strategy is applied with a k constant value of 50. The current has a THD of 4.67% and the voltage at the PCC is 3.42%. The results corroborate the behavior expected by expression (5.2), where a high k value ($k \gg Z_{Sh}$) reduces the harmonic content of the source current.

On the other hand, the voltage at the PCC is practically sinusoidal. Expression (5.4) establishes a dependence on the source voltage and the voltage at the load side. If it is taken into account that the voltage at the source side is sinusoidal, the voltage at the common connection point will also be sinusoidal, if the k choice ensures that the APF eliminates voltage harmonics caused by the load. Therefore, this control strategy makes possible the elimination of the harmonic components of the voltage at the PCC produced by the load. Figure 5.6 shows the harmonic spectra of source current and voltage at the PCC, before and after the connection of the SAF.

In this compensation strategy it is possible to predict what k value will get the lowest THD. The Figure 5.7 shows a graphical representation of current THD for different values of the proportionality constant, k. It can be observed how from a k value of about 70, the THD deteriorates slightly, which is due to the appearance of a higher ripple on the current signal. This depends on the values of L and C of the ripple filter, which is connected to the output of the inverter. *LC* filter design becomes an important issue for the SAF. A detailed analysis will be left for a later section.

Another important issue is to analyze how the presence of voltage harmonics at the source side effects the compensation. To study this situation, a third harmonic of 8% with an initial phase of π rad and fifth harmonic of 5% with an initial phase of 0 rad have been added. So the applied voltage is

$$v(t) = V_p \left[\sin(\omega_1 t) + 0.08 \sin(3\omega_1 t + \pi) + 0.05 \sin(5\omega_1 t) \right] \tag{5.5}$$

This waveform has been chosen as one of those included in the IEC 61000 standard, in its section on immunity to voltage harmonics.

Figure 5.8 shows the waveforms of the source current and voltage at the PCC for phase "a," when the active filter is not connected. In this case, note the presence of a third harmonic in the source current before compensation, this is due to the presence of this harmonic in the source voltage.

When the active filter is connected with $k = 50$, the source current THD improves significantly; decreasing from 15.22% to 5.61%. As regards the voltage at the PCC, the THD decreases from 17.32% to 11.75%. This minor reduction compared to the case of sinusoidal supply voltage is because the active filter can only mitigate the harmonics generated by the load and not the harmonics present at the PCC due to the supply voltage. The waveforms of interest are shown in Figure 5.9.

Figure 5.10 shows the frequency spectra of the current and voltage before and after the filter connection, which allows comparison of the harmonic content in both situations. As was already indicated, SAF "isolates" the PCC of voltage harmonics produced by the load.

On the other hand, the configuration shown in Figure 5.1 is not appropriate in the case of HCS-type loads. Indeed, there is a series circuit where the source current is always equal to the load current and therefore will not be able to cancel the harmonic currents. Figure 5.11 shows the equivalent single-phase circuit of Figure 5.2 in which the HVS load is replaced by an HCS-type load, modeled by a current source of value I_{Lh}. The circuit shows that this configuration cannot eliminate current harmonics. Respect to the voltage at the PCC, it is given by

$$V_{PCCh} = V_{Sh} - Z_{Sh} I_{Lh} \tag{5.6}$$

So, it demonstrates that this configuration does not allow compensation for this type of load.

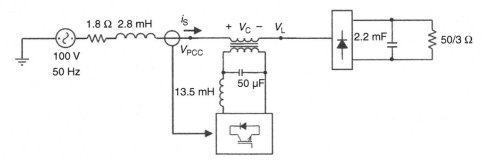

FIGURE 5.3 Power circuit. Example 5.1.

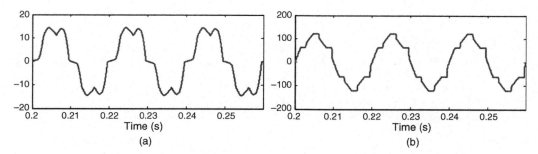

FIGURE 5.4 Example 5.1, waveforms without active filter. (a) Source current; (b) PCC voltage.

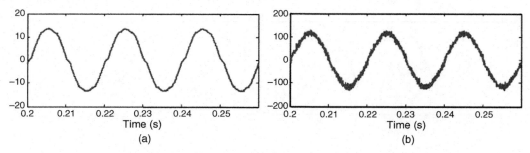

FIGURE 5.5 Example 5.1, waveforms with active filter. (a) Source current; (b) PCC voltage.

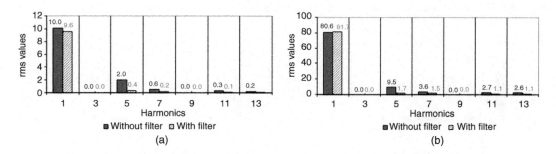

FIGURE 5.6 Harmonic spectrum. (a) Source current; (b) PCC voltage.

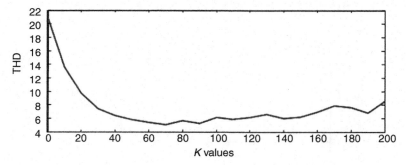

FIGURE 5.7 THD source current versus *k* constant.

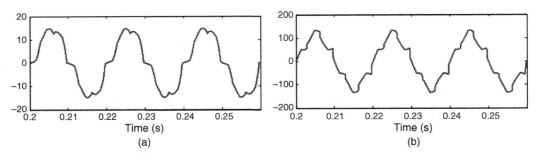

FIGURE 5.8 Example 5.1, waveforms without filter, distorted voltage supply. (a) Source current; (b) PCC voltage.

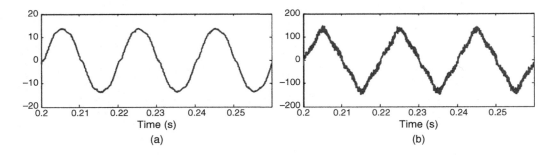

FIGURE 5.9 Example 5.1, waveforms with filter, distorted voltage supply. (a) Source current; (b) PCC voltage.

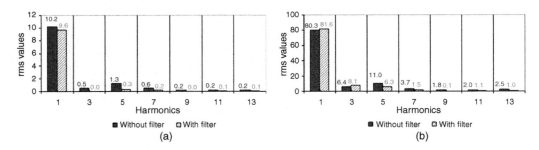

FIGURE 5.10 Example 5.1, harmonic spectrum with distorted voltage supply. (a) Source current; (b) PCC voltage.

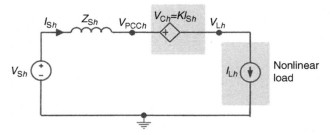

FIGURE 5.11 Single-phase circuit of a system with SAF and HCS load, control strategy: $V_{Ch} = kI_{Sh}$.

5.2.2 Load Voltage Detection Control

This is a control strategy that detects the load voltage and generates a waveform, which contains the load voltage harmonics to counterphase. Thus if the voltage at the load terminals is v_{Lh}, (where h is the harmonic order), for all harmonics different from h ($h\neq1$) it should satisfy

$$v_{Ch} = -v_{Lh}$$

(5.7)

For a HVS-type load, the voltage at the PCC due to a harmonic h in the load is zero, $V_{PCCh} = 0$, as it is shown in Figure 5.12.

However, the detection of the load voltage harmonics is generally dependent upon the sensitivity of the instrumentation, as well, for in this compensation strategy, the voltage generated by the APF can be expressed in the form

$$v_{Ch} = -k_v \, v_{Lh}$$

(5.8)

Where k_v represents the relationship between the value detected by the control circuit and the harmonic values of the load voltage signal. Generically, the k_v shows a frequency dependence, though here, in a first approximation it is considered constant.

Accordingly, the voltage at the PCC for any harmonic h is given by the expression

$$V_{PCCh} = V_{Lh}(1 - k_v)$$

(5.9)

For the condition, $k_v = 1$, at the PCC there will be no distortion due to the load. This represents a condition of ideality that from a practical point of view it would be impossible to achieve. Furthermore, the source current is given by the expression.

$$I_{Sh} = \frac{1}{Z_{Sh}} V_{Sh} - \frac{(1-k_v)}{Z_{Sh}} V_{Lh}$$

(5.10)

Thus, if the source voltage is sinusoidal, (i.e., $V_{Sh} = 0$, $\forall h\neq1$) the source current harmonics can also be mitigated when the condition $k_v = 1$ is fulfilled. However, it is clear that if the source voltage contains harmonics of order h, they will also be present in the spectrum of the source current due to the presence of the first term in (5.10).

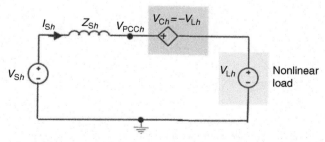

FIGURE 5.12 Single-phase circuit of a system with SAF, control strategy: $V_{Ch} = -V_{Lh}$.

EXAMPLE 5.2

Here, Example 5.1 will be analyzed according to the control strategy, $v_{Ch} = -v_{Lh}$.

Figure 5 shows the waveform before compensation. Figure 5.13 shows current and voltage waveforms, once the active filter is connected. For simulation it was considered that $k_v = 0.95$. The voltage THD and current THD are 2.79% and 3.47%, respectively. When the supply voltage is sinusoidal, the source current is practically sinusoidal also.

If it is considered that the voltage source is distorted and it is defined by the expression

$$v(t) = V_p \left[\sin(\omega_1 t) + 0.08 \sin(3\omega_1 t + \pi) + 0.05 \sin(5\omega_1 t)\right]$$

The current source and the voltage at the PCC have the waveforms shown in Figure 5.14. The current THD is 5.61% and the voltage THD is 11.98%. Clearly there is significant improvement in current THD relative to the situation before the compensation, that was 15.22%, not so with the voltage THD where the reduction is less. This is because these harmonics are present in the supply voltage, thus cannot be eliminated by the active filter according to (5.10).

Figure 5.15 shows harmonic spectra before and after connecting the SAF.

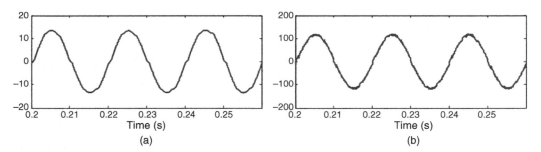

FIGURE 5.13 Example 5.2, waveforms with active filter. (a) Source current; (b) PCC voltage.

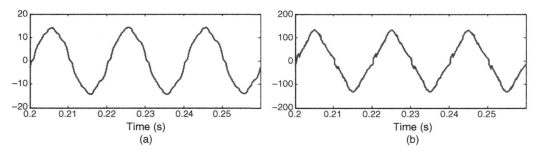

FIGURE 5.14 Example 5.2, waveform with SAF connected and source voltage distorted with 3rd and 5th order harmonics. (a) Source current; (b) PCC voltage.

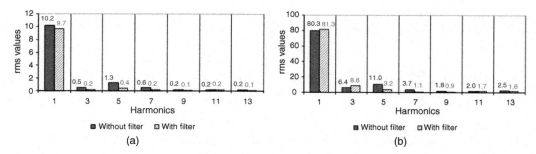

FIGURE 5.15 Example 5.2, harmonic spectrum with distorted voltage supply. (a) Source current; (b) PCC voltage.

5.2.3 Combined Control

After the above analysis of the two compensation strategies was applied to the SAF, the following conclusions could be drawn:

- Control strategy by detection of the source current.
  - It is possible to mitigate source current harmonics when the condition $k \gg Z_S$ is satisfied regardless of voltage harmonics present in the power supply and those generated by the load.
  - Voltage harmonics present at the PCC are reduced when $k \gg Z_S$, and therefore, the SAF "isolates" the PCC from load voltage harmonics.
  - It is impossible to determine the optimal value of k because it depends on the source impedance, as this parameter is usually variable and difficult to know.
- Control strategy by detection of the load voltage.
  - It mitigates the source current harmonics in the case that the voltage at the PCC is not distorted.
  - It is possible to reduce the voltage harmonics at the PCC produced by the load.

Hereinafter, a hybrid strategy will be developed. It is conceivable that this approach will improve the filtering characteristics of the series filter with respect to previous strategies, since it combines both methods in a single expression. So, the voltage generated by the active filter is given by the expression:

$$v_{Ch} = k\, i_{Sh} - k_v\, v_{Lh} \tag{5.11}$$

For the circuit shown in Figure 5.16, the h order harmonic of the source current is given by

$$I_{Sh} = \frac{1}{(Z_{Sh}+k)} V_{Sh} + \frac{(1-k_v)}{(Z_{Sh}+k)} V_{Lh} \tag{5.12}$$

According to (5.12) the harmonic currents are due primarily by voltage harmonics present in the supply voltage (V_{Sh}), and secondly by the harmonics generated by the load voltage (V_{Lh}). In the power system, the supply voltage is generally considered slightly

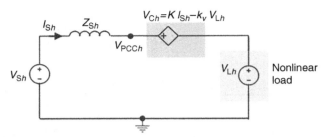

FIGURE 5.16 Single-phase circuit of a system with SAF, combined control strategy: $V_{Ch} = kI_{Sh} - k_v V_{Lh}$.

distorted, so the first term could be reduced to a constant k with a not excessively high value. Furthermore, the second term depends on the voltage harmonics present at the load side. However, the simultaneous presence in the numerator of the $(1-k_v)$ factor and in the denominator of the $(Z_S + k)$ term cause a reduction, first, in the error influence of the instrumentation for load voltage harmonic detection, and on the other, allow the use of a lower k value compared with the control strategy by detection of the source current.

As can be seen from Figure 5.16, the V_{PCC} voltage to an h harmonic is given by the expression

$$V_{PCCh} = \frac{k}{(Z_{Sh}+k)} V_{Sh} + \frac{Z_{Sh}(1-k_v)}{(Z_{Sh}+k)} V_{Lh} \qquad (5.13)$$

Analysis of expression (5.13) allows us to conclude that it is not possible to cancel the voltage harmonics at the PCC due to the supply network, but it is possible to avoid the presence of harmonics produced by the load with a k value reduced, as well as, decrease the influence of the error in the detection of the load harmonic.

EXAMPLE 5.3

Same Example 5.1, where the hybrid strategy is applied to the filter, this is $v_{Ch} = ki_{Sh} - k_v v_{Lh}$.

First a sinusoidal supply voltage is considered. In the detection of load voltage harmonic, an error will be taken into account, for which is chosen a $k_V = 0.95$. Furthermore, the value of k has been set at 10. Waveforms when the filter is connected are shown in Figure 5.17. The current THD is reduced from 20.95% without filter to 0.95% with filter. Regarding the voltage at the PCC, the THD decreases from 13.59% to 0.90%. It is verified that a significant improvement is achieved in the THD of the current and voltage respect to the strategies by detection of the current source and by detection of the load voltage.

Current and voltage spectrum before and after connecting the active filter are shown in Figure 5.18.

Another test of interest is to consider that the supply voltage is distorted. A supply voltage defined by expression (5.5) is assumed. Figure 5.19 shows the waveforms after the SAF connection. The source current THD passes from 15.22% to 0.94%. This result verifies the behavior deduced from the expression (5.12). Therefore, it is possible to reduce the harmonic content of the source current with a *k* value smaller than with the strategy by detection of the source current. Furthermore, the voltage THD at the PCC decreases from 17.32% to 11.38%. This value is close to that of the source voltage defined by (5.5), which is 9.43%. The difference is due to the existence of high-frequency harmonics as a result of switching of the electronic devices, which makes the voltage waveform present a ripple, as shown in Figure 5.19b. The ripple reduction depends on the inductance and capacitance parameters, which constitute the inverter output filter.

Figure 5.20 shows harmonic spectrum before and after connecting the active filter.

Finally, in Table 5.1 the results of the voltage and current THD for the three compensation strategies and different source voltage conditions are summarized. This allows for easy comparison of the results obtained and consideration of the theoretical analysis.

In conclusion, from the point of view of compensation, the hybrid strategy performs better than strategies that use detection of the source current or detection of the load voltage. However, from a practical point of view it has the disadvantage of requiring the measurement six variables (three voltages and three currents) compared to the other strategies that only need three variables. As a consequence, the hybrid strategy requires a larger number of sensors. It is also necessary to determine harmonics of a higher number of variables, which also increases the complexity of the control.

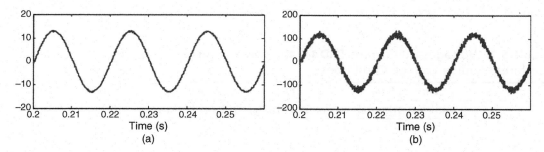

FIGURE 5.17 Example 5.3, SAF connected with combined control strategy. (a) Source current; (b) PCC voltage.

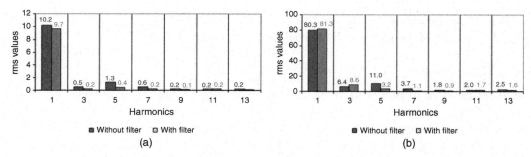

FIGURE 5.18 Example 5.3, harmonic spectrum. (a) Source current; (b) PCC voltage.

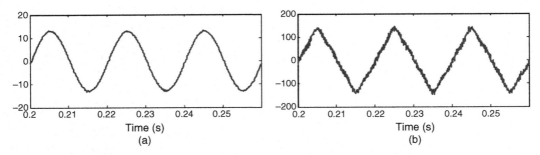

FIGURE 5.19 Example 5.3, waveforms with SAF connected and distorted voltage supply. (a) Source current; (b) PCC voltage.

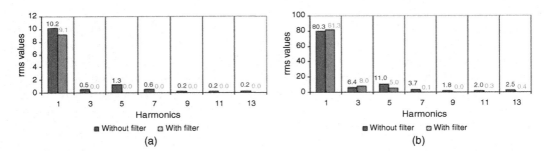

FIGURE 5.20 Example 5.3, harmonic spectrum when supply voltage is distorted: (a) Source current; (b) PCC voltage.

Table 5.1   Voltage and Current THD for the Three Compensation Strategies and Different Source Voltage Conditions

	Without filter		$V_{Ch} = 50\,I_{Sh}$		$V_{Ch} = -0.95\,V_{Lh}$		$V_{Ch} = 10\,I_{Sh} - 0.95\,V_{Lh}$	
	THDv (%)	THDi (%)	THDv (%)	THDi (%)	THDv (%)	THDi (%)	THDv (%)	THDi (%)
Sinusoidal, V_S	13.59	20.95	3.42	4.67	2.79	3.47	0.90	0.95
Nonsinusoidal, V_S	17.32	15.22	11.75	5.61	11.98	5.61	11.38	0.94

5.3 Design of SAF from State Space

In this section the model in the state space of a SAF will be obtained. This will permit a dynamic analysis of the system for each of the proposed control strategies.

5.3.1 State Space

Use of series APFs in power systems can eliminate the harmonics caused by nonlinear loads. Different control strategies have so far been proposed to achieve this end. However, the establishment of each of the proposals has been based on the equivalent circuit analysis and results obtained in laboratory tests rather than the formal analysis control structure [10,17,18].

The representation in the state space of systems has been widely used in control theory since 1960. During that decade the so-called "modern control theory" was developed. The state space is another way to describe a dynamic model of a system. It is applicable to both linear systems and nonlinear systems. This representation is always referred to as internal description model because the internal state variables are fully defined in the model representation.

A system with p inputs $u_i(t)$, $(i = 1, ..., p)$, q outputs $y_j(t)$, $(j = 1, ..., q)$ with n states, can be defined by a vector of state variables as

$$x = [x_1, x_2, ..., x_n] \tag{5.14}$$

The general expression in the state space of a system can be written in the form

$$\begin{aligned} \dot{x}_i &= f_i(x_1, x_2, ..., x_n, u_1, ..., u_p), & i &= 1, ..., n \\ y_j &= g_j(x_1, x_2, ..., x_n, u_1, ..., u_p), & j &= 1, ..., q \end{aligned} \tag{5.15}$$

Where $f_i(\cdot)$ and $g_j(\cdot)$ may be nonlinear functions. The first expression of (5.15) is called the equation of state and the second expression is known as output equation. In linear time invariant (LTI), the state space function is usually written in the form

$$\begin{aligned} \dot{\mathbf{x}}(t) &= \mathbf{A}\,\mathbf{x}(t) + \mathbf{B}\,u(t) \\ \mathbf{y}(t) &= \mathbf{C}\,\mathbf{x}(t) + \mathbf{D}\,u(t) \end{aligned} \tag{5.16}$$

Where $\mathbf{u}(t)$ is the vector of inputs defined by

$$\mathbf{u}(t) = \left[u_1, ..., u_p \right]^T \tag{5.17}$$

With $\mathbf{y}(t)$ as a vector of outputs given by

$$\mathbf{y}(t) = \left[y_1, ..., y_q \right]^T \tag{5.18}$$

The matrix $\mathbf{A}$ has dimension $n \times n$, B is a matrix of dimension $n \times p$, and the matrix $\mathbf{D}$ is of dimension $q \times p$.

When the Laplace transform is applied to the state model defined in (5.16) for zero initial conditions, we get the expression

$$\begin{aligned} s\,\mathbf{I}\,\mathbf{X}(s) &= \mathbf{A}\,\mathbf{X}(s) + \mathbf{B}\,\mathbf{U}(s) \\ \mathbf{Y}(s) &= \mathbf{C}\,\mathbf{X}(s) + \mathbf{D}\,\mathbf{U}(s) \end{aligned} \tag{5.19}$$

Where $\mathbf{I}$ is the identity matrix of the same dimension as the matrix $\mathbf{A}$. Thus, the first term of (5.19), is obtained

$$\mathbf{X}(s) = (s\,\mathbf{I} - \mathbf{A})^{-1}\,\mathbf{B}\,\mathbf{U}(s) \tag{5.20}$$

From the second term of (5.19), the transfer function of the system is obtained

$$\mathbf{G}(s) = \mathbf{Y}(s)\mathbf{U}^{-1}(s) = \mathbf{C}(s\,\mathbf{I} - \mathbf{A})^{-1}\,\mathbf{B}\,\mathbf{U}(s) + \mathbf{D} \tag{5.21}$$

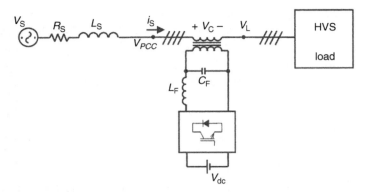

FIGURE 5.21 Scheme of a SAF with HVS load.

A system is stable if its transfer function $\mathbf{G}(s)$ has its poles in the left semiplane of "s" plane. Therefore, the problem of analysis of the stability of a system is reduced to the analysis of the denominator polynomial roots of (5.21).

5.3.2 SAF State Model

Figure 5.21 shows a three-phase system consisting of a nonlinear load, HVS type and a SAF connected to a voltage source with R_S resistance and L_S inductance.

To represent the system by state variables [19] the circuit model shown in Figure 5.22 is used. This is the single-phase equivalent of the network shown in Figure 5.21, for any h order harmonic different from the fundamental. The active filter is modeled by a controlled voltage source of $\mathbf{u}$ value. This value depends on the applied control strategy and will generally be expressed in terms of the system state variables. On the other hand, the load is represented by its Norton equivalent [20] whose parameters were described in Chapter 1. This load will be characterized by resistor R_L connected in parallel with inductance L_L and a current source, i_L, which will have a nonzero value for harmonics different from the fundamental.

The system allows a representation [15] in the state space given by

$$\dot{x} = \mathbf{A}\,x + \mathbf{B}_1\,\mathbf{u} + \mathbf{B}_2\,\mathbf{v}$$
$$y = \mathbf{C}\,x + \mathbf{D}_1\,\mathbf{u} + \mathbf{D}_2\,\mathbf{v}$$

(5.22)

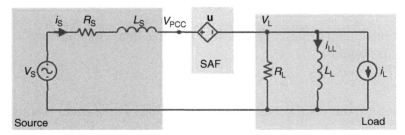

FIGURE 5.22 Equivalent single-phase circuit of the system shown in Figure 5.21.

In expression (5.22), the state vector is

$$\mathbf{x} = \begin{bmatrix} i_S & i_{LL} \end{bmatrix}^T \tag{5.23}$$

The system input vector has been divided into two terms, first the voltage signal of the source controlled, **u**, that is, the control signal, and on the other hand, the source vector, **v**, which includes the values of the source voltage, v_S, and the source current, i_L, defined by

$$\mathbf{v} = \begin{bmatrix} v_S & i_L \end{bmatrix}^T \tag{5.24}$$

The system matrix, **A**, is given by

$$\mathbf{A} = \begin{bmatrix} -\dfrac{(R_s + R_L)}{L_s} & \dfrac{R_L}{L_s} \\ \dfrac{R_L}{L_L} & -\dfrac{R_L}{L_L} \end{bmatrix} \tag{5.25}$$

$\mathbf{B_1}$ is a vector defined by

$$\mathbf{B}_1 = \begin{bmatrix} -\dfrac{1}{L_S} & 0 \end{bmatrix}^T \tag{5.26}$$

And finally, the matrix $\mathbf{B_2}$ is

$$\mathbf{B}_2 = \begin{bmatrix} \dfrac{1}{L_S} & \dfrac{R_L}{L_S} \\ 0 & -\dfrac{R_L}{L_L} \end{bmatrix} \tag{5.27}$$

When source current, i_S, is chosen as output variable, the matrix **C** is defined by

$$\mathbf{C} = \begin{bmatrix} 1 & 0 \end{bmatrix} \tag{5.28}$$

And for the output equation (5.22), $\mathbf{D_1} = [0]$ and $\mathbf{D_2} = [0 \ 0]$
If the voltage at the PCC, v_{PCC}, is selected as output signal, the matrix **C** is

$$\mathbf{C} = \begin{bmatrix} R_L & -R_L \end{bmatrix} \tag{5.29}$$

And for the output equation

$$\mathbf{D}_1 = [1]$$
$$\mathbf{D}_2 = \begin{bmatrix} 0 & -R_L \end{bmatrix} \tag{5.30}$$

Figure 5.23 shows the block diagram representing the state equation defined by (5.22).

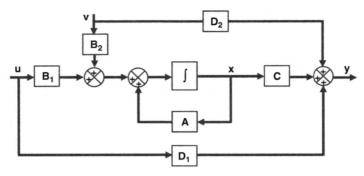

FIGURE 5.23 Block diagram for the system defined in (5.22).

5.3.3 Control Strategies

This section describes the system state models for the three proposed control strategies. This will allow to us carry out a dynamic analysis of the system behavior and its stability.

The first strategy which arises, is by detection of the source current. Here, the control signal applied to the system is proportional to the harmonics of the source current. This is

$$u = k\, i_S \tag{5.31}$$

If the above equation is expressed in terms of the state variables as was defined in (5.22), it results in

$$\mathbf{u} = \begin{bmatrix} k & 0 \end{bmatrix} \begin{bmatrix} i_S \\ i_{LL} \end{bmatrix} \tag{5.32}$$

Where k is the proportionality constant.

According to (5.32), in this control strategy the state vector is fed back through a gain matrix **K** of the form

$$\mathbf{K} = \begin{bmatrix} k & 0 \end{bmatrix} \tag{5.33}$$

Now, the estate equation can be expressed by means of

$$\dot{\mathbf{x}} = (\mathbf{A} + \mathbf{B}_1\mathbf{K})\mathbf{x} + \mathbf{B}_2\, \mathbf{v} \tag{5.34}$$

So, the new system matrix is given by

$$\mathbf{A'} = (\mathbf{A} + \mathbf{B}_1\mathbf{K}) = \begin{bmatrix} -\dfrac{(R_s + R_L + k)}{L_s} & \dfrac{R_L}{L_s} \\ \dfrac{R_L}{L_L} & -\dfrac{R_L}{L_L} \end{bmatrix} \tag{5.35}$$

Figure 5.24 shows the block diagram of the state equation (5.34).

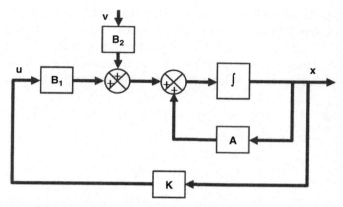

FIGURE 5.24 Block diagram of the state equation for control strategy $u = ki_s$.

The system characteristic polynomial is obtained from (5.36)

$$\varphi(s) = |s\mathbf{I} - \mathbf{A}'|$$

(5.36)

Where $\mathbf{I}$ is the identity matrix. The polynomial obtained is

$$\varphi(s) = \frac{s^2 L_S L_L + s\left(R_L L_S + R_S L_L + R_L L_L + k L_L\right) + R_L\left(R_S + k\right)}{L_S L_L}$$

(5.37)

The roots of the characteristic polynomial are the system poles. Therefore, taking into account (5.37), these will be located in the left half plane provided that the following condition is fulfilled

$$R_L L_S + R_S L_L + R_L L_L + k L_L \geq 0$$

(5.38)

Thus, to make the system stable, k value must satisfy the condition

$$k \geq -\frac{\left(R_L L_S + R_S L_L + R_L L_L\right)}{L_L}$$

(5.39)

Equation (5.39) establishes the minimum value of the proportionality constant, k, which makes the system stable. As noted, this value depends on the resistance and inductance of source, which are difficult to estimate in a power system. However, the term at the right of the inequality is always less than zero, so that a positive value of k will always ensure system stability, regardless of the parameter values of the network impedance.

On the other hand, when source current is taken as an output signal, the state model is defined by

$$\dot{x} = (\mathbf{A} + \mathbf{B}_I \mathbf{K})\,x + \mathbf{B}_2\,v = \mathbf{A}'x + \mathbf{B}_2 v$$
$$y = \mathbf{C}x$$

(5.40)

Where $\mathbf{C}$ is defined according to (5.28). Therefore, the proportionality constant k changes only the matrix $\mathbf{A'}$ of the system. Taking into account the transfer function given by expression (5.21), k adjustment affects the denominator of the transfer function, that is, the system poles. Thus, k value must choose such that it provides the adequate gain, at the frequencies of interest, to achieve a given attenuation of the source current harmonics.

When the voltage at the PCC is taken as the output signal, the transfer function is given by

$$\begin{aligned}
\dot{\mathbf{x}} &= (\mathbf{A} + \mathbf{B}_1 \mathbf{K})\mathbf{x} + \mathbf{B}_2\,\mathbf{v} = \mathbf{A'}\mathbf{x} + \mathbf{B}_2\mathbf{v} \\
\mathbf{y} &= (\mathbf{C} + \mathbf{D}_1\mathbf{K})\mathbf{x} + \mathbf{D}_2\mathbf{v} = \mathbf{C'}\mathbf{c} + \mathbf{D}_2\mathbf{v}
\end{aligned} \tag{5.41}$$

Here, $\mathbf{C}$ matrix is defined according to (5.29) and $\mathbf{D}_1$, $\mathbf{D}_2$ according to (5.30). In this case, the k constant affects either poles as zeros of the system. The criterion for selecting the k value will be based on achieving a given target attenuation of the voltage harmonics at the PCC.

Next, the strategy of detecting the load voltage will be analyzed. In this case the voltage generated by the active filter is given by the expression:

$$u = -k_v\, v_L \tag{5.42}$$

Where k_v represents a term which models the instrumentation sensitivity and v_L the voltage at the load side. Considering the equivalent circuit of Figure 5.22, the control signal, u, can be expressed as

$$u = -k_v R_L\, i_S + k_v R_L\, i_{LL} + k_v R_L\, i_L \tag{5.43}$$

Equation (5.43) can be expressed in matrix form in terms of the state vector and the system input vector, this is

$$\mathbf{u} = \mathbf{K}_1\, \mathbf{x} + \mathbf{K}_2\, \mathbf{v} \tag{5.44}$$

Where,

$$\mathbf{K}_1 = \begin{bmatrix} -k_v R_L & k_v R_L \end{bmatrix} \tag{5.45}$$

$$\mathbf{K}_2 = \begin{bmatrix} 0 & k_v R_L \end{bmatrix} \tag{5.46}$$

The state equation can be written in the form

$$\dot{\mathbf{x}} = (\mathbf{A} + \mathbf{B}_1\mathbf{K}_1)\mathbf{x} + (\mathbf{B}_2 + \mathbf{B}_1\mathbf{K}_2)\mathbf{v} \tag{5.47}$$

Thus the new system matrix is defined by

$$\mathbf{A'} = (\mathbf{A} + \mathbf{B}_1\mathbf{K}_1) = \begin{bmatrix} -\dfrac{(R_s + R_L(1 - k_v))}{L_s} & \dfrac{R_L(1 - k_v)}{L_s} \\[2mm] \dfrac{R_L}{L_L} & -\dfrac{R_L}{L_L} \end{bmatrix} \tag{5.48}$$

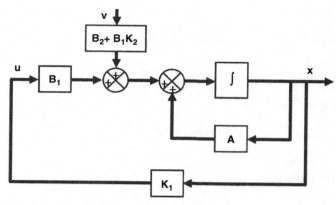

FIGURE 5.25 Block diagram of the state equation, control strategy $u = -k_v v_L$.

The block diagram of the state equation when this control strategy is applied, is shown in Figure 5.25.

Moreover, the characteristic equation is given by

$$\varphi(s) = |s\mathbf{I} - \mathbf{A}'| = \frac{L_S L_L \, s^2 + s\left[R_L L_S + R_S L_L + R_L L_L \left(1 - k_v\right)\right] + R_S R_L}{L_S L_L} \tag{5.49}$$

From this, it follows that the system is stable as long as the condition is fulfilled.

$$R_L L_S + R_S L_L + R_L L_L \left(1 - k_v\right) > 0 \tag{5.50}$$

This criterion ensures that the system poles are located in the left half plane. Otherwise,

$$k_v < \frac{R_L L_S + R_S L_L + R_L L_L}{R_L L_L} \tag{5.51}$$

Generally, $k_v \approx 1$ so that the condition given in (5.51) will always be satisfied. In the particular case where the source impedance is negligible ($R_S \approx L_S \approx 0$) of (5.51), it follows that values $k_v > 1$ can destabilize the system.

Finally, hybrid control strategy, which combines detection of the source current and detection of the load voltage is analyzed. Thus, the active filter generates a voltage waveform defined by the expression

$$u = k \, i_S - k_v \, v_L \tag{5.52}$$

Considering the circuit shown in Figure 5.22, the control signal is of the form

$$u = \left(k - k_v R_L\right) i_S + k_v R_L \, i_{LL} + k_v R_L \, i_L \tag{5.53}$$

Equation (5.53) can be rewritten in matrix form as

$$\mathbf{u} = \mathbf{K}_1 \, \mathbf{x} + \mathbf{K}_2 \, \mathbf{v} \tag{5.54}$$

Where

$$\mathbf{K}_1 = \begin{bmatrix} k - k_v R_L & k_v R_L \end{bmatrix} \tag{5.55}$$

$$\mathbf{K}_2 = \begin{bmatrix} 0 & k_v R_L \end{bmatrix} \tag{5.56}$$

The resulting state equation is

$$\dot{\mathbf{x}} = (\mathbf{A} + \mathbf{B}_1 \mathbf{K}_1)\mathbf{x} + (\mathbf{B}_2 + \mathbf{B}_1 \mathbf{K}_2)\mathbf{v} \tag{5.57}$$

Its block diagram is shown in Figure 5.26.
The new system matrix is

$$\mathbf{A}' = (\mathbf{A} + \mathbf{B}_1 \mathbf{K}_1) = \begin{bmatrix} -\dfrac{(R_s + k + R_L(1 - k_v))}{L_s} & \dfrac{R_L(1 - k_v)}{L_s} \\ \dfrac{R_L}{L_L} & -\dfrac{R_L}{L_L} \end{bmatrix} \tag{5.58}$$

This matrix allows the characteristic polynomial be determined,

$$\varphi(s) = |s\mathbf{I} - \mathbf{A}'| = \frac{L_s L_L\, s^2 + s\left[R_L L_s + R_s L_L + kL_L + R_L L_L(1 - k_v)\right] + R_L[R_s + k]}{L_s L_L} \tag{5.59}$$

The system will be stable as long as the condition is fulfilled

$$R_L L_s + R_s L_L + kL_L + R_L L_L(1 - k_v) > 0 \tag{5.60}$$

This is,

$$kL_L - k_v R_L L_L > -R_L L_s - R_s L_L - R_L L_L \tag{5.61}$$

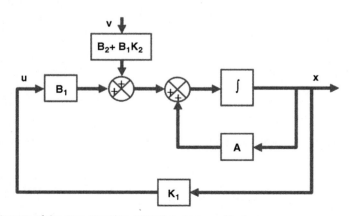

FIGURE 5.26 Block diagram of the state equation, control strategy $u = ki_s - k_v\, v_L$.

Assuming $k_v \approx 1$, this condition will be satisfied whenever

$$kL_L > -R_L L_S - R_S L_L \tag{5.62}$$

Which will be ensured when $k > 0$ since the term on the right of (5.62) is always less than zero, therefore it will be assumed as a general design rule.

On the other hand, for $k > 0$, the k_v parameter must satisfy the condition

$$k_v < \frac{R_L L_S + R_S L_L + R_L L_L + kL_L}{R_L L_L} \tag{5.63}$$

According to (5.63), a high k value provides more robustness to the system because it increases the second term of the inequality by reducing its dependence on variations of R_L, L_L, R_S, y L_S. Furthermore, this term is greater than one even in the case that the load is resistive, because at the limit when $L_L = 0$, it holds that

$$\lim_{L_L \to 0} \frac{R_L L_S + R_S L_L + R_L L_L + kL_L}{R_L L_L} = \frac{R_S + R_L + k}{R_L} \tag{5.64}$$

EXAMPLE 5.4

Figure 5.27 shows a SAF connected to an HVS-type load. The load consists of an uncontrolled three-phase rectifier with a capacitor of 2200 μF and a resistor of 50/3 Ω in parallel at the dc side. The source is sinusoidal, with an rms value of the phase voltage of 100 V and a frequency of 50 Hz. A resistance of 1.8 Ω and an inductance of 2.8 mH are included to model the equivalent impedance from connection point. The active filter is connected to the system by means of three-phase transformers with ratio 1:1, an LC ripple filter with inductance of 0.15 mH and a capacity and 50 μF in order to reduce the ripple of the output voltage. The inverter is a three-phase IGBT Bridge, the dc side has a connected constant source of 100 V. The simulation has been done in MATLAB–Simulink models with devices SymPowerSystem library.

To obtain the system model, the load without active filter is simulated and active and reactive power [21] are determined [22]. The values obtained by simulation are: 2125 W and 454 var, respectively. Considering that the rms value of the fundamental component of the load voltage across is 82.7 V, a value for the load resistor is obtained, $R_L = 9.65 \, \Omega$ and load inductance $L_L = 144$ mH. These values allow us to define the single-phase equivalent circuit model shown in Figure 5.22.

In the state model defined by the expression (5.22), matrix system without active filter is given by

$$\mathbf{A} = \begin{bmatrix} -4089 & 3446 \\ 66.5 & -66.5 \end{bmatrix}$$

$\mathbf{B_1}$ vector is given by

$$\mathbf{B_1} = \begin{bmatrix} -357 & 0 \end{bmatrix}^T$$

$\mathbf{B_2}$ matrix is

$$\mathbf{B_2} = \begin{bmatrix} 357 & 3446 \\ 0 & -66.5 \end{bmatrix}$$

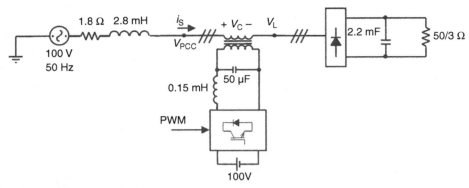

FIGURE 5.27 Circuit of the Example 5.4.

When the current source is chosen as output signal, the matrix $\mathbf{C}$ is

$$\mathbf{C} = \begin{bmatrix} 1 & 0 \end{bmatrix}$$

With $\mathbf{D}_1 = [0]$ and $\mathbf{D}_2 = [0\ 0]$
Thereby state vector is formed

$$\dot{x} = \mathbf{A}\,x + \mathbf{B}_1\,u + \mathbf{B}_2\,v$$
$$y = \mathbf{C}\,x + \mathbf{D}_1\,u + \mathbf{D}_2\,v$$

Figure 5.28 shows the pole diagram of the uncompensated system, two real poles located at $s = -10.3$ and $s = -4150$ are observed.

On the other hand, Figure 5.29 shows the Bode magnitude when source current is considered as an output signal. When the input signal is the supply voltage harmonics, the system has a gain of -21.9 dB for a frequency of 250 Hz, corresponding to harmonic 5. Considering the distortion of the load current as input, the gain at that frequency is higher, -2.19 dB. This enables us to ensure that with this load type, the source current distortion is mainly due to the load rather than the supply voltage.

Subsequently, the active filter is connected to the system. First, the control strategy by detection of source current is applied. The new equation of state is defined by (5.34). The gain of the state vector is set to $k = 20$, which is obtained

$$\mathbf{K} = \begin{bmatrix} 20 & 0 \end{bmatrix}$$

The new system matrix is defined in the form

$$\mathbf{A} + \mathbf{B}_1\,\mathbf{K} = \begin{bmatrix} -11229 & 3446 \\ 66.5 & -66.5 \end{bmatrix}$$

Figure 5.30 shows the new pole diagram. The fed back system has two real poles located at $s = -46$ and $s = -11200$. As predicted, from the theoretical point of view, the system is stable for values of $k > 0$. Furthermore, the poles are moved away off the origin, allowing to provide greater system robustness.

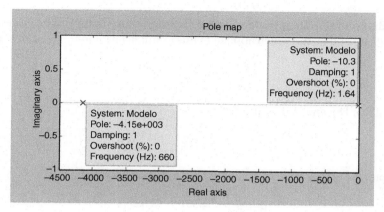

FIGURE 5.28 Example 5.4, pole map without SAF.

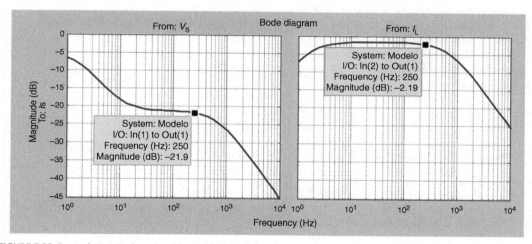

FIGURE 5.29 Example 5.4, Bode gain without SAF.

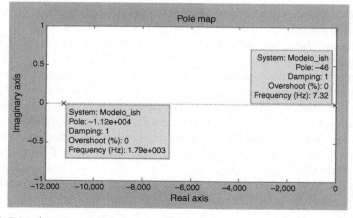

FIGURE 5.30 Example 5.4, pole map control strategy, $u = ki_{sh}$.

Secondly, the control strategy by detection of the load voltage is applied. Matrices $\mathbf{K}_1$ and $\mathbf{K}_2$ as defined in (5.45) and (5.46), which applied to this example result in:

$$\mathbf{K}_1 = \begin{bmatrix} -9.18 & 9.18 \end{bmatrix}$$
$$\mathbf{K}_2 = \begin{bmatrix} 0 & 9.18 \end{bmatrix}$$

Where $k_v = 0,95$. The new system matrix is

$$\mathbf{A} + \mathbf{B}_1\,\mathbf{K}_1 = \begin{bmatrix} -816.2 & 173.2 \\ 66.5 & -66.5 \end{bmatrix}$$

Furthermore, the matrix to be multiplied by the input vector is given by

$$\mathbf{B}_2 + \mathbf{B}_1\mathbf{K}_2 = \begin{bmatrix} 357 & 173.2 \\ 0 & -66.5 \end{bmatrix}$$

In this case the system has the pole diagram shown in Figure 5.31. Two real poles located at $s = -51.4$ and $s = -831$ are observed. The two poles are located in the left half plane, which ensures system stability.

Finally, the hybrid control strategy is applied, which is a combination of the two previous approaches. It is considered $k_v = 0.95$ and $k = 20$, so that according to (5.55) and (5.56), the following vectors are obtained

$$\mathbf{K'}_1 = \begin{bmatrix} 10.83 & 9.16 \end{bmatrix}$$
$$\mathbf{K'}_2 = \begin{bmatrix} 0 & 9.16 \end{bmatrix}$$

Therefore, the system matrix is

$$\mathbf{A} + \mathbf{B}_1\,\mathbf{K'}_1 = \begin{bmatrix} -7956 & 173.2 \\ 66.5 & -66.5 \end{bmatrix}$$

The matrix that multiplies the input vector is

$$\mathbf{B}_2 + \mathbf{B}_1\mathbf{K'}_2 = \begin{bmatrix} 357 & 173.2 \\ 0 & -66.5 \end{bmatrix}$$

This result shows how this matrix is the same as the previously obtained, with strategy by detection of the load voltage.

Figure 5.32 presents the system poles diagram when this control strategy is applied. Two real poles appear in the left half plane located at $s = -65$ and $s = -7960$.

From the point of view of the system gain is of interest for analyzing the behavior of the three strategies for the two harmonic input sources: the load distortion and source voltage. When the supply voltage is distorted, the strategy that presents a lower gain is that based on the detection of the current source, Figure 5.33. This is almost constant with a value of about -30 dB in a frequency range between 50 Hz and 1 kHz (this involves considering the first 20 harmonics). By contrast, the strategy for detection of the load voltage has highest gain. For the same range of frequencies, the gain varies from -8 dB to -30 dB. These values are higher than those obtained when the system has no active filter. Therefore, when the supply voltage is distorted, strategy by detection of the load voltage amplifies source current harmonics.

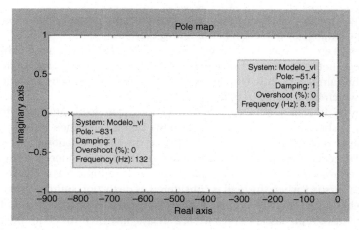

FIGURE 5.31 Example 5.4, pole map control strategy, $u = -k_v v_{\text{Lh}}$.

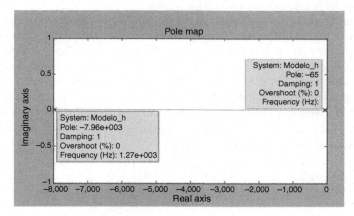

FIGURE 5.32 Example 5.4, pole map control strategy, $u = ki_{\text{sh}} - k_v v_{\text{Lh}}$.

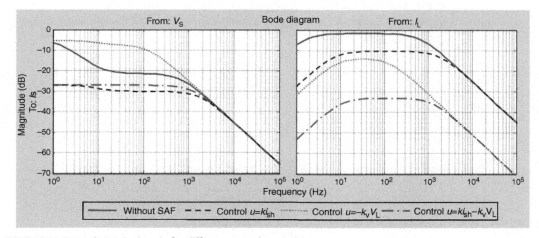

FIGURE 5.33 Example 5.4, Bode gain for different control strategies.

Control hybrid strategy presents a practically constant gain, though slightly higher than that achieved with the strategy by detection of the source current.

When harmonics are generated by the load (Figure 5.33), the lowest gain is achieved by the hybrid strategy. In the frequency range of the most significant harmonics of the system (100–1000 Hz) the gain is -33.4 dB. On the opposite side is the strategy by detection of the source current, which has a gain of -10 dB.

In the strategy by detection of the current source, the system gain value depends on the k value. Figure 5.34 shows how the gain varies with the k constant. When k increases the gain decreases. However, this variation is not linear because the gain reduction is not proportional to the k constant. This occurs regardless of the input source that is considered. As design criterion, a k value, which allows obtaining a desired gain value can be adopted. The Figure 5.35 shows the absolute gain versus k when the load is considered as distortion source. Thus, when the goal is to have a gain of 0.1, the value of k must be 90. However, this is a cost of the control signal, which is in many cases unnecessary, since we will always have an error about the tension filter output due to the switching of the power devices and the measurement system itself. Thereby, it can produce acceptable results with lower gain values.

The hybrid strategy requires less gain from the point of view of the harmonics elimination of source current. In regard to the most appropriate k value, with smaller k values in strategy by detection of the source current, it is possible to obtain smaller gain. Figure 5.36 shows the gain for different k values when the hybrid strategy is applied. In the case that the supply voltage is distorted, k must be high, since this strategy has a gain slightly greater than the strategy by detection of the source current. If the supply voltage is sinusoidal with a value of $k = 10$ it reaches -28.2 dB at a frequency of 250 Hz, which means reducing this harmonic at 95.3%.

In what follows, the system when the output is the voltage at the PCC to the network is analyzed. In this situation, the matrix $\mathbf{C}$ is

$$\mathbf{C} = \begin{bmatrix} 9.65 & -9.65 \end{bmatrix}$$

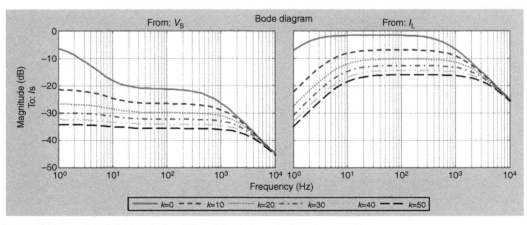

FIGURE 5.34 Example 5.4, Bode gain for different k values, control strategy, $u = ki_{sh}$.

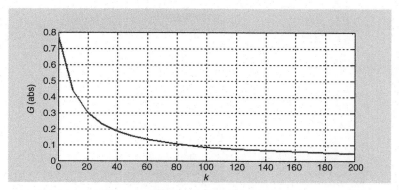

FIGURE 5.35 Example 5.4, 5th harmonic gain for different k values. Input signal, i_L.

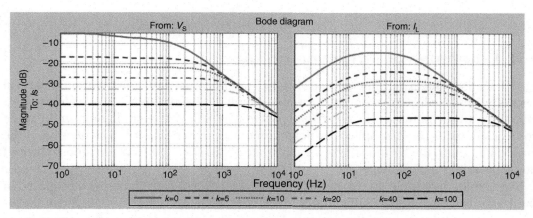

FIGURE 5.36 Example 5.4, Bode gain for different k values, combined control strategy.

In addition, $\mathbf{D_1} = [1]$ and $\mathbf{D_2}=[0\ -9.65]$

Figure 5.37 shows the gains for different strategies when the output is the voltage at the PCC. When the input are the supply voltage harmonics, the strategy by detection of the current source and the hybrid strategy have a gain of the same order as that of the system without the active filter. Therefore, these control strategies do not improve the THD of this voltage. Moreover, the strategy of detecting the load voltage shows a similar gain to the system without compensating, so that the voltage at the PCC remains the same total harmonic distortion.

Finally, when the distortion of the load source current is taken into account, the control strategy with lower gain is the hybrid strategy. At the same time, the voltage at the PCC is less influenced by the harmonic distortion produced by the load.

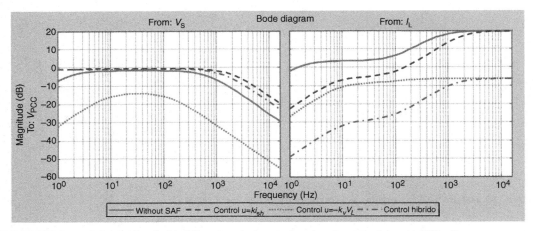

FIGURE 5.37 Example 5.4, Bode gain for different control strategies when the output signal is PCC voltage.

5.4 Experimental Prototype of SAF

In this section, the experimental platform that has been used to contrast the theoretical results is presented. For this purpose, just one circuit has been assembled in which an HVS type three-phase nonlinear load is compensated by a SAF. In this configuration the three control strategies can be applied: by detection of the source current, by detection of the load voltage, and the hybrid strategy. Each strategy has been subjected to sinusoidal and nonsinusoidal supply voltage.

For eliminating ripple due to the switching of the power devices have been proposed in the literature with different configurations of passive filters [23]. Due to its simplicity, the filter most used is the LC topology shown in Figure 5.38. More complex configurations manage to reach a greater attenuation to the switching frequency, but these have proved to be impractical. The LC filter design is based on the fulfillment of some of the following points:

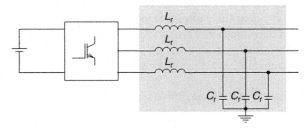

FIGURE 5.38 Ripple filter scheme.

- That the cutoff frequency is less than the switching frequency. For this we must take into account that

$$L_r C_r = \frac{1}{\omega_0^2}$$

(5.65)

Where L_r is the coils inductance, C_r the capacitor capacitance and ω_0 the resonance frequency of the ripple filter.

- The filter volume or weight of the reactive elements should be minimized. Taking into account that the energy densities are different for capacitors and coils, the next relationship is obtained:

$$\frac{Q_L}{Q_C} = \frac{1}{\left[3 \cdot \omega_0 \cdot V_n^2\right]^2} \cdot \left(\frac{P_n}{C_r}\right)^2$$

(5.66)

For equation (5.66):

V_n, converter rating voltage; I_n, rated current; P_n, active power for the converter is designed, considering unity power factor, as well $P_n \cong 3 \cdot V_n \cdot I_n$; Q_L, reactive power of the reactor, it can be obtained as $Q_L = \omega_n L_r I_n^2$; Q_C, reactive power of the capacitor, $Q_C = \omega_n C_r V_n^2$; ω_n, network fundamental frequency.

- Minimize the voltage drop across the filter inductance to the nominal current in order to provide a high-voltage ratio. That is

$$\frac{\Delta V}{V_n} = 1 - \sqrt{1 - (\omega_n \cdot L_r)^2 \cdot \left(\frac{I_n}{V_n}\right)^2}$$

(5.67)

Where ΔV is the voltage drop in the filter inductance.

For the prototype developed, a cut-off frequency of 2 kHz and a design power of 4500 W were considered adequate. A commercial capacitor of 50 μF and 230 V was chosen. The coil was constructed from a standard iron core. According to expression (5.65) and after different simulations its value is set at 0.13 mH.

The three control strategies are based on the determination of the load voltage harmonics, the source current harmonics, or both. These harmonics are obtained by measurement of a v signal. Then, this signal is multiplied first by $\sin \omega t$ and second by $\cos \omega t$, where ω is the system pulsation at the fundamental frequency. The mean value of both products can be obtained by

$$\frac{1}{T}\int_0^T v \sin \omega t \, dt = \frac{V_f \sqrt{2}}{2} \cos \varphi$$

$$\frac{1}{T}\int_0^T v \cos \omega t \, dt = \frac{V_f \sqrt{2}}{2} \sin \varphi$$

(5.68)

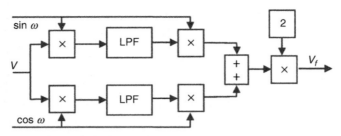

FIGURE 5.39 Block diagram to obtain the fundamental component.

In (5.68), V_f is the rms value of the fundamental component of v and φ is its phase angle. If the mean values are multiplied again by the same sine and cosine functions, the results are added and multiplied by the instantaneous value of the fundamental harmonic, v_f, is obtained,

$$v_f = V_f \sqrt{2}\left(\cos\varphi \sin\omega t + \sin\varphi \cos\omega t\right) = V_f \sqrt{2}\sin\left(\omega t + \varphi\right) \tag{5.69}$$

Harmonics can be determined by the difference between the measured signal and the instantaneous value of the fundamental harmonic,

$$v_h = v - v_f \tag{5.70}$$

Figure 5.39 shows the block diagram with which this calculation method is developed. This block diagram is modeled in Simulink. The system fundamental frequency is fixed at 50 Hz, which defines the pulsation ω. On the other hand, two lowpass filters (LPF) allow the mean values to be determined. For these filters a Simulink block has been used, which models a second order filter. Parameters defining this block are the cutoff frequency, which is set at 100 Hz and the damping factor that is set to 0.707.

5.4.1 Results of Practical Cases

Figure 5.40 shows the scheme of the experimental prototype of the SAF. The load is a three-phase noncontrolled rectifier, concretely the 36MT60 from International Rectifier, to which it is connected at the dc side with a capacitor of 2200 μF in parallel with a resistance of 16.7 Ω. This load is connected to the 4500-iL three-phase programmable source from California Instruments, with a sinusoidal voltage of 90 V and frequency of 50 Hz. Figure 5.40 also shows the position of the current and voltage sensors.

When the active filter is not connected, the current waveform and voltage at the common connection point, recorded by a 424 Wavesurfer oscilloscope from LECROY are like those shown in Figure 5.41.

Furthermore, these signals are measured with a three-phase network analyzer, model 434 from Fluke. These results showed for the voltage a THD of 10.0% and for the current a THD of 18.1%, for the first phase. Figure 5.42 shows the harmonic spectra of both signals.

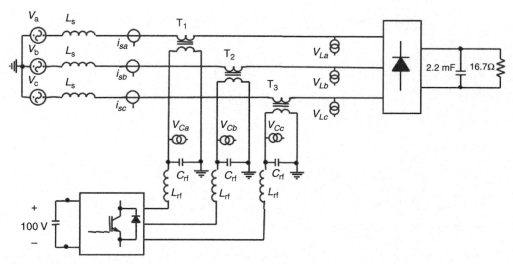

FIGURE 5.40 SAF experimental prototype scheme.

Only odd harmonics are observed, except those multiples of three. In the voltage and current, these harmonics are nearly zero.

Similarly the active, reactive, and apparent power per phase are measured, resulting in 0.66 kW, 0.19 kvar, and 0.69 kVA, respectively. The power factor is 0.96 inductive.

Once submitted and analyzed waveforms of the voltage at the PCC and system source current, results are analyzed when the SAF is connected and the control three strategies are applied. The first analysis strategy was by detection of the source current. The proportionality constant is set to $k = 50$. The source current and voltage PCC are shown in Figure 5.41. The THDs are 5.3% and 3.8%, respectively. The active, reactive, and apparent power per phase are: $P = 0.65$ kW, $Q = 0.09$ kvar, and $S = 0.66$ kVA. The power factor obtained is 0.99 inductive.

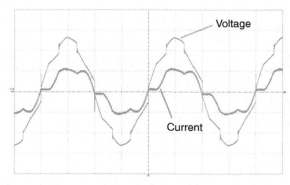

FIGURE 5.41 Voltage waveform (48V/div) and current (10A/div), without SAF.

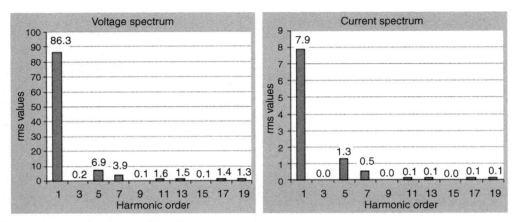

FIGURE 5.42 Harmonic spectrum of the voltage and current.

When the control strategy by detection of the load voltage is applied, the voltage and current waveforms shown in Figure 5.44 are obtained. With this compensation strategy the voltage THD is reduced to 2.9% and the source current THD to 3.8%. Comparison with the previous strategy shows a slight improvement in the voltage and current THDs. The measured powers by phase are: $P = 0.65$ kW, $Q = 0.08$ kvar, and $S = 0.66$ kVA. The power factor is 0.99 inductive (Figure 5.43).

The hybrid strategy presents the waveforms shown in Figure 5.45. It has been considered a proportionality constant $k = 20$. The voltage THD is 2.9% and the current THD is 3.1%. The measured powers by phase are: $P = 0.66$ kW, $Q = 0.06$ kvar, and $S = 0.66$ kVA. The measured power factor is 1.00. It must be borne in mind that the analyzer used is only capable of appreciating the hundredth.

Table 5.2 summarizes the measured values of THDs, most significant harmonics, power, and power factors when each of the compensation strategies are applied.

The same tests were carried out for a nonsinusoidal voltage source. In this experiment, the voltage source is programmed so that the voltage waveform includes 5th order

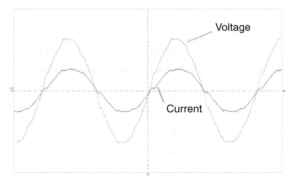

FIGURE 5.43 Source current detection control strategy, voltage waveform (48V/div) current waveform (10A/div).

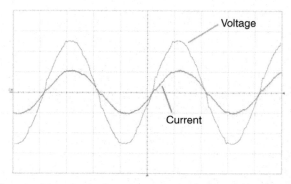

FIGURE 5.44 Load voltage detection control strategy, voltage waveform (48V/div) and current waveform (10A/div).

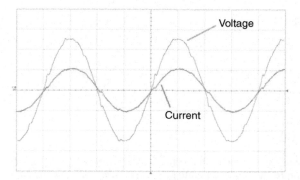

FIGURE 5.45 Combined control strategy, voltage waveform (48V/div) and current waveform (10A/div).

Table 5.2 Results of the Practical Case of SAF Filter: Sinusoidal Source

		THD (%)	RMS	H1	H3	H5	H7	H9	H11	H13	P (kW)	Q (kvar)	S (kVA)	PF
Without filter	V	10.0	86.7	86.3	0.2	6.9	3.9	0.1	1.6	1.5	0.66	0.19	0.69	0.96
	I	18.1	8	7.9	0.0	1.3	0.5	0.0	0.1	0.1				Ind.
SAF detection source current k = 50	V	3.8	87.2	87.1	0.1	1.6	1.4	0.1	1.4	1.1	0.65	0.08	0.66	0.99
	I	5.3	7.6	7.5	0.0	0.3	0.2	0.0	0.1	0.1				Ind.
SAF detection load voltage	V	2.9	87.1	87.1	0.2	1.1	0.8	0.1	0.8	0.7	0.65	0.08	0.66	0.99
	I	3.8	7.6	7.5	0.0	0.2	0.1	0.0	0.1	0.0				Ind.
SAF hybrid control k = 20	V	2.9	87.2	87.2	0.2	0.8	0.8	0.1	1.0	0.9	0.66	0.06	0.66	1.00
	I	3.1	7.6	7.6	0.0	0.1	0.1	0.0	0.1	0.0				Ind.

harmonic with an amplitude of 12% of the fundamental component and with the same initial phase angle.

Figure 5.46 shows the voltage waveforms at the PCC and the source current waveforms: (a) without filter, (b) with the strategy by detection of the source current with $k = 50$, (c) with the strategy by detection of the load voltage, and (d) the hybrid strategy with $k = 20$. The most significant result should be noted that the strategy by detection of

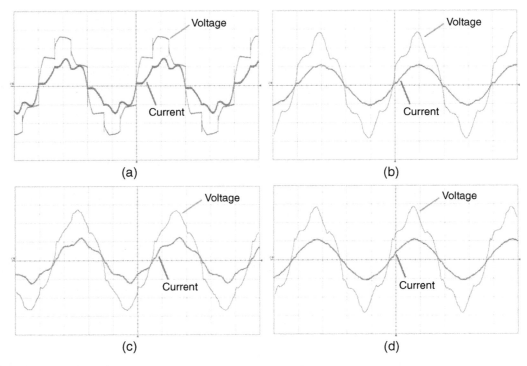

FIGURE 5.46 SAF behavior when source voltage is no sinusoidal. (a) Without SAF; (b) source current detection control; (c) load voltage detection control; (d) combined strategy, Voltage 48V/div and current 10A/div.

the load voltage is not possible to eliminate the current source harmonic, with a 5 order harmonic as most significant. On the contrary, the strategy by detection of the source current and hybrid can eliminate the harmonic distortion of the current source due to the distortion of the supply voltage. Table 5.3 shows the most significant numerical results for this test.

Table 5.3   Results of the Practical Case of SAF filter: Nonsinusoidal Source

		THD (%)	RMS	H1	H3	H5	H7	H9	H11	H13	P (kW)	Q (kvar)	S (kVA)	PF
Without filter	V	17.2	86.9	85.2	0.1	12.8	5.6	0.0	3.0	2.2	0.64	0.18	0.67	0.96
	I	13.9	7.7	7.6	0.0	0.6	0.8	0.0	0.3	0.2				Ind.
SAF, detection source current $k = 50$	V	15.7	87.2	86.1	0.2	13.2	1.4	0.1	1.4	1.2	0.64	0.13	0.65	0.98
	I	4.7	7.5	7.5	0.0	0.2	0.2	0.0	0.1	0.1				Ind.
SAF, detection load voltage	V	9.7	86.8	86.4	0.3	7.6	1.4	0.1	1.4	1.4	0.65	0.10	0.66	0.99
	I	13.1	7.5	7.5	0.0	0.9	0.2	0.0	0.1	0.0				Ind.
SAF, hybrid control, $k = 20$	V	14.3	87.1	86.2	0.2	12	0.6	0.1	0.8	0.9	0.65	0.11	0.66	0.99
	I	3.2	7.5	7.5	0.0	0.2	0.1	0.0	0.1	0.1				Ind.

5.5 Summary

In this chapter the steady state behavior of SAFs was analyzed. Three different control strategies for the active filter have been studied: by detection of the source current, by detection of the load voltage, and a hybrid control that includes a combination of the previous two strategies. Since the single-phase equivalent circuit, the expressions of the two variables of greatest interest from the viewpoint of harmonic filtering have been obtained, that is, the voltage at the PCC and the source current. Once obtained the equations in steady state, the set filter-load has been subjected to different situations in order to compare their behavior. Thus, the set has been fed to a distorted supply voltage and nonlinear loads at the PCC have been connected.

The SAF has proven effective in compensating for the type loads of harmonic tension source, HVS. Of the strategies applied to the active filter, the hybrid strategy has proven to be the most effective from the point of view of the elimination of harmonics in both the case of distorted and undistorted voltage.

Furthermore, the state space model of an active filter series configuration, SAF, has been set up and analyzed for the three proposed compensation strategies. This has allowed the establishment of different design rules, from the point of view of stability and system gain. So, for the SAF topology the following conclusions can be established:

- Strategy by detection of the source current.
  As a general design rule, the condition $k > 0$ must be satisfied for the constant of proportionality. This ensures the stability of the closed-loop system. When acting on the k constant is possible to change the location of the system poles. A high k value will require the poles be placed further away from the origin. Adjustment of the k proportionality constant can act on the system gain, so its value should be set in order to achieve a given attenuation of harmonics.
- By detection of the load voltage.
  System stability is guaranteed if the condition $k_v \approx 1$ is fulfilled. This strategy changes the position of the poles and zeros of the system, because it acts on the matrices **B, C,** and **D** of the system. Harmonic attenuation maximum is reached when $k_v = 1$.
- Hybrid strategy
  The system is stable when it holds that $k > 0$. This provides greater robustness to the system because the source impedance variations and load changes made less sensitive to errors in detecting the load voltage. This strategy changes the position of the poles and zeros of the system, because it acts on the matrices **B, C,** and **D** states space model.

When the supply voltage is distorted, the strategy that has a higher attenuation is that which detects the current source. Detection of the load voltage produces less attenuation, and may even amplify the source current harmonics.

When the harmonic source is the load, the smallest gain is achieved with a hybrid strategy.

The design of an experimental prototype has allowed to verify the performance of series APF and contrast these results with the theoretical results presented.

References

[1] Emanuel E. Harmonics in the early years of electrical engineering: a brief review of events, people and documents, In: Proceedings of the 9th International Conference on harmonics and quality of power, vol. 1. October 2000. pp. 1–7.

[2] Owen EL. A history of harmonics in power systems. IEEE Ind Appl Mag 1998;4(1):6–12.

[3] Nabae A.1; Enviromental issues and power electronics. In: Proceedings of the IEEE ISIE'2000; 2000. pp. PL1–PL5.

[4] Holmes DG, Lipo TA. Pulse width modulation for power converters. New York: IEEE Press; 2003.

[5] Orts-Grau S, Gimeno-Sales FJ, Abellan-Garcia A, Segui-Chilet S, Alfonso-Gil JC. Improved shunt active power compensator for IEEE Standard 1459 compliance. IEEE Trans Power Deliv 2010;25(4):2692–701.

[6] Litrán SP, Salmerón P, Prieto J, Vázquez JR. Aplicación de los filtros en la mejora de la calidad de la potencia eléctrica. In: Proceedings of the Spanish Portuguese Congress on Electrical Engineering, CHLIE, Marbella; 2005.

[7] Prieto J, Salmerón P, Litrán SP. "Acondicionamiento de cargas no lineales con tensiones distorsionadas". In: Proceedings of the Spanish Portuguese Congress on Electrical Engineering, CHLIE. Marbella; 2005.

[8] Salmerón P, Montaño JC, Vazquez JR, Prieto J, Perez A. Compensation in non-sinusoidal, unbalanced three-phase four-wire systems with active power-line conditioner. IEEE Trans Power Deliv 2004;19(4):1968–2197.

[9] da Silva SAO, Donoso-Garcia PF, Cortizo PC, Seixas PF. A three-phase line interactive UPS system implementation with series-parallel active power-line conditioning capabilities. IEEE Trans Ind Appl 2003;38:1581–90.

[10] Shu Z, Guo Y, Lian J. Steady-state and dynamic study of active power filter with efficient fpga-based control algorithm. IEEE Trans Ind Electr 2008;55(4):1527–36.

[11] Tanaka T, Koshio N, Akagi H, Nabae A. Reducing supply current harmonics. IEEE Ind App Mag 1998;4:31–7.

[12] Wang Z, Wang Q, Yao W, Liu J. A series active power filter adopting hybrid control approach. IEEE Trans Power Electr 2001;16(3):301–10.

[13] Rivas D, Moran L, Dixon JW, Espinoza JR. Improving passive filter compensation performance with active techniques. IEEE Trans Ind Electr 2003;50(1):161–70.

[14] Salmerón P, Litrán SP. Improvement of the electric power quality using series active and shunt passive filters. IEEE Trans Power Deliv 2010;25(2):1058–67.

[15] Litrán SP, Salmerón P, Herrera RS, Vázquez JR. "Control of series active power filter by state feedback". In: Proceedings of the International Conference on Renewable Energy and Power Quality, ICREPQ 08. Santander; 2008.

[16] Peng FZ, Akagi H, Nabae A. A new approach to harmonic compensation in power systems-a combined system of shunt passive and series active filters. IEEE Trans Ind Appl 1990;26(6):983–90.

[17] Yingying L, Yonghai X, Xiangning X, Yongqiang Z, Chunlin G. A stages compensation and control strategy for series power-quality regulator. IEEE Trans Power Deliv 2010;25(4):2807–13.

[18] Zhong H, Chen P, Lu Z, Qian Z, Ma H. Three-phase four-wire series hybrid active power filter with fundamental current bypass channel". In: Proceedings of the Industrial Electronics Society, IECON, vol. 1. November 2004. pp. 536–39.

[19] Litrán SP, Salmerón P. Analysis and design of different control strategies of hybrid active power filter based on the state model. IET Power Electr 2012;5(8):1341–50.

[20] IEEE Std. 1531-2003, IEEE Guide for Application and Specification of Harmonic Filters. November 2003.

[21] Vázquez JR, Salmerón P, Herrera RS,Litrán SP. Los términos de potencia en el marco del Standard IEEE 1459–2000. In: Proceedings of the XVII Reunión de :grupos de investigación en Ingeniería Eléctrica. Sevilla; 2007.

[22] Vázquez JR, Salmerón P, Herrera RS, Litrán SP. "Discusión sobre la potencia reactiva y la potencia aparente en sistemas con desequilibrios y distorsión". In: Proceedings of the XVII Reunión de grupos de investigación en Ingeniería Eléctrica. Sevilla; 2007.

[23] Rashid Muhammad H. Power electronics handbook. London: Academic Press; 2001.

Hybrid Filters: Series Active Power Filters and Shunt Passive Filters

CHAPTER OUTLINE

In this chapter, three control strategies for a combination of series active filters (SAFs) and parallel passive filters (SAPPF) are analyzed.

- Control by source current detection.
- Control by load voltage detection.
- Hybrid control combining source current and load voltage detection.

First of all, these strategies are analyzed in steady state with a single-phase equivalent circuit. The theoretical analysis is made considering two types of nonlinear loads with a dual distorting behavior: harmonic current source loads and harmonic voltage source loads. Furthermore, this analysis is carried out while considering the distortion produced by the load and the distortion present in the system due to another nonlinearities different from the own load.

By the other side, a state space model for the hybrid filter configuration, SAPPF, is obtained. With this model, the behavior of the filter set is analyzed considering the three compensation strategies; and some design rules are established for each control strategy.

P. Salmerón Revuelta, S.P. Litrán, and J.P. Thomas: Active Power Line Conditioners. http://dx.doi.org/10.1016/B978-0-12-803216-9.00006-7

The theoretical analysis has been contrasted with the results obtained with various simulation cases; a MATLAB–Simulink simulation platform has been developed for that purpose. These results have allowed us to establish, from the harmonics cancellation point of view, which is the most suitable active filter topology and control strategy for each type of nonlinear load.

6.1 Introduction

Active power filters (APF) have proved to be effective for compensating nonlinear loads [1–3]. Shunt configuration has been the most studied topology, in which the APF is connected parallel to the load. Its traditional use is the elimination of current harmonics produced by loads generating disturbances, known as harmonic current source (HCS) loads. However, parallel APF is not effective in situations where the load generates voltage harmonics, which are called harmonic voltage source (HVS) loads [4–7]. In this case, a series connection APF configuration has been proposed and different control strategies have been tried out [8]. In any event, compensation systems composed only of an APF, whether in parallel or in series, do not completely solve the problem of harmonic elimination for all load types. To this end, other configurations have been proposed [9–16] that are combinations of series and parallel topologies with active and passive filters. These are called hybrid topologies.

The aim of using a hybrid topology is to enhance the passive filter performance and power-rating reduction of the active filter. Two configurations have mostly been adopted: an active filter connected in series with a shunt passive filter, and a SAF combined with shunt passive filter (series active-parallel passive filter, SAPPF). Both topologies are useful to compensate HCS load types. However, when the load also generates voltage harmonics, the second topology is the most appropriate [17].

Different control strategies have been proposed for SAPPF configurations. One of them originated in the early 1990s, and the control objective of this configuration was based on generating a voltage proportional to the source current harmonics through a proportionality constant k [18]. In this instance, APF allowed the filtering features of the parallel connection passive filter to be improved. The functionality of the new strategy was analyzed using a steady state model. However, this theoretical development is not helpful for determining the proportionality constant value. Besides, this is not the proper way to study the system stability.

Subsequently, other control strategies for series APFs appeared. Thus, in [8] three control strategies are analyzed from the point of view of performance, although this analysis does not specify the design criteria for each compensation strategy.

This chapter studies the hybrid filter configuration composed of a combination of a series APF and various tuned passive filters, for harmonic compensation of systems with nonlinear loads. The application of each of these components has been systematized, with a practical point of view; when they are used to compensate type load HVS or HCS. Thus, the different compensation strategies (source current detection, load voltage detection, and hybrid control which combines the two previous) have been analyzed; and it has allowed to summarize the common uses of this type of compensation equipment collected in the scientific literature.

6.2 Series Active Filters and Shunt Passive Filters

There are many hybrid filter configurations that combine active and passive filters [1]. The topology that will be analyzed here is composed of an active filter connected in series with the supply and a passive filter shunt connected with the load, SAPPF. Figure 6.1 shows the scheme of this configuration.

The shunt passive filter is tuned at the frequency of the most significant current harmonics. There are various passive filter topologies, depending on the number of energy storage elements [19], although from a practical point of view some of them are difficult to implement [20]. Thus, Figure 6.2 just show the configurations most used in power systems.

The band pass filter (Figure 6.2a) is the simplest of them all and the most commonly used. Its main advantage is that it has a substantially zero impedance at the resonance frequency, which makes it an almost perfect sink at that frequency. However it suffers from the risk of producing resonances at frequencies below the tuning point. This may involve the amplification of other frequencies, which could cause a new harmonic problem. Furthermore, this configuration filters the harmonics above the tuning point poorly.

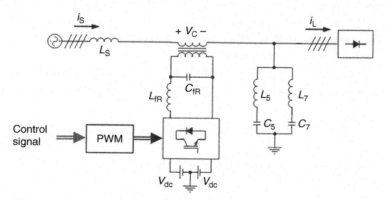

FIGURE 6.1 Topology and connection of a hybrid filter SAPPF.

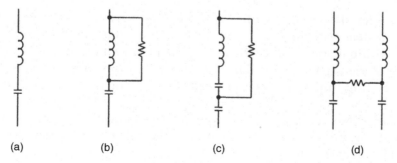

FIGURE 6.2 Passive filter types. (a) Band pass filter; (b) first order high pass filter; (c) C filter; (d) Pi filter.

The high-pass filter (Figure 6.2b) allows harmonics above the cut-off frequency to be eliminated. It is usually tuned to frequencies above the 7th or 11th harmonics. Such filters may also be suitable for removing high-frequency oscillations. The resistance can be adjusted to eliminate low-order parallel resonances. Due to the power consumption of the resistance, it is not desirable to apply it below the 5th harmonic.

The C-type harmonic filter (Figure 6.2c) has performance characteristics similar to the high-pass filter, with the advantage that the resistance does not consume power at the fundamental frequency. This is because it is designed so that the inductors and capacitors that are connected in parallel with the resistor, and are tuned for the fundamental frequency. It is mainly used when a tuned filter is required below the 5th harmonic and is often used in arc furnaces or cycloconverter applications to avoid amplifications of low-order harmonics and interharmonics.

The Pi-type filter (Figure 6.2d) is composed of two band pass filters connected in the middle point through a resistor. This configuration allows the tuning of each LC branch to a different frequency. With this topology the filtering feature is improved for both frequencies. The resistor may have a lower rated power, compared with the above filters because the current that flows through it is reduced.

For the analysis of the SAPPF configuration the band pass filter setting has been chosen (Figure 6.2a). For the design of these filters [21–23], the first step is to choose the capacitor. The rated voltage of the capacitor must be between 10% and 25% of the supply voltage of the network. On the other hand, the reactive power required by the load is also taken into account [24]. This power is usually determined from the measurement of the load power, its power factor, and displacement factor.

After selecting the capacitor, the inductance of the coil is determined, considering that the resonance frequency is given by

$$f_0 = \frac{1}{2\pi\sqrt{LC}}$$

(6.1)

Therefore, to eliminate the harmonic order h, the inductance is set to

$$L = \frac{1}{(2\pi h f)^2 C}$$

(6.2)

Generally, when designing the filter, it is usual to consider a somewhat lower harmonic than the ideal resonance (e.g., if you want to tune the filter for the 5th harmonic, the calculations are made for $h = 4.7$). This is a common practice that is motivated by the reduction of the capacity with aging, and the tolerances in the construction of the coils. Moreover, fine-tuning makes it more prone to overload due to nearby harmonic sources.

However, for a proper choice of the components, it should be checked with simulation; so it is necessary to choose a model as accurate as possible for the system behavior at the point of connection of the filter.

EXAMPLE 6.1

Figure 6.3 shows a system with two nonlinear loads connected to a busbar. This system has been simulated with MATLAB–Simulink, where a sinusoidal 230 V terminal voltage at the source is assumed, a source impedance of L_S = 2 mH, and L_1 = L_2 = 0.5 mH for each load line. Regarding the current consumption of the loads, Table 6.1 shows the rms values of each of the current harmonics present in the system.

In this situation the source current waveform is shown in Figure 6.4a, with a total harmonic distortion, THD, of 15.8%. The harmonic spectrum is shown in Figure 6.4b, where it can be seen that the current harmonics present are 5th, 7th, and 9th order.

The voltage at the point of common coupling (PCC) is distorted due to the nonsinusoidal supply current, as can be seen in the waveform and harmonic spectrum shown in Figure 6.5. The measured THD is 14.5%. Figure 6.5 also shows the presence of harmonics with the same order as those of the source current.

Next, a passive filter is connected to eliminate the 5th harmonic and to compensate the reactive power of load 2, as shown in Figure 6.6. The values of the filter components are 50 μF for the capacity and 8 mH for the inductance.

The resultant harmonic spectra of the currents i_S, i_1, and i_2 are shown in Figure 6.7. Being the passive filter tuned to the frequency of the harmonic of order 5, in principle one would expect a reduction in the harmonic intensity i_2; however, its value is greater than the circulation before connecting the passive filter. This is explained by the fact that the passive filter has a low-impedance path not only for the load but also for nearby current harmonic producing loads; namely, the passive filter acts as a harmonic "sink" for the rest of the installation, with the consequent overload of the passive filter. This behavior is one of the disadvantages of the use of passive filters. Another consequence is that the harmonic content of the current source is reduced, as can be seen in Figure 6.7, due to this "sink" effect that the passive filter produces in the whole system.

Another drawback to consider with respect to the passive filter is the potential risk of series and parallel resonances. Power systems are basically inductive at the fundamental frequency, usually despising the capacitive effects of the distribution lines. However, it is usual to connect capacitor banks, either at the consumer side to correct the power factor, or in the substation busbar to control the voltage level; therefore the short circuit impedance becomes a decisive parameter in the analysis of the frequency response of the resulting system. As is well known, circuits that contain multiple capacities and inductances have more than one natural resonance frequency. When one of these resonant frequencies coincides with one of the frequencies of the harmonic voltage or current, it can produce a state of resonance in the power system; and the voltages and currents at this frequency can reach dangerously high values. Thus, the parallel resonance occurs when, for a given frequency, the parallel combination of the impedances of the system has an infinite value. If the resonant frequency matches that of a harmonic current, this leads to a high voltage at this point. On the other side, an inductor and a capacitor connected in series at a point in the network can lead to a low-impedance path to one of the harmonics current. As a consequence, a high voltage appears across the terminals of the capacitor.

Considering the situation of Example 6.1, Figure 6.8 shows the variation with the frequency of the impedance seen from the load 2 terminals. The existence of an unexpected parallel resonance, particularly at 220 Hz, can be seen. That is, if there is any source current harmonic at this frequency it could be amplified and as a result cause an overvoltage at this point.

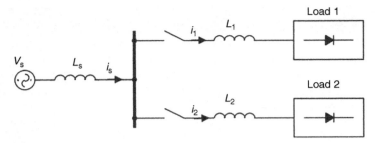

FIGURE 6.3 System with two nonlinear loads, Example 6.1.

Table 6.1 Load Harmonic Currents of Example 6.1

| Harmonic Order | rms Values (A) | |
	Load 1	Load 2
1	40	20
5	5	3
7	3	2
9	1	–

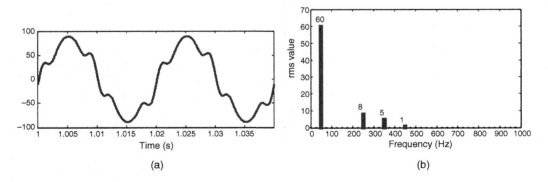

(a)

(b)

FIGURE 6.4 Source current, Example 6.1. (a) Waveform; (b) harmonic spectrum.

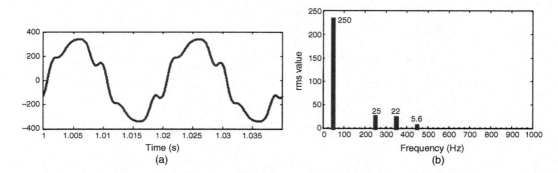

FIGURE 6.5 Voltage at the PCC, Example 6.1. (a) Waveform; (b) harmonic spectrum.

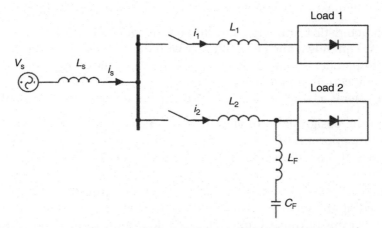

FIGURE 6.6 Example 6.1, connection of a passive filter to eliminate the harmonics of load 2.

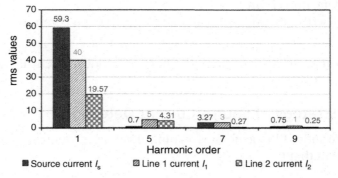

FIGURE 6.7 Harmonic spectrum when the passive filter is installed, Example 6.1.

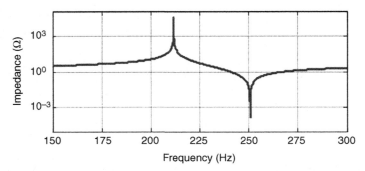

FIGURE 6.8 Resonances that appear in the system when the passive filter is connected, Example 6.1.

These passive filters are easy to design and have a low cost. However, they have some drawbacks [25] that make them inadvisable in many cases. The most prominent of these are:

- Its filter characteristic is influenced by the source impedance, which in most applications is unknown and can even vary with the system configuration.
- Series and/or parallel resonances may appear between the source impedance and the passive filter at certain frequencies.
- The passive filter is a "sink" for the harmonics generated in other points of the system.

Finally from the design point of view, it is necessary to know the current harmonics in order to choose appropriate values of L and C to be tuned at the desired frequencies. Therefore, these settings will be limited to systems that do not exhibit random variations in the load current.

The filtering characteristics of the passive filter are improved with the connection of a SAF, thus avoiding the inconveniences indicated when they are connected to a power system [26,27]. For this purpose, initially, they could use the same control strategies as in the case of SAF, namely:

- Source current detection control.
- Load voltage detection control.
- Hybrid control, where the APF generates a voltage that is a combination of the above two.

Hereinafter, a detailed analysis with the help of various practical cases will be performed.

6.2.1 Control Strategy of Source Current Detection

In this strategy, the active filter generates an output voltage proportional to the harmonics of the source current that is

$$v_{Ch} = k\,i_{Sh} \tag{6.3}$$

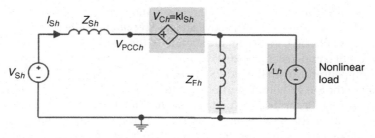

FIGURE 6.9 Single-phase equivalent circuit of a system with a SAPPF, HVS-type load, and control $V_{Ch} = kI_{Sh}$.

In a first case, an ideal load of voltage harmonics source type, HVS, is considered. Figure 6.9 shows the single-phase equivalent circuit for a certain harmonic order h. At this frequency the passive filter has an impedance value Z_{Fh}.

The source current will be given by

$$I_{Sh} = \frac{V_{Sh}}{(Z_{Sh} + k)} - \frac{V_{Lh}}{(Z_{Sh} + k)} \tag{6.4}$$

It can be seen that this current is independent of the value of the passive filter impedance, its connection or disconnection does not alter the source current waveform.

The voltage at the PCC is given by the expression

$$V_{PCCh} = \frac{k}{(Z_{Sh} + k)} V_{Sh} + \frac{Z_{Sh}}{(Z_{Sh} + k)} V_{Lh} \tag{6.5}$$

As with the source current, the voltage at the PCC does not include the term Z_{Fh}. Thus, the filtering feature of the source current and the voltage at the PCC only depend on the source impedance and the proportionality constant k. Values of k such that $k \gg Z_S$ mitigate the harmonics of both network variables.

EXAMPLE 6.2

Figure 6.10 shows a nonlinear load composed of a three-phase noncontrolled rectifier with a 2200 µF capacitor in the dc side and a 50/3 Ω resistor in parallel, which is a typical HVS load. It is powered by a three-phase sinusoidal source, 100 V rms, 50 Hz, with grounded neutral. The equivalent series impedance of the power supply network is modeled by a 1.8 Ω resistor and a 2.8 mH inductance. The active filter is connected through three 1:1 transformers, with a LC ripple filter of 13.5 mH inductance and 50 µF capacity. The inverter is a three-phase IGBT bridge with a 100 V constant source in the dc side. The passive LC filter is composed of three branches tuned to 250 Hz, with 13.5 mH inductance and 30 µF capacity, connected in star with the neutral of the network. The simulation has been done in MATLAB–Simulink using the device models of the SymPowerSystem library.

The active filter generates a voltage signal proportional to the source current harmonics. The proportionality constant used has a value of 50. As a simulation result the harmonic spectrum of the source current is obtained for three different situations: before compensation, only with the compensation of SAF, and with both active and passive filter connected.

With both kinds of filtering, the harmonic content of the source current is reduced. The current THD without compensation is 20.95%, and decreases to 4.24% when the active filter is connected. Compensating with the SAPPF the THD decreases to 2.2%. This is because the passive filter of the SAPPF eliminates the 5th harmonic, as it is shown in the harmonic spectrum of Figure 6.11. Regarding the fundamental harmonic, a reduction in the source current is observed when the active filter is connected. This is due to the voltage drop produced at the primary side of the coupling transformer, which reduces the voltage level at the load terminals and thus the load current. In the simulation, the transformer model includes the winding resistance and leakage reactance.

Regarding the voltage at the PCC, Figure 6.12 shows its harmonic spectrum. With the active filter the voltage THD is reduced, starting from a value of 13.59% before compensation it decreases to 3.00% with the SAF connection and improves slightly to 2.55% with the SAPPF compensation, in a similar way as the source current does.

The circuit model shown in Figure 6.9 has a voltage source V_{Lh} (that models a HVS load) in parallel with the LC branch. If the load generates a voltage harmonic at the resonance frequency of the passive filter, high values of current through the LC branch may be reached, theoretically an infinite value. In Example 6.2, the load has a voltage harmonic of 5th order of 9.5 V rms (Figure 6.12). However, the 5th harmonic of the current through the passive filter has a value of 2.98 A rms. Therefore, one would think that the load does not have the behavior of a HVS load, maybe because the capacity of the dc side of the rectifier is not high enough (2200 μF). It should be noted that the ripple factor[1] in the dc side is 0.39%. If the capacity is increased in a factor of 10, the ripple value reaches 0.039%; almost insignificant. Its behavior should be nearer to an ideal HVS load. However, the 5th order harmonic of the current in the LC branch is then 3.00 A. That is, it is almost unchanged although the capacity has increased ten times. A similar analysis can be done for other harmonics resulting in similar values for the current through the LC branch, with dc capacitors of 2200 μF and above. It can be concluded that the ideal model of HVS load is an approach that allows a comprehensive analysis of the system performance, but it is not possible to make a complete analysis of the system behavior. Later, an analysis will be made with state variables formulation, and using a more complete load model.

---

[1]In a three-phase rectifier with a capacitor in the dc side, the ripple factor is given by $r = 1 / (8\sqrt{3} f C R_L)$ being f the network frequency, C the capacity, and R_L the resistance in the dc side.

The previous analysis shows that the passive filter, applied to HVS loads, does not substantially improve the waveform of the voltage at the PCC or the source current. Once this control strategy has been analyzed for the case of a HVS type load, the same procedure is applied to a HCS type load. Figure 6.13 shows the equivalent single-phase circuit, where the source current models the nonlinear load. The source current is defined by

$$I_{Sh} = \frac{Z_{Fh}}{(Z_{Sh} + Z_{Fh} + k)} I_{Lh} + \frac{1}{(Z_{Sh} + Z_{Fh} + k)} V_{Sh} \tag{6.6}$$

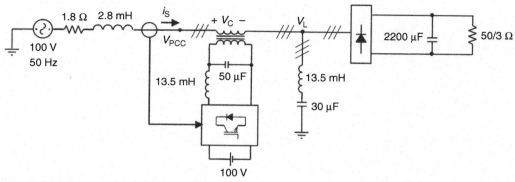

FIGURE 6.10 Circuit of Example 6.2.

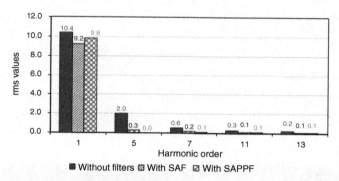

FIGURE 6.11 Harmonic spectrum of the source current. Example 6.2.

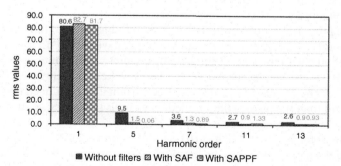

FIGURE 6.12 Harmonic spectrum of the voltage at the PCC. Example 6.2.

The expression (6.6) has two terms, one related to the harmonics of the load current and the other dependent on the harmonics of the supply voltage. In the same way as for the case of a HVS-type load, current harmonics can be mitigated if k takes a sufficiently high value. The main difference from the previous case is that now the passive filter impedance

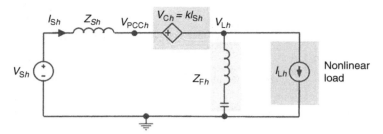

FIGURE 6.13 Single-phase equivalent circuit of a system with a SAPPF, HCS-type load, and control $V_{Ch} = kI_{Sh}$.

appears (Z_{Fh}) in the expression. Therefore, to mitigate the harmonics of the source current, considering the expression (6.6), the condition that must be fulfilled is

$$k \gg Z_{Fh}; Z_{Sh} \tag{6.7}$$

On the other side, the voltage at the PCC is given by

$$V_{PCCh} = \frac{Z_{Fh} + k}{(Z_{Sh} + Z_{Fh} + k)} V_{Sh} - \frac{Z_{Sh} Z_{Fh}}{(Z_{Sh} + Z_{Fh} + k)} I_{Lh} \tag{6.8}$$

A value of k such that $k \gg Z_{Fh}; Z_{Sh}$ will cause voltage harmonics at the connection point to be due only to the distortion of the supply voltage, as the second term of (6.8) is close to zero.

EXAMPLE 6.3

The same circuit of Example 6.2 is considered, with a new load composed by an uncontrolled three-phase rectifier with a 55 mH inductance in series with a $50/3\ \Omega$ load resistor at the dc side, as shown in Figure 6.14.

In this case the proportionality constant k is set to 50 Ω. Figure 6.15 shows the harmonic spectrum of the source current, without compensation and with the SAPPF. The 5th order harmonic is eliminated with the passive filter. Being the most significant harmonic, this results in an improved THD of the source current, from 18.09% without filtering to 2.27% when the SAPPF is connected. The 7th harmonic of the source current is also reduced because the constant k is greater than the passive filter impedance at this frequency, so the first term of (6.6) is practically zero.

In respect to the voltage at the PCC, its THD is reduced from 12.60% to 2.15%, when the SAPPF is connected. Figure 6.16 shows the harmonic spectrum in both situations. In view of the expression (6.8), if the supply voltage has no harmonics, the first term is zero. Regarding the second term, when condition (6.7) is fulfilled, the load current is multiplied by a factor less than one, so that the harmonics values of the voltage at the PCC are attenuated.

A slight increase in the rms value of the voltage at the fundamental frequency is observed in Figure 6.16 because the current drawn by the load with SAPPF is also lower, and thus the voltage drop across the source impedance is reduced.

Next, a distorted supply voltage is assumed, with a voltage waveform defined by

$$v(t) = V_p \left[\sin(\omega_1 t) + 0.08 \sin(3\omega_1 t + \pi) + 0.05 \sin(5\omega_1 t) \right] \tag{6.9}$$

In this case, the source current has a THD value of 3%, which is a slightly higher value than in the case of sinusoidal supply voltage. Since V_{Sh} is not zero, the second term on expression (6.6) is neither, although its value is reduced when condition (6.7) holds.

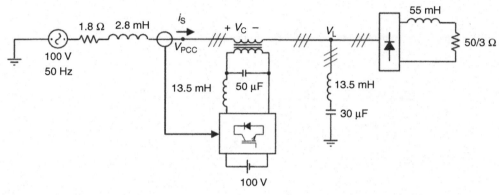

FIGURE 6.14 Circuit of Example 6.3.

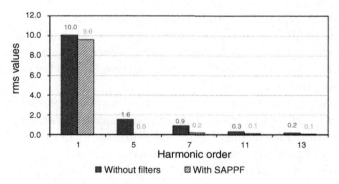

FIGURE 6.15 Harmonic spectrum of the source current. Example 6.3.

Another interesting issue in this analysis is to determine how the presence of a HCS load connected to the PCC can affect the set of active and passive filters, as shown in Figure 6.17, where an ideal current source has been connected to the PCC. The current that flows through the SAPPF + load (I_{Sh1}) is given by

$$I_{Sh1} = \frac{Z_{Fh}}{(Z_{Sh} + Z_{Fh} + k)} I_{Lh} + \frac{Z_{Sh}}{(Z_{Sh} + Z_{Fh} + k)} I_{Lh2} + \frac{1}{(Z_{Sh} + Z_{Fh} + k)} V_{Sh} \tag{6.10}$$

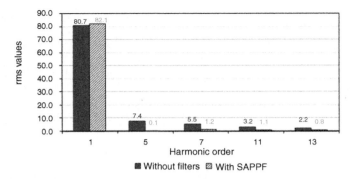

FIGURE 6.16 Harmonic spectrum of the voltage at the PCC. Example 6.3.

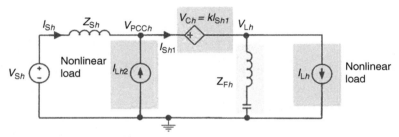

FIGURE 6.17 Circuit with a SAPPF filter, and a HCS load connected to the PCC.

For values of k that accomplish condition (6.7), this current is not affected by the connection of a nearby harmonics-producing load, since the addend term that includes I_{Lh2} tends to zero for values of $k \gg Z_{Sh}$; Z_{Fh}. Thus, the presence of the active filter in the compensator equipment avoids the possibility of the passive filter to becoming a sink for current harmonics due to other nonlinear loads.

The behavior of the passive filter can be improved if instead of only a LC branch tuned to the 5th harmonic, that for this load is the most significant harmonic, one more branch is connected that is tuned to the next higher harmonic value; in this case the harmonic of order 7. This new configuration allows us to enhance the source current THD with lower k values. Furthermore, the compensation voltage generated by the active filter is reduced and so is the filter power.

6.2.2 Control Strategy of Load Voltage Detection

The active filter of the SAPPF, with the control strategy of load voltage detection generates a voltage equal but opposite in sign to the voltage harmonics at the load side. Thus, for a harmonic of order "h" different from the fundamental, the compensation voltage of the active filter is

$$V_{Ch} = -k_v V_{Lh} \tag{6.11}$$

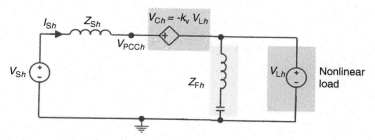

FIGURE 6.18 Equivalent circuit of a single-phase system with a SAPPF, HVS load type, and control $V_{Ch} = -k_v V_{Sh}$.

Where k_v is a parameter value close to 1, which takes into account the errors in the load voltage estimation and in the filter voltage generation.

Figure 6.18 shows the hybrid filter connected to an ideal load of HVS type, the latter modeled by an ideal voltage source of value V_{Lh}. In this circuit the source current I_{Sh} is given by

$$I_{Sh} = \frac{1-k_v}{Z_{Sh}} V_{Lh} + \frac{1}{Z_{Sh}} V_{Sh} \qquad (6.12)$$

In this expression two terms appear. The first depends on the load voltage and constant k_v. This term is close to zero when k_v is close to one, that is, the higher the accuracy of the instrumentation. The second term includes the voltage of the power supply and the network impedance. On this second term the active filter does not act, so when the supply voltage is distorted, the current source will also be distorted.

Therefore, for a load HVS type and sinusoidal voltage supply, the source current is free from harmonics. As above, for the same load the passive filter does not contribute to improving the source current, so that disconnection does not affect operation of the system; this is similar to the situation analyzed in the previous section.

The voltage at the PCC, is given by the expression

$$V_{PCCh} = (1-k_v) V_{Lh} \qquad (6.13)$$

So the voltage at the PCC is free of harmonics produced by the load when $k_v \approx 1$, regardless of connection or not the passive filter.

For HCS-type load, the equivalent single-phase circuit shown in Figure 6.19 allows to analyze the operation of SAPPF. The source current is given by an expression of the form

$$I_{Sh} = \frac{Z_{Fh}(1-k_v)}{(Z_{Sh} + Z_{Fh}(1-k_v))} I_{Lh} + \frac{1}{(Z_{Sh} + Z_{Fh}(1-k_v))} V_{Sh} \qquad (6.14)$$

In the ideal case $k_v = 1$, the first term of (6.14) is zero. This is not dependent on the value of the passive filter, which Z_{Fh} not have to vanish for any particular harmonic. However, in general $k_v \neq 1$, whereby for "h" harmonic (it would be best that "h" was the most significant)

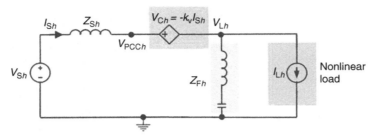

FIGURE 6.19 Equivalent circuit of a single-phase system with a SAPPF, HVS load type and control $V_{Ch} = -k_v V_{Sh}$.

must be satisfied that $Z_{Fh} \ll Z_{Sh}$ to get the first term is negligible. This condition is possible when the passive filter is tuned to the frequency of the harmonic order "h".

Regarding the second term, when the supply voltage (V_{Sh}) presents a harmonic voltage, source current includes harmonics of the same order. Its magnitude depends on the source impedance value.

Relative to the voltage at the PCC is given by

$$V_{PCCh} = \frac{Z_{Fh}(1-k_v)}{\left(Z_{Sh}+Z_{Fh}(1-k_v)\right)} V_{Sh} - \frac{Z_{Fh}\, Z_{Sh}\,(1-k_v)}{\left(Z_{Sh}+Z_{Fh}(1-k_v)\right)} I_{Lh} \qquad (6.15)$$

Therefore, the voltage at the PCC when $k_v \approx 1$ is free of harmonics, even when the supply voltage is distorted.

EXAMPLE 6.4

The control by load voltage detection to the circuit described in Example 6.2 is applied, and is reproduced in Figure 6.20.

Figure 6.21 shows the harmonic spectrum of the source current. A value of $k_v = 0.95$ for the simulations is considered. With the hybrid filter can improve the source current waveform because the THD decreases from 18.09% to 2.03%. This is justified as the first term of the expression (6.14) is almost zero, and on the other hand, since the supply voltage is sinusoidal, the second term of (6.14) is zero.

Relative to the voltage at the PCC, Figure 6.22 shows the frequency spectrum. The hybrid filter THD is improved, passing from 12.60% to 1.89%, which justifies their behavior according to (6.15).

Then it is considered that the supply voltage is distorted. Thus, the waveform of the voltage source is defined by the expression (6.9). This includes two voltage harmonics: one of order 3 of 8% and other of order 5 of 5%. Figure 6.23 shows the harmonic spectrum of source current before and after connecting the hybrid filter. The presence of a harmonic of order 5 is observed to be nonzero. This is because the second term of (6.14) is nonzero V_{Sh} also. A harmonic that was not present in the previous case also appears, this is the harmonic of order 3. Due to this harmonic component, THD increases going from 16.14% to 27.84%.

With respect to the voltage at the PCC, the most significant order harmonics are the 3rd and 5th order (Figure 6.24). When the SAPPF is connected, voltage THD decreases from 19.78% to 3.19%. From expression (6.15), the voltage harmonic content at the PCC depends on the k_v value and source impedance Z_{Sh}. As $k_v \neq 1$, the two terms tend to decrease depending on the value of the source impedance, which justifies the reduction of THD.

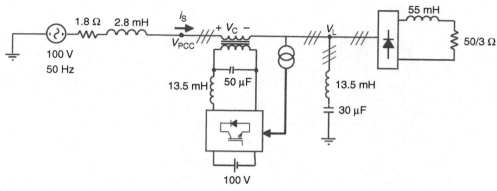

FIGURE 6.20 Circuit of Example 6.4.

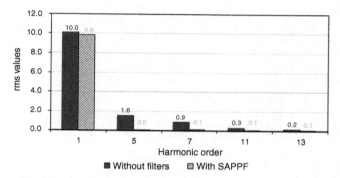

FIGURE 6.21 Harmonic spectral of current source. Example 6.4.

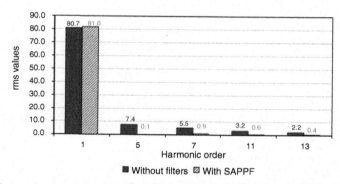

FIGURE 6.22 Voltage harmonic spectrum at the connection point to the network. Example 6.4.

Another important aspect is to analyze the behavior of the set SAPPF-load when there is a load close to a current harmonic source, as performed in previous compensation strategy. Figure 6.25 shows the equivalent phase circuit. The current drawn from the supply, I_{Sh1} is given by the expression

$$I_{Sh1} = \frac{Z_{Fh}(1-k_v)}{(Z_{Sh}+Z_{Fh}(1-k_v))}I_{Lh} + \frac{1}{(Z_{Sh}+Z_{Fh}(1-k_v))}V_{Sh} + \frac{Z_{Sh}}{(Z_{Sh}+Z_{Fh}(1-k_v))}I_{Lh2} \qquad (6.16)$$

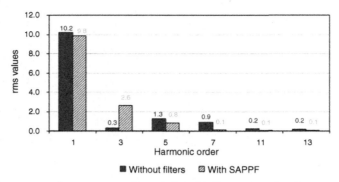

FIGURE 6.23 Harmonic spectrum of source current. Example 6.4 with distorted supply voltage.

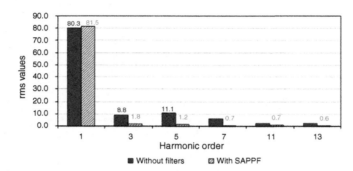

FIGURE 6.24 Voltage harmonic spectrum at the PCC. Example 6.4 with distorted supply voltage.

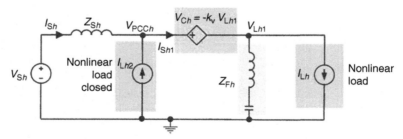

FIGURE 6.25 Circuit with a SAPPF and with additional HCS load connected at the PCC.

In equation (6.16), the source current has an additional term that depends on the nonlinear load current connected to the PCC. When $k_v \approx 1$, SAPPF-load set becomes a sink current harmonics of this nonlinear load. This is because when k_v is close to one, the 3rd term in (6.16) is equal to the load current, I_{Lh2}.

The voltage at the PCC is given by

$$V_{PCCh} = \frac{Z_{Fh}(1-k_v)}{(Z_{Sh}+Z_{Fh}(1-k_v))}V_{Sh} - \frac{Z_{Fh}\,Z_{Sh}\,(1-k_v)}{(Z_{Sh}+Z_{Fh}(1-k_v))}I_{Lh} + \frac{Z_{Fh}(1-k_v)}{(Z_{Sh}+Z_{Fh}(1-k_v))}I_{Lh2} \qquad (6.17)$$

This expression differs from (6.15) in that it includes a new term that depends on the nonlinear load current. This term is attenuated in proportion to the term that includes the supply voltage.

Ultimately, with this strategy it is possible to improve the waveform when the supply voltage is sinusoidal and there is not connected to the PCC additional nonlinear loads. With respect to the voltage at the PCC, in any situation improves THD.

6.2.3 Combined Control

From the analysis of the control strategy of source current detection when $k \gg Z_{Fh}$; Z_{Sh}, for $h \neq 1$ we can draw the following conclusions:

- The harmonic content of the source current after compensation does not depend on the distortion of the supply voltage and the load current harmonics.
- The voltage at the PCC is only distorted by the harmonics from the supply voltage and not affected by harmonic distortion produced by the load.
- It avoids the compensator-load system becomes a sink of harmonic currents from nearby sources.

Regarding the control strategy by load voltage detection, its main features are:

- Its operation depends on the error in the measurement of the voltage on the load side.
- Harmonic content of the source current depends on the distortion of the supply voltage and the source impedance.
- The voltage distortion at the PCC produced by the supply voltage or the load current is attenuated depending on k_v.
- When HCS-type loads are connected to the PCC, the SAPPF do not prevent the passive filter to behave as a sink of harmonics.

From previous conclusions, it is evident that the control by source current detection seems more appropriate versus control by load voltage detection. The analysis assumed that the constant k is greater than the source impedance and the passive filter impedance. Such impedance increases with frequency, and therefore the k value should be high if it is to eliminate higher order harmonics. If Example 6.4 is analyzed, the filter impedance and the source impedance are seen to vary, as shown in Figure 6.26. The passive filter impedance has a value of 50 Ω at the frequency of about 650 Hz (harmonic order 13). This value of 50 is the same as was taken as the constant k, which for this frequency, the relationship $k \gg Z_{Fh}$ is not satisfied, therefore, the reduction of a harmonic of this order may not be very significant (0.2–0.1 A). However, it should be noted that in practice, high order harmonics are less significant. The other premise $k \gg Z_{Sh}$ it is possible to achieve because in an electrical system, Z_{Sh}, usually has a reduced value, as shown in Figure 6.26.

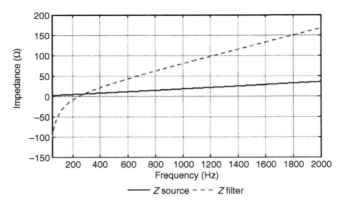

FIGURE 6.26 Variation of the source impedance and the passive filter with frequency. Example 6.4.

In this section a compensation strategy that combines the two previous methods will be addressed. Thus, the hybrid control of SAPPF includes a control by source current detection and by load voltage detection. The generated voltage waveform by the active filter for harmonic of order h is given by the expression

$$v_{Ch} = k\,i_{Sh} - k_v\,v_{Lh} \tag{6.18}$$

For single-phase equivalent circuit shown in Figure 6.27, the current source is

$$I_{Sh} = \frac{Z_{Fh}(1-k_v)}{\left(Z_{Sh}+k+Z_{Fh}(1-k_v)\right)}I_{Lh} + \frac{1}{\left(Z_{Sh}+k+Z_{Fh}(1-k_v)\right)}V_{Sh} \tag{6.19}$$

It can be seen that when $k_v \approx 1$, the first term is close to zero, regardless of the value of k and the impedance of the passive filter. Therefore, the harmonics of the source current are less influenced by the harmonics of load current. Regarding the second term, its value depends on k, the larger the k, the smaller will be the current harmonics due to the distortion of the supply voltage.

Furthermore, the voltage at the PCC is given by the expression

$$V_{PCCh} = \frac{Z_{Fh}(1-k_v)+k}{\left(Z_{Sh}+Z_{Fh}(1-k_v)+k\right)}V_{Sh} - \frac{Z_{Sh}\,Z_{Fh}(1-k_v)}{\left(Z_{Sh}+Z_{Fh}(1-k_v)+k\right)}I_{Lh} \tag{6.20}$$

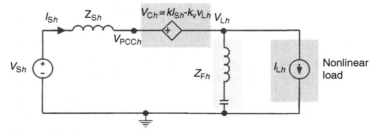

FIGURE 6.27 Single-phase equivalent circuit of a system with a SAPPF and a load HCS type; control: $V_{Ch} = k i_{Sh} - k_v v_{Lh}$.

It is observed that the second term of (6.20) mitigates the harmonic voltage due to load current. Another important aspect is that the effect of a high impedance passive filter can be minimized because $k_v \approx 1$.

EXAMPLE 6.5

A hybrid control shall be applied to the circuit presented in Example 6.4.

In this case it has been considered a proportionality constant of 50 ($k = 50$) for the current source term and $k_v = 0.95$ to the corresponding term of the load voltage. Figure 6.28 shows the frequency spectrum of the source current before and after connecting the compensation system. A significant improvement of the source current waveform is observed, as its THD is reduced from 18.19% to 1.79%. This confirms the behavior according to expression (6.19).

The voltage at the PCC before and after connecting the SAPPF also shows an improvement in its THD, as it is seen in the harmonic spectrum that is shown in Figure 6.29. Its value is reduced from 12.60% to 2.30%.

Then in order to verify the functionality of SAPPF in presence of harmonics in the supply voltage, a source voltage waveform defined by expression (6.9) is considered. The distortion of the source current depends on the value of k selected (6.19), the higher value does lower value of the harmonics due to the supply voltage; however, as a result, the rated power of the filter increases. Relative to the voltage at the PCC, according to expression (6.20), the harmonic content will be virtually the same as the supply voltage.

When the hybrid strategy is applied, another aspect that deserves to be analyzed is the operating SAPPF-load system in the presence of harmonic currents produced by additional loads connected at the PCC. The current source value I_{Sh2} in Figure 6.25, represents this nonlinear load. For the hybrid control, the filtered current (I_{Sh1}) is given by

$$I_{Sh1} = \frac{Z_{Fh}(1-k_v)}{(Z_{Sh}+k+Z_{Fh}(1-k_v))}I_{Lh} + \frac{1}{(Z_{Sh}+k+Z_{Fh}(1-k_v))}V_{Sh} + \frac{Z_{Sh}}{(Z_{Sh}+k+Z_{Fh}(1-k_v))}I_{Lh2} \quad (6.21)$$

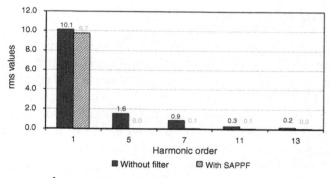

FIGURE 6.28 Harmonic spectrum of source current. Example 6.5.

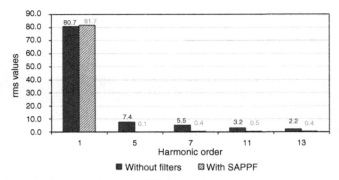

FIGURE 6.29 Harmonic spectrum of the voltage at the PCC. Example 6.5.

Table 6.2 THD of Voltage and Current for the Three Compensation Strategies and Different Supply Conditions

	Sin compensar		$V_{Ch} = 50\ I_{Sh}$		$V_{Ch} = -0.95\ V_{Lh}$		$V_{Ch} = 50\ I_{Sh} - 0.95\ V_{Lh}$	
	THDv (%)	THDi (%)	THDv (%)	THDi (%)	THDv (%)	THDi (%)	THDv (%)	THDi (%)
V_S sinusoidal	12.60	18.19	2.15	2.27	1.89	2.03	2.30	1.79
V_S nonsinusoidal	19.78	16.14	11.35	3.00	3.19	27.84	11.30	2.83

In (6.21), a new term that includes the current of the nonlinear load I_{Lh2} appears. The effect of this current is reduced such that the condition $k \gg Z_{Sh}$ is met.

Table 6.2 summarizes the value of THD in the different situations analyzed after connecting SAPPF. For a sinusoidal supply voltage three strategies provide low values of the current THDs while the hybrid approach is the one with the lowest value. In the case of nonsinusoidal supply voltage, the hybrid approach is that the lower source current THD presents. Instead, the strategy of load voltage detection may deteriorate source current THD over the uncompensated system.

6.3 State Model of SAPPF

In this section, the state-variable analysis of a SAPPF topology will be addressed [28,29]. The equations of state will be obtained when the following control strategies are applied: current source detection, load voltage detection, and the hybrid control. The analysis was performed on one HCS-type load because this filter topology is suitable for harmonic elimination of such loads, as demonstrated in the previous sections of this chapter.

Figure 6.30 shows a three-phase circuit with a nonlinear HCS load type and compensator SAPPF. That is, a SAF connected to a voltage source with resistance R_S and inductance L_S, and a passive filter connected in parallel with the load comprising two branches LC tuned to the most significant harmonic currents.

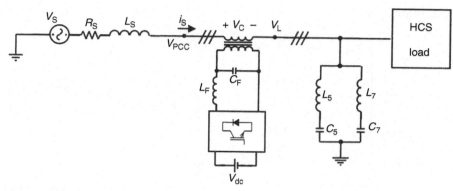

FIGURE 6.30 SAPPF filter system and load type HCS.

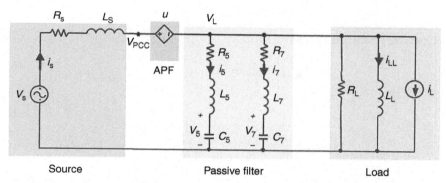

FIGURE 6.31 Single-phase equivalent circuit of the system of Figure 6.18.

For the representation of the system by state variables, the circuit model shown in Figure 6.31 is used. This represents the single-phase equivalent of the network of Figure 6.30 for any harmonic order "h" different from the fundamental. The active filter is modeled by a controlled voltage source u. For the load, a Norton equivalent circuit model consisting of a parallel combination of resistor R_L, inductance L_L, and current source i_L is used [30]. This current source will be zero for the fundamental harmonic. Passive LC filters consist of two branches tuned to frequencies of the most significant harmonics. Each LC branch includes a resistor, which models the resistive effect of its own coils because in practice this is not negligible.

In the circuit of Figure 6.31, a state equation can be formulated as follows:

$$\dot{x} = A\,x + B_1\,u + B_2\,v$$
$$y = C\,x + D_1\,u + D_2\,v \tag{6.22}$$

For the state variables and the references given in Figure 6.31 the state vector is defined

$$x = \begin{bmatrix} i_S & i_5 & i_7 & i_{LL} & v_5 & v_7 \end{bmatrix}^T \tag{6.23}$$

The systematic application of Kirchhoff's laws to determine the matrix of the system, that is,

$$
A = \begin{bmatrix}
-\dfrac{(R_S + R_L)}{L_S} & \dfrac{R_L}{L_S} & \dfrac{R_L}{L_S} & \dfrac{R_L}{L_S} & 0 & 0 \\[2mm]
\dfrac{R_L}{L_5} & -\dfrac{(R_L + R_5)}{L_5} & -\dfrac{R_L}{L_5} & -\dfrac{R_L}{L_5} & -\dfrac{1}{L_5} & 0 \\[2mm]
\dfrac{R_L}{L_7} & -\dfrac{R_L}{L_7} & -\dfrac{(R_L + R_7)}{L_7} & -\dfrac{R_L}{L_7} & 0 & -\dfrac{1}{L_7} \\[2mm]
\dfrac{R_L}{L_L} & -\dfrac{R_L}{L_L} & -\dfrac{R_L}{L_L} & -\dfrac{R_L}{L_L} & 0 & 0 \\[2mm]
0 & \dfrac{1}{C_5} & 0 & 0 & 0 & 0 \\[2mm]
0 & 0 & \dfrac{1}{C_7} & 0 & 0 & 0
\end{bmatrix}
\tag{6.24}
$$

In this case the vector $\mathbf{B_1}$ takes the form

$$
B_1 = \begin{bmatrix} -\dfrac{1}{L_S} & 0 & 0 & 0 & 0 & 0 \end{bmatrix}^T
\tag{6.25}
$$

$\mathbf{B_2}$ is

$$
B_2 = \begin{bmatrix}
\dfrac{1}{L_S} & 0 & 0 & 0 & 0 & 0 \\[2mm]
\dfrac{R_L}{L_S} & -\dfrac{R_L}{L_5} & -\dfrac{R_L}{L_7} & -\dfrac{R_L}{L_L} & 0 & 0
\end{bmatrix}^T
\tag{6.26}
$$

Which multiplies the input vector defined by,

$$
v = \begin{bmatrix} v_S & i_L \end{bmatrix}^T
\tag{6.27}
$$

The matrices $\mathbf{C}$, $\mathbf{D_1}$, and $\mathbf{D_2}$, depending on the chosen output. If the source current i_S, is taken as the output signal, it holds that

$$
C = \begin{bmatrix} 1 & 0 & 0 & 0 & 0 & 0 \end{bmatrix}
\tag{6.28}
$$

$$
\begin{aligned}
D_1 &= [0] \\
D_2 &= \begin{bmatrix} 0 & 0 \end{bmatrix}
\end{aligned}
\tag{6.29}
$$

6.3.1 Source Current Detection

When a control value proportional to the source current harmonics is applied, the control signal takes the form

$$
u = \begin{bmatrix} k & 0 & 0 & 0 & 0 & 0 \end{bmatrix} \begin{bmatrix} i_S & i_5 & i_7 & i_{LL} & v_5 & v_7 \end{bmatrix}^T
\tag{6.30}
$$

Where k is the proportionality constant.

As in the case of filter SAF, this strategy is based on the feedback of the state vector through a gain matrix **K** defined by

$$K = \begin{bmatrix} k & 0 & 0 & 0 & 0 & 0 \end{bmatrix} \tag{6.31}$$

With this control the state equation can be expressed by

$$\dot{x} = (A + B_1 K) x + B_2 v \tag{6.32}$$

The matrix system is defined by

$$A' = (A + B_1 K) = \begin{bmatrix} -\dfrac{(R_S + R_L + k)}{L_S} & \dfrac{R_L}{L_S} & \dfrac{R_L}{L_S} & \dfrac{R_L}{L_S} & 0 & 0 \\[2mm] \dfrac{R_L}{L_5} & -\dfrac{(R_L + R_5)}{L_5} & -\dfrac{R_L}{L_5} & \dfrac{R_L}{L_5} & -\dfrac{1}{L_5} & 0 \\[2mm] \dfrac{R_L}{L_7} & -\dfrac{R_L}{L_7} & -\dfrac{(R_L + R_7)}{L_7} & \dfrac{R_L}{L_7} & 0 & -\dfrac{1}{L_7} \\[2mm] \dfrac{R_L}{L_L} & -\dfrac{R_L}{L_L} & -\dfrac{R_L}{L_L} & -\dfrac{R_L}{L_L} & 0 & 0 \\[2mm] 0 & \dfrac{1}{C_5} & 0 & 0 & 0 & 0 \\[2mm] 0 & 0 & \dfrac{1}{C_7} & 0 & 0 & 0 \end{bmatrix} \tag{6.33}$$

This matrix (6.33) determines the dynamics of the control strategy by source current detection from the constant of proportionality k.

6.3.2 Load Voltage Detection

In this control strategy the voltage generated by the active filter is given by the expression

$$u = -k_v v_L \tag{6.34}$$

Where k_v represents the parameter which models the instrumentation sensitivity. For the circuit of Figure 6.31, the voltage v_L is

$$v_L = R_L i_S - R_L i_5 - R_L i_7 - R_L i_{LL} - R_L i_L \tag{6.35}$$

Thus the control signal u can be expressed by

$$u = -k_v R_L i_S + k_v R_L i_5 + k_v R_L i_7 + k_v R_L i_{LL} + k_v R_L i_L \tag{6.36}$$

Equation (6.36) can be written in matrix form as a function of the input vector v and system state vector x, that is,

$$u = K_1 x + K_2 v \tag{6.37}$$

Where

$$K_1 = \left[\begin{array}{cccccc} -k_v R_L & k_v R_L & k_v R_L & k_v R_L & 0 & 0 \end{array} \right] \tag{6.38}$$

And

$$K_2 = \left[\begin{array}{cc} 0 & k_v R_L \end{array} \right] \tag{6.39}$$

From (6.22) and (6.37) the state equation of the system is defined as

$$\dot{x} = (A + B_1 K_1) x + (B_2 + B_1 K_2) v \tag{6.40}$$

The system matrix **A'** is given by

$$A' = (A + B_1 K_1) = \left[\begin{array}{cccccc} -\dfrac{(R_S + R_L(1-k_V))}{L_S} & \dfrac{R_L(1-k_V)}{L_S} & \dfrac{R_L(1-k_V)}{L_S} & \dfrac{R_L(1-k_V)}{L_S} & 0 & 0 \\[2mm] \dfrac{R_L}{L_5} & -\dfrac{(R_L + R_5)}{L_5} & -\dfrac{R_L}{L_5} & \dfrac{R_L}{L_5} & -\dfrac{1}{L_5} & 0 \\[2mm] \dfrac{R_L}{L_7} & -\dfrac{R_L}{L_7} & -\dfrac{(R_L + R_7)}{L_7} & \dfrac{R_L}{L_7} & 0 & -\dfrac{1}{L_7} \\[2mm] \dfrac{R_L}{L_L} & -\dfrac{R_L}{L_L} & -\dfrac{R_L}{L_L} & -\dfrac{R_L}{L_L} & 0 & 0 \\[2mm] 0 & \dfrac{1}{C_5} & 0 & 0 & 0 & 0 \\[2mm] 0 & 0 & \dfrac{1}{C_7} & 0 & 0 & 0 \end{array} \right] \tag{6.41}$$

Therefore, with this control strategy, the dynamics of the system are affected by the parameter k_v.

6.3.3 Hybrid Control

This strategy combines the strategies by source current detection and by load voltage detection. Thus, when the hybrid strategy is used, the active filter must generate a voltage waveform defined by

$$u = k i_S - k_v v_L \tag{6.42}$$

Considering the circuit of Figure 6.31, the control signal takes the form

$$u = (k - k_v R_L) i_S + k_v R_L i_5 + k_v R_L i_7 + k_v R_L i_{LL} + k_v R_L i_L \tag{6.43}$$

Expressed in matrix form which is given by

$$u = K_1 x + K_2 v \tag{6.44}$$

Where the gain matrices are

$$K_1 = \left[\begin{array}{cccccc} (k - k_v R_L) & k_v R_L & k_v R_L & k_v R_L & 0 & 0 \end{array} \right] \tag{6.45}$$

$$K_2 = \begin{bmatrix} 0 & k_v R_L \end{bmatrix}$$ (6.46)

In this way the state equation is defined by

$$\dot{x} = (A + B_1 K_1) x + (B_2 + B_1 K_2) v$$ (6.47)

Now, the system matrix is

$$A' = (A + B_1 K_1) = \begin{bmatrix} -\dfrac{(R_S + R_L(1-k_v)+k)}{L_S} & \dfrac{R_L(1-k_v)}{L_S} & \dfrac{R_L(1-k_v)}{L_S} & \dfrac{R_L(1-k_v)}{L_S} & 0 & 0 \\[2mm] \dfrac{R_L}{L_5} & -\dfrac{(R_L+R_5)}{L_5} & -\dfrac{R_L}{L_5} & -\dfrac{R_L}{L_5} & -\dfrac{1}{L_5} & 0 \\[2mm] \dfrac{R_L}{L_7} & -\dfrac{R_L}{L_7} & -\dfrac{(R_L+R_7)}{L_7} & -\dfrac{R_L}{L_7} & 0 & -\dfrac{1}{L_7} \\[2mm] \dfrac{R_L}{L_L} & -\dfrac{R_L}{L_L} & -\dfrac{R_L}{L_L} & -\dfrac{R_L}{L_L} & 0 & 0 \\[2mm] 0 & \dfrac{1}{C_5} & 0 & 0 & 0 & 0 \\[2mm] 0 & 0 & \dfrac{1}{C_7} & 0 & 0 & 0 \end{bmatrix}$$ (6.48)

In this strategy, the dynamics of the system are affected by both the proportionality constant k as the parameter k_v.

EXAMPLE 6.6

Figure 6.32 shows a sinusoidal voltage source with 1.8 Ω resistance and 2.8 mH inductance. The load is formed by an uncontrolled three-phase rectifier with a resistance of 50/3 Ω connected in series with an inductance of 55 mH. This load is connected in parallel with a passive filter comprised of two LC branches tuned to the 5th and 7th harmonic. Resistors of 2.1 Ω and 1.1 Ω model the resistive effect of the two passive filter coils. The active filter is connected in series with the voltage source through three coupling transformers of ratio 1:1. A dc voltage source of 100 V is connected at the dc side of the inverter.

Figure 6.33 shows the system pole diagram without an active filter. The system has two real poles and four complex poles. All poles are located on the left semi plane.

On the other hand, Figure 6.34 shows the Bode gain when the current source is considered as an output signal. In the left diagram, where the input signal is the supply voltage, two peaks appear with gains of: −12.1 dB at 225 Hz and −14.6 dB at 323 Hz. These peaks represent two series resonances between the network and the passive filter. They are close to the resonance frequency of the LC branches, and depend on the impedance of the network from the PCC. In the ideal case of zero impedance, network will correspond to the tuning frequency of the LC branches.

In the second diagram of Figure 6.34, the input signal is the current source, which represents the load Norton equivalent. In this situation, the gain decreases to −10.1 dB at 249 Hz and −17.3 dB at 352 Hz, therefore, the system has two resonances at these frequencies. This does not occur exactly at 250 Hz and 350 Hz, due to tolerance of passive elements in the harmonic

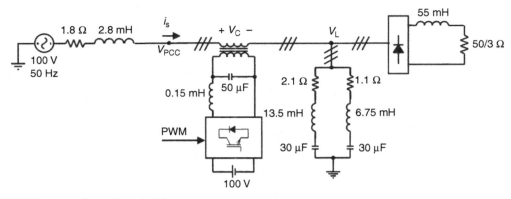

FIGURE 6.32 Power circuit. Example 6.6.

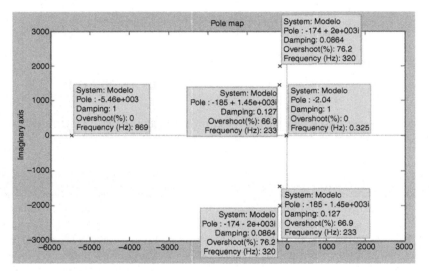

FIGURE 6.33 Pole map without active filter. Example 6.6.

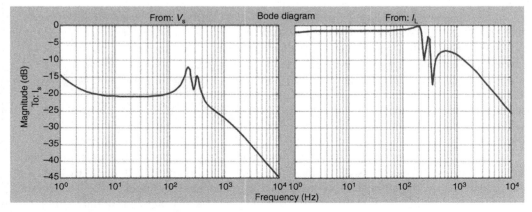

FIGURE 6.34 Bode gain without active filter. Example 6.6.

filter. Moreover, it shows that the gain value is not too low for a situation of resonance, however, must take into account that the coil resistive effect is modeled by a resistance of 2.1 Ω for tuned branch to the 5th harmonic, and 1.1 Ω for tuned branch to the 7th harmonic.

When the control strategy of source current detection is applied, with $k = 40$, the system pole position is modified. Figure 6.35 shows the new location of the poles. When the results shown in Figure 6.33 and Figure 6.35 are compared, a shift of the system poles on the left semiplane is observed. The implementation of this strategy causes the system poles to move away from the imaginary axis so that system stability is ensured.

To check what k values could destabilize the system, the Routh–Hurwitz criterion is used. The system characteristic polynomial with the connected active filter is analyzed. When the control strategy of source current detection is applied the obtained polynomial is

$$\varphi(s) = s^6 + (6314.10 + 357.10\,k)s^5 + (1.03 \times 10^7 + 8.50 \times 10^5\,k)s^4 + (3.75 \times 10^{10} + 2.77 \times 10^9\,k)s^3 + \\ + (2.11 \times 10^{13} + 2.86 \times 10^{12}\,k)s^2 + (4.82 \times 10^{16} + 4.35 \times 10^{15}\,k)s + (9.87 \times 10^{16} + 5.48 \times 10^{16}\,k)$$

For $k = -1.8$, the resulting table is

s^6	1	8.78×10^6	1.6×10^{13}	1.04×10^3
s^5	5.67×10^3	3.25×10^{10}	4.03×10^{16}	0
s^4	3.05×10^6	8.89×10^{12}	1.04×10^3	
s^3	1.59×10^{10}	4.03×10^{16}		
s^2	1.18×10^{12}	1.04×10^3		
s^1	4.03×10^{16}			
s^0	1.04×10^3			

The absence of any change of sign in the first column indicates that there are poles with negative real part, therefore, all are located in the left semiplane.

For a value of $k = -1.85$, the application of the stability criterion of Routh–Hurwitz allows obtaining the next table

s^6	1	8.74×10^6	1.59×10^{13}	-2.74×10^{15}
s^5	5.65×10^3	3.24×10^{10}	4.01×10^{16}	0
s^4	3.01×10^6	8.76×10^{12}	-2.74×10^{15}	
s^3	1.59×10^{10}	4.01×10^{16}		
s^2	1.17×10^{12}	-2.74×10^{15}		
s^1	4.02×10^{16}			
s^0	-2.74×10^{15}			

As can be seen, there is a sign change in the first column, which indicates the existence of a positive real pole, therefore, the system is unstable.

Figure 6.36 shows the graphical representation of the real parts of the system poles for k values between -100 and 100. The system will be stable for values of $k > -1.8$.

On the other hand, it is appropriate to establish the most suitable k value from the point of view of harmonic mitigation. The goal is that the system presents a low-input gain for a given harmonic content. Figure 6.37 represents the Bode diagram magnitude for $k = 0, 10, 20, 40, 80,$ and 120. Regardless of the input or output variable chosen, a larger value of k means that the gain is less. This fact is evident when the current source is taken as output signal and the gain is plotted for k value different. Highlights the fact that the increase in k is not proportional to the decrease in the gain.

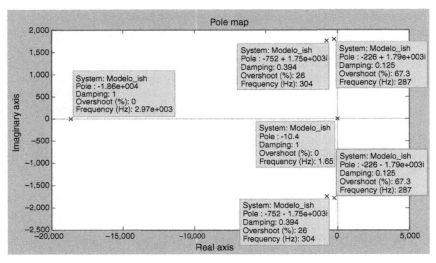

FIGURE 6.35 Pole map with active filter, control strategy of source current detection (Example 6.6).

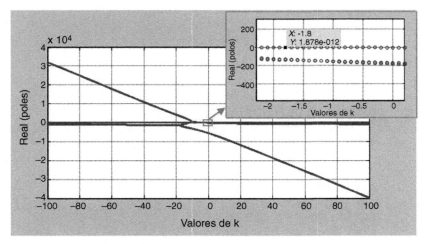

FIGURE 6.36 Real part of the poles versus k. Example 6.6.

Figure 6.38 reflects more clearly this nonproportionality between gain and k constant. This shows the gain at the frequency of 250 Hz for different k values. A gain of -30 dB (about 0.030 in absolute value) implies that the magnitude of 5 order harmonic drops to 3%. So, when the distortion source is the supply voltage, a value of $k = 25$ will be sufficient, whereas if it is considered that the distortion source is the load, it is necessary that $k = 50$.

Subsequently, the strategy control of load voltage detection is applied. The system matrix changes and therefore the system poles. The stability criterion application of Routh–Hurwitz for k_v values between -2 and 2 allows to represent the graph shown in Figure 6.39. This shows the actual values of the system poles in terms of k_v. It is noted that when the condition $k_v \geq 1.198$ is accomplished the system has a pole in the right semiplane.

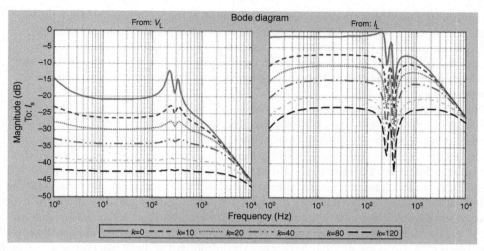

FIGURE 6.37 Bode gain, strategy source current detection. Example 6.6.

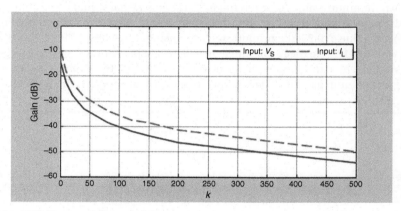

FIGURE 6.38 Gain at 250 Hz versus k. Example 6.6.

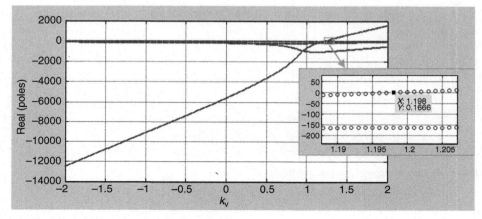

FIGURE 6.39 Real part of the poles versus k_v. Example 6.6.

Figure 6.40 shows the system gain for different values of the k_v parameter. When the load harmonic source is considered as input signal (Figure 6.40 right) the minimum value of the gain is obtained for $k_v = 1$. This does not happen when the input signal is the supply voltage, as it is shown in the Bode diagram in the Figure 6.40 left. In this situation it is clear that the implementation of this strategy can even increase the harmonic distortion of the current source.

Finally, results are analyzed when the control strategy that combines the previous two are applied. A stability analysis must take into account two parameters: k_v and k. In a first analysis, it is considered that $k_v = 1$. The stability criterion application of Routh–Hurwitz allows us to determine the minimum value of k for which the system is stable. This occurs for $k > -1.8$. This value is the same as obtained with the control strategy of source current detection. If the value of k_v is set at 1.2, the system is stable whenever $k > 0$. It must be remembered that the strategy of load voltage detection, with $k_v = 1.2$ the system was unstable. Therefore, the hybrid strategy provides a more robust system. For $k_v < 1$, the stability limit is located at values of $k < -1.8$. Therefore, it can be concluded that if k_v has a value close to the unit, the system will be stable as long as k is greater than zero. Thus the fulfillment of these two premises will be the design goal.

With known minimum values of k, it remains to establish what would be its appropriate value from the point of view of system gain. When set to target the low-frequency gain is -30 dB, with a k_v value of 0.95, the proportionality constant must be $k = 29$. As shown in Figure 6.41, with this k value the maximum gain will be 30 dB, regardless of the input signal. In case of a system where the supply voltage is slightly distorted, the k value can be reduced to 12, having a maximum gain of -30dB at low frequencies, as shown in the Bode plot on the Figure 6.41 right.

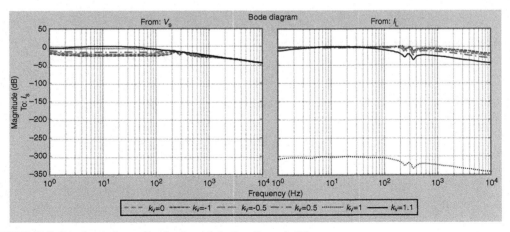

FIGURE 6.40 Bode gain, strategy of load voltage detection. Example 6.6.

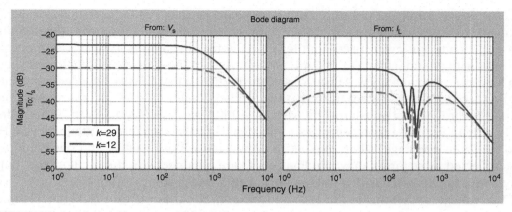

FIGURE 6.41 Bode gain, hybrid strategy with $k_v = 0.95$. Example 6.6.

6.4 Experimental Prototype of SAPPF

In this section, the experimental platform that has allowed the verification of the theoretical results is presented. It consists of a nonlinear load, HCS type, a parallel passive filter and a SAF. In this configuration the three control strategies are implemented: source current detection, load voltage detection, and hybrid control strategy. Experimental results allow the operation of each compensation strategy to be verified.

Figure 6.42 shows the schematic of the experimental prototype of SAPPF, which was designed in the laboratory. The load used is composed of a three-phase noncontrolled

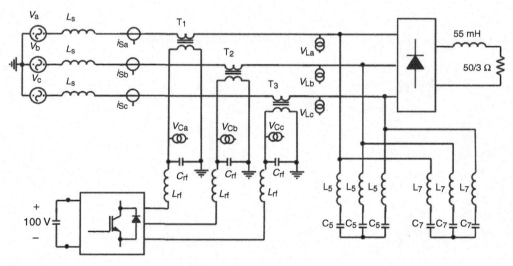

FIGURE 6.42 Experimental prototype of hybrid filter SAPPF.

rectifier. It was implemented with a three-phase bridge 36MT60 IR. At the dc side, a coil of 55 mH in series with a resistor of 50/3 Ω was connected. Two LC branches tuned to the 5th and 7th harmonic were used. For the 5th harmonic filter a capacity of 30 μF connected in series with an inductance of 13.5 mH were chosen. In the case of the filter tuned to the 7th harmonic the capacity was of 30 μF and the inductance of 6.7 mH. The element values of the LC branches were chosen so that the fundamental frequency reactive power consumed by the load and transferred by the passive filter were approximately the same order.

In a first test, the source generates a sinusoidal voltage waveform of frequency 50 Hz and 90 V phase voltage. When the load is connected with the active and passive filters disconnected, waveforms shown in Figure 6.43 are obtained with the oscilloscope. Both waveforms are distorted with a voltage THD of 9.6% and current THD of 17.2%.

The voltage waveforms at the PCC and the source current shown in Figure 6.43 are characterized by having odd harmonics with the exception of the harmonic of 3rd order and its multiples that are practically zero. Figure 6.44 shows the spectrum of harmonics of

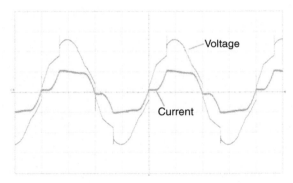

FIGURE 6.43 Voltage waveform, 48V/div and current waveform, 10A/div, without filters.

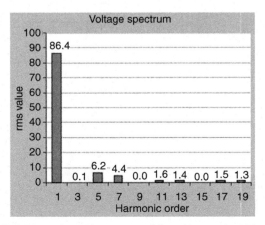
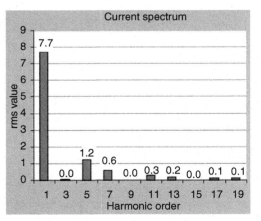

FIGURE 6.44 Harmonic spectrum of the voltage at the PCC and source current without filters.

both. The voltage rms value is 86.8 V and the current rms value of 7.7 A. The most significant harmonics are of 5th and 7th order.

The active power, reactive power, and apparent power per phase are: $P = 0.66$ kW, $Q = 0.18$ kvar, and $S = 0.68$ kVA. The measured power factor is 0.96 inductive.

When the passive filter is connected, the waveforms shown in Figure 6.45 are obtained. An improvement of the two waveforms is observed. The voltage THD is 5.1% and current THD 7.9%.

Figure 6.46 shows the harmonic spectrum of the source current and voltage at the PCC. Highlights the decrease of the rms values of 5th and 7th harmonics, however, they do not become zero because in practice, the LC branches do not present a zero impedance at tuning frequencies. The voltage rms value is 88.1 V and the current is 9.6 A.

With regard to the power consumed by the set, their values per phase are: $P = 0.71$ kW, $Q = 0.07$ kvar, and $S = 0.71$ kVA. The power factor improves to 0.99. There is a small increase

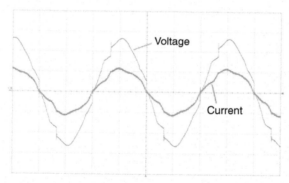

FIGURE 6.45 Voltage waveform, 48V/div and current waveform, 10A/div, with passive filter.

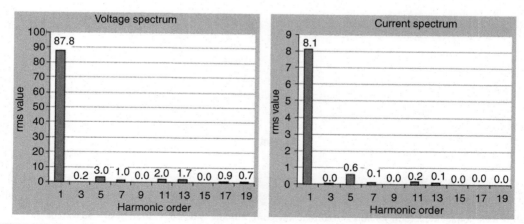

FIGURE 6.46 Harmonic spectrum of the voltage at the PCC and source current with passive filter.

in active power, which is justified by the resistive effect of the LC branch coils and a decrease of reactive power also due to passive filter.

In the first test with active filter, the strategy of source current detection is applied. The proportionality constant is set to 50. Source current waveform and voltage waveform at the PCC is shown in Figure 6.47. The voltage THD measured is 2% and 2.3% for the current. Therefore, there is a clear reduction in THD, compared to the system without active filter. With respect to the voltage and current rms values, these are 87.6 V and 7.9 A, respectively.

Moreover, the active power, reactive power, and apparent power measured per phase are: $P = 0.69$ kW, $Q = 0.03$ kvar, and $S = 0.70$ kVA. The power factor is 0.99.

When the strategy of load voltage detection is applied to the active filter, the waveforms shown in Figure 6.48 are obtained. For voltage, the measured THD is 2.1% and 3.5% for the current. A slight deterioration in terms of source current THD is observed. The voltage THD value is maintained, while the voltage and current rms values are 87.6 V and 8.0 A. In addition, active power, reactive power, and apparent power per phase are: $P = 0.70$ kW, $Q = 0.04$ kvar, and $S = 0.70$ kVA. The power factor is 0.99.

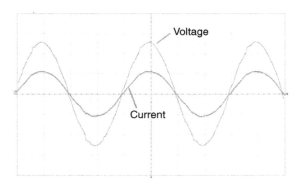

FIGURE 6.47 Voltage waveform, 48V/div and current waveform, 10A/div, with SAPPF. Strategy of source current detection.

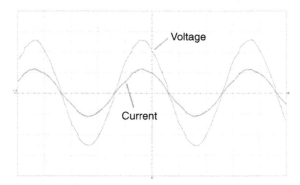

FIGURE 6.48 Voltage waveform, 48V/div and current waveform, 10A/div, with SAPPF. Strategy of load voltage detection.

Finally, the hybrid strategy is applied. The proportionality constant of current source is $k = 20$. The waveforms obtained are shown in Figure 6.49. The voltage THD is 1.7%, and the current THD is 2.1%. Therefore, the THD decreases compared to previous strategies. The measured rms values are 87.6 V for voltage and 7.9 A for current. The measured powers at the PCC are: $P = 0.70$ kW, $Q = 0.03$ kvar, and $S = 0.70$ kVA, being the power factor 0.99 capacitive.

Table 6.3 has been built with the results obtained using the three strategies. Here, the measured data in each of the experiments are summarized. Emphasize that from the point of view of the harmonic elimination, when the supply voltage is sinusoidal the results obtained with the different strategies are similar.

Finally, the behavior of compensation equipment is analyzed when the supply voltage is distorted. For this, the power source is programmed to generate the fundamental component

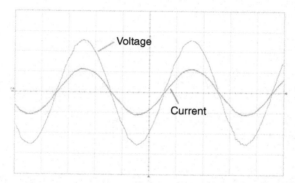

FIGURE 6.49 Voltage waveform, 48V/div and current waveform, 10A/div, with SAPPF. Hybrid strategy.

Table 6.3   Results from the Practical Case SAPPF Filter. Source Sinusoidal

		THD (%)	rms	H1	H3	H5	H7	H9	H11	H13	P (kW)	Q (kvar)	S (kVA)	PF
Without filters	V	9.6	86.8	86.4	0.1	6.2	4.4	0.0	1.6	1.4	0.78	0.25	0.82	0.95 Ind.
	I	17.2	7.8	7.7	0.0	1.2	0.6	0.1	0.3	0.2				
Strategy: Source current detection $k = 50$	V	2.0	87.6	87.5	0.2	0.8	0.3	0.1	0.8	0.7	0.69	0.03	0.70	0.99 Cap.
	I	2.3	7.9	7.9	0.0	0.1	0.0	0.0	0.1	0.1				
Strategy: Load voltage detection	V	2.1	87.6	87.6	0.3	1.1	0.3	0.1	0.8	0.6	0.70	0.04	0.70	0.99 Cap.
	I	3.5	8.0	8.0	0.1	0.2	0.0	0.0	0.1	0.0				
Hybrid strategy, $k = 20$	V	2.0	87.6	87.5	0.2	0.8	0.3	0.1	0.8	0.7	0.69	0.03	0.70	0.99 Cap.
	I	2.3	7.9	7.9	0.0	0.1	0.0	0.0	0.1	0.0				

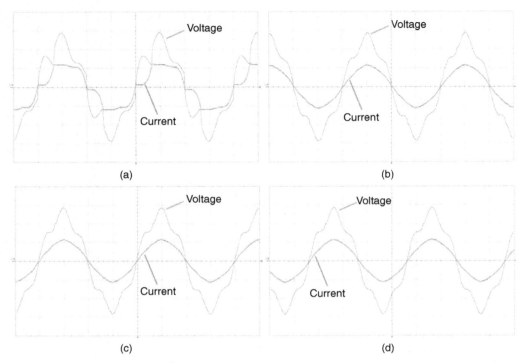

FIGURE 6.50 SAPPF behavior for nonsinusoidal source voltage. (a) Without filters; (b) source current detection; (c) load voltage detection; (d) hybrid strategy. Tensión, 48 V/div, intensidad, 10 per div.

and a 5th order harmonic of 12% on the fundamental harmonic. Figure 6.50 shows the source current waveform and voltage at the PCC before and after connecting the active filter and for the three compensation strategies analyzed. For the strategy of source current detection the proportionality constant, k, is 50 and for hybrid strategy the k constant is 20.

Table 6.4 summarizes the most significant measured values: THD, harmonics, powers, and power factor. The strategies for source current detection and hybrid allows significantly reducing the source current THD. The voltage at the PCC is distorted because of the harmonics present in the supply voltage, but they are not due to the load current harmonics.

6.5 Summary

In this chapter, the steady state behavior of a hybrid filter formed by an SAPPF was analyzed. Three different control strategies for the active filter were studied: current source detection, load voltage detection, and a hybrid control that includes a combination of these two strategies. From the equivalent single-phase circuit, expressions of the two variables of interest from the point of view of harmonic filtering were obtained, these are: the voltage at the PCC and source current. After obtaining the steady state equations, the set

Table 6.4   Results from the Practical Case SAPPF Filter. Source Nonsinusoidal

		THD (%)	rms	H1	H3	H5	H7	H9	H11	H13	P (kW)	Q (kvar)	S (kVA)	PF
Without filters	V	19.1	87.1	85.5	0.2	15.4	3.2	0.1	2.4	2.1	0.63	0.20	0.66	0.95 Ind.
	I	13.3	7.6	7.5	0.0	0.8	0.5	0.0	0.2	0.2				
Strategy:	V	13.4	87.5	86.6	0.2	11.5	0.2	0.1	0.8	0.7	0.68	0.09	0.68	0.99 Cap.
source current	I	2.5	7.8	7.8	0.0	0.2	0.0	0.0	0.1	0.1				
detection														
$k = 50$														
Strategy:	V	9.0	87.0	86.6	0.3	7.7	0.2	0.1	0.5	0.4	0.68	0.06	0.68	0.99 Ind.
load voltage	I	10.3	7.8	7.8	0.1	0.8	0.0	0.0	0.0	0.0				
detection														
Strategy:	V	13.4	87.5	86.7	0.2	11.5	0.2	0.1	0.8	0.7	0.68	0.09	0.68	0.99 Ind.
hybrid,	I	2.5	7.8	7.8	0.0	0.2	0.0	0.0	0.1	0.1				
$k = 20$														

load-filter was subjected to different situations in order to compare their performance. Thus, the set was connected to a voltage supply distorted with different nonlinear loads at the PCC.

The SAPPF configuration was adequate for removing source current harmonics when the loads are HCS type. In the case of sinusoidal supply voltage, the three control strategies allowed a reduction of the same order in the THD source current and voltage THD at the PCC. However, the strategy of load voltage detection did not avoid the passive filter to become a drain from harmonics of nearby load to passive filter. When the supply voltage is distorted, the strategies of source current detection and hybrid can reduce current harmonics, not showing a clear difference between the two for the same value of k. In contrast, the control strategy of load voltage detection was adequate to improve the THD voltage at the PCC; however, when the supply voltage is distorted the THD source current increases respect to the uncompensated system.

Moreover, the state-variable model of the hybrid active filter SAPPF was obtained. With the state model, it was possible to perform a dynamic analysis of the three control strategies presented. This allowed establishing different design rules from the point of view of stability and the system gain. Thus, these conclusions could be established:

- Source current detection
  The stability is ensured when design criteria $k > 0$ is chosen. The higher the k value, the lower the system gain. The k value is adjusted according to the objective sought harmonic attenuation.
- Load voltage detection
  The minimum value of gain is obtained when $k_v = 1$, that is, when the error in the measurement of the load voltage is null. Furthermore, a value of $k_v > 1$ can destabilize the system.

- Hybrid strategy
  If k_v has a value close to unity and k is greater than zero, the system will be stable. This will provide robustness to the system, and also allows the error in measurement of the load voltage be increased.

From the point of view of the system gain when the supply voltage is distorted, strategies of source current detection and hybrid achieve a lower value. Furthermore, when the supply voltage is distorted, the greater attenuation of the source current harmonics is given to the strategy of load voltage detection with $k_v = 1$.

Finally, a laboratory prototype was designed and subjected to two different situations of voltage network: sinusoidal and nonsinusoidal voltage. The obtained results allowed the theoretical development to be verified.

References

[1] Peng FZ, Adams DJ. Harmonics sources and filtering approaches. Proc Ind Appl Conf 1999;1:448–55.

[2] Akagi H. Active harmonic filters. Proc IEEE 2005;93(12):2128–41.

[3] Singh B, Verma V, Chandra A, Al-Haddad K. Hybrid filters for power quality improvement. IEE Proc Gen Trans Distrib 2005;152(3):365–78.

[4] Leong Hua Tey, Ping Lam So, Yun Chung Chu. Adaptive neural network control of active filters. IEEE Trans Power Electr 2005;74(1):37–56.

[5] Montero MIM, Cadaval ER, Gonzalez FB. Comparison of control strategies for shunt active power filters in three-phase four-wire systems. IEEE Trans Power Electr 2007;22(1):229–36.

[6] Uyyuru KR, Mishra MK, Ghosh A. An optimization-based algorithm for shunt active filter under distorted supply voltages. IEEE Trans Power Electr 2009;24(5):1223–32.

[7] Herrera RS, Salmerón P. Instantaneous reactive power theory: a reference in the nonlinear loads compensation. IEEE Trans Ind Electr 2009;56(6):2015–22.

[8] Wang Z, Wang Q, Yao W, Liu J. A series active power filter adopting hybrid control approach. IEEE Trans Power Electr 2001;16(3):301–10.

[9] Jou H-L, Wu K-D, Wu J-C, Li C-H, Huang MS. Novel power converter topology for three-phase four-wire hybrid power filter. IET Power Electr 2008;1(1):164–73.

[10] Shuai Z, Luo A, Tu C, Liu D. New control method of injection-type hybrid active power filter. IET Power Electr 2011;4(9):1051–7.

[11] Milanés-Montero MI, Romero-Cadaval E, Barrero-González F. Hybrid multiconverter conditioner topology for high-power applications. IEEE Trans Ind Electr 2011;58(6):2283–92.

[12] Akagi H, Kondo R. A transformerless hybrid active filter using a three-level pulsewidth modulation (PWM) converter for a medium-voltage motor drive. IEEE Trans Power Electr 2010;25(6):1365–74.

[13] An Luo, Wei Zhao, Xia Deng, Shen ZJ, Jian-Chun Peng. Dividing frequency control of hybrid active power filter with multi-injection branches using improved i_p–i_q algorithm. IEEE Trans Power Electr 2009;24(10):2396–405.

[14] Rodriguez P, Candela JI, Luna A, Asiminoaei L. Current harmonics cancellation in three-phase four-wire systems by using a four-branch star filtering topology. IEEE Trans Power Electr 2009;24(8):1939–50.

[15] Zhao W, Luo A, Shen ZJ, Wu C. Injection-type hybrid active power filter in high-power grid with background harmonic voltage. IET Power Electr 2011;4(1):63–71.

[16] Rivas D, Moran L, Dixon J, Espinoza J. A simple control scheme for hybrid active power filter. IEEE Proc Gen Transm Distrib 2002;149(4):485–90.

[17] Salmerón P, Litrán SP. Improvement of the electric power quality using series active and shunt passive filters. IEEE Trans Power Deliv 2010;25(2):1058–67.

[18] Peng FZ, Akagi H, Nabae A. A new approach to harmonic compensation in power systems-a combined system of shunt passive and series active filters. IEEE Trans Ind Appl 1990;26(6):983–90.

[19] IEEE Std. 1531-2003, IEEE Guide for Application and Specification of Harmonic Filters, November 2003.

[20] Ginn HL, Czarnecki LS. An optimization based method for selection of resonant harmonic filter branch parameters. IEEE Trans Power Deliv 2006;21(3):1445–51.

[21] Peng FZ, Su GJ. A series LC filter for harmonic compensation of ac drives. In: Proceedings of the IEEE Power Electronics Specialists Conference; 1999 PESC'99. p. 213–8.

[22] Pomilio JA, Deckmann SM. Characterization and compensation of harmonics and reactive power of residential and commercial loads. IEEE Trans on Power Deliv 2007;22(2):1049–55.

[23] Zobaa AF. The optimal passive filters to minimize voltage harmonic distortion at a load bus. IEEE Trans Power Deliv 2005;20(2 Part 2):1592–7.

[24] González DA, McCall JC. Design of filters to reduce harmonic distortion in industrial power systems. IEEE Trans Ind Appl 1987;IA-23:504–12.

[25] Das JC. Passive filters-potentialities and limitations. IEEE Trans Ind Appl 2004;40(1):232–41.

[26] Vázquez JR, Juan L, Flores P, Salmerón SP, Litrán. A hybrid approach to compensate nonlinear loads in electrical power systems. In: Proceedings of the International Conference on Renewable Energy and Power Quality; 2005 ICREPQ 05. Zaragoza.

[27] Vázquez JR, Flores JL, Salmerón P, Litrán SP. Diseño de filtros pasivos, activos e híbridos para la compensación armónica de cargas trifásicas no lineales. In: Proceedings of the Spanish Portuguese Congress on Electrical Engineering; 2005 CHLIE. Marbella .

[28] Litrán SP, Salmerón P, Vázquez JR. Analysis by state equation of a control strategy for hybrid filter. In: Proceedings of the International Conference on Renewable Energy and Power Quality; 2009 ICREPQ 09. Valencia.

[29] Litrán SP, Salmerón P. Analysis and design of different control strategies of hybrid active power filter based on the state model. IET Power Electr 2012;5(8):1341–50.

[30] Thunberg E, Soder L. A Norton approach to distribution network modeling for harmonic studies. IEEE Trans Power Deliv 1999;14(1):272–7.

7

Combined Shunt and Series Active Power Filters

This chapter describes the main features that can be achieved by using combined series and shunt active power filters. The first section focuses on the concept of load conditioning and power quality. It studies the different strategies to provide balanced, sinusoidal, and regulated voltages at the load terminals, as well as to compensate the harmonic, unbalance, and reactive components of the load currents to obtain supply currents with the same waveform and in phase with the fundamental direct-sequence component of the supply voltages. These targets allow us to define the structure and control calculations of a unified power quality conditioner (UPQC) to provide this load conditioning and to enhance the electric power quality (EPQ) in the point of common coupling (PCC).

Analysis of the state-space model of the power circuit and the control implementation will show the filtering characteristics of the conditioner, so that the values of the passive components around both converters can be selected, and the control parameters tuned.

The theoretical analysis is contrasted with the results obtained in simulation cases, developed in a platform based on MATLAB-Simulink; as well as with the results of an experimental laboratory prototype.

The last section of the chapter focuses on another interesting application of the series-shunt power conditioner: power flow control and the voltage regulation in electric power systems with the unified power flow controller (UPFC). The control techniques of this device are compatible with the tasks of the UPQC and can be combined to design a universal active power line conditioner (UPLC) with both functions. The complete control implementation is now defined, and a practical case is developed to show the behavior of the combined conditioner.

7.1 Introduction

The increasing sophistication of APFs has allowed them to be used in an increasing number of compensating functions, and the experience obtained from their use has shaped the objectives of the conditioners, depending on their situation in the electrical networks [1,2]. Thus, the main objective of an active filters installed by individual users would be the harmonic and unbalance compensation of its own specific nonlinear loads. On the other hand, the main objective of active filters installed by the electrical companies would be the compensation of the supplied voltages, as well as a harmonic dumping between distribution systems.

Depending on the specific objectives, different appropriate configurations for each application can be selected:

1. Shunt active filter (Figure 7.1): This is the simplest configuration and the one now most commonly used. Its main purpose is to compensate for nonlinear loads, injecting in parallel the harmonic components of the load current. Thus, its control strategy would be:

$$i_C = i_{Lh} \tag{7.1}$$

   Further applications assigned them a more complete compensation of the load current, including the reactive components (i_{LfQ}) as well as the unbalance components (i_{LfU}) at the fundamental frequency:

$$i_C = i_{Lh} + i_{LfU} + i_{LfQ} \tag{7.2}$$

   This increase in performance does not imply necessarily higher intensity peak values for the electronic devices, although they may need a higher capacity in the dc side to compensate for higher oscillating instantaneous power.

   Finally it is worth mentioning the ability of the shunt APF to dump harmonic resonances between existing passive components in the electrical networks [3], acting as a harmonic resistance:

$$i_C = G \cdot v_{Lh} \tag{7.3}$$

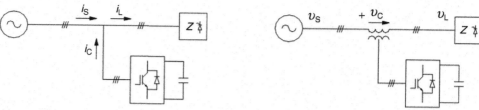

FIGURE 7.1 Shunt active filter.   **FIGURE 7.2** Series active filter.

2. Series active filter (Figure 7.2): This configuration is specifically appropriate for individual HVS loads [4] where this load acts as a source of harmonic voltages, instead of harmonic currents. In this case, the compensation strategy should be:

$$v_C = -v_{Lh} \qquad (7.4)$$

Another more general application is to compensate for the unbalances and voltage harmonics of the mains supply [5], in order to provide an enhanced voltage to the installation. In this case, the strategy should be:

$$v_C = v_{Sh} + v_{SfU} \qquad (7.5)$$

It is usual to include, in both applications, an additional harmonic current compensating term $R \cdot i_{Sh}$; since in both cases there may be harmonic voltage sources on the side of the circuit that is intended to be compensated. This term has also a dumping effect against transients and resonances, giving the conditioner a more robust behavior.

3. Hybrid filters (Figures 7.3–7.5) [2]: The cost, complexity, and energy efficiency of active filters, compared to passive filters of equal power, has promoted the development of hybrid solutions that use the advantages of both elements. The basic idea is to minimize the rated power of the active components, assigning them a tuning and damping behavior of the passive filters.

The combination of parallel passive and active filters of Figure 7.3 allows us to use the active filter as a damper for the passive filter against possible resonances. (The passive filter generically includes tuned filters and high-pass filter.) The series combination of passive filters and active filter in Figure 7.4 tries to achieve similar objectives.

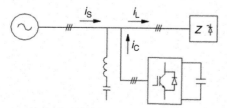

FIGURE 7.3 Parallel passive, parallel active hybrid filter.

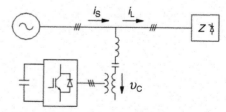

FIGURE 7.4 Parallel active–passive hybrid filter.

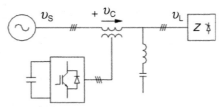

FIGURE 7.5 Series active, parallel passive hybrid filter.

The configuration in Figure 7.5 follows a slightly different principle, in which the series active filter – acting as a harmonic resistance – aims to force the flow of the load harmonics through the passive filters, providing also a harmonic damping between source and load.

4. Combined series–parallel active filters (Figure 7.6): The dual properties of series and shunt active filters has led to combined designs, which have been generically called UPQC [6–10].

The configuration shown in Figure 7.6 is the most commonly used of the different options. Both converters share a common dc link, which is regulated through the shunt converter. The series converter usually compensates the supply magnitudes (v_S, i_S) and the shunt converter is mainly focused on the load side (v_L, i_L). There is another usual topology that is symmetrical to this one, where the series APF is at the load side of the circuit [7]. This topology is usually intended for specific loads and each converter compensates only one of the target magnitudes. The configuration of Figure 7.6 is particularly suitable for distribution systems and combines the functions of both filters to compensate the distortion and unbalances of the supply voltages, as well as the load currents. The control strategy would be:

$$v_C = v_{Sh} + v_{SfU} + R \cdot i_{Sh} \tag{7.6}$$

$$i_C = i_{Lh} + i_{LfU} + i_{LfQ} \tag{7.7}$$

Thus, the series filter compensates the supply voltages and provides harmonic damping between source and load; and the parallel filter compensates the load currents. Furthermore, the topology of this conditioner allows to each converter to

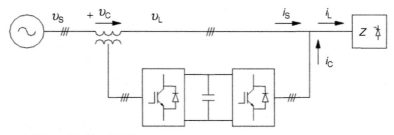

FIGURE 7.6 Series–parallel combined active filter.

control both target variables – the load voltages and the supply currents – and to design a global control strategy that enhances the general behavior of the conditioner.

7.2 Unified Power Quality Conditioner

This section explains and develops the design and control implementation of a UPQC to compensate a broad range of load types, improving the EPQ in the point of common coupling (PCC). This conditioner is called load compensation active conditioner (LCAC) and has a specific design to enhance the voltages applied to the load and the currents delivered by the electrical supply.

The general target is the compensation of the distortion and unbalance produced by nonlinear loads, and the supply network itself. With an adequate compensation strategy it is possible to obtain balanced sinusoidal supply currents in phase with the direct-sequence component of the supply voltages; and regulated and balanced sinusoidal voltages applied to the load terminals. From this point of view, the combination of series and parallel active filters is seen as one of the most appropriate configurations [1], and the topology shown in Figure 7.6 has the capability to achieve these requirements.

In the first stage, the series converter should apply a controlled voltage equal to the harmonic and unbalance components of the supply voltages (equation (7.8)) to isolate the load from this source of distortion; as well as a direct-sequence fundamental voltage to regulate the voltage level at the load side. On the other hand, the shunt converter should draw the harmonic, unbalance, and reactive components of the current required by the load (equation (7.9)) to avoid that they flow through the whole supply network. And the other task of the shunt converter is to maintain the voltages of the dc link, through an active component, i_{adcreg}, to compensate the internal losses of the conditioner.

$$v_C^* = v_{Sh} + v_{SfU} + \Delta v_{Sf+} \tag{7.8}$$

$$i_C^* = i_{Lh} + i_{LfU} + i_{LfQ} - i_{adcreg} \tag{7.9}$$

In the second stage, it is advisable to consider the inclusion of the current harmonic dumping term ($R \cdot i_{Sh}$) in the series converter due to the reasons explained in the previous section. This way, the presence of HVS loads in the generic distribution line is mitigated, and the additional control term provides damping against transients and possible resonances. At this point, it is important to state that pure HVS loads should have their own specific compensation methods, like active filters or smoothing reactors. Otherwise they could not be connected to pure sinusoidal voltage sources, when the resulting currents exceed the capabilities of their own electronic power devices. And when they are in the same line with another loads, strong voltage distortions and unbalance can be produced.

It is also convenient to include a corrective term for the load voltages in the reference current of the shunt converter ($G \cdot v_{Lh}$). When the conditioner is applied to a strong HVS load or the changes in the load current are faster than the tracking capability of the shunt converter, the correcting term of the series filter reach significant values and therefore the load voltages differ substantially from their reference values. If the shunt converter can provide this requested current as fast as possible, the voltage gap will be minimized. Otherwise it must be assumed that the load voltages will remain distorted (it is a voltage distortion caused by the own load and remains only in the load side) and that the dumping effort will flow through the series converter and the supply network. With the additional term in the shunt converter, the stabilization efforts of the conditioner are shared between the converters and the part of these transients that flow through the supply is reduced.

$$v_C^* = v_{Sh} + v_{SfU} + \Delta v_{Sf+} + R \cdot i_{Sh} \qquad (7.10)$$

$$i_C^* = i_{Lh} + i_{LfU} + i_{LfQ} - i_{adcreg} - G \cdot v_{Lh} \qquad (7.11)$$

These correcting terms, in both converters, should be limited in the control implementation to maintain the voltages and currents of the conditioner in affordable values. In any case, they act as a feedback of the general compensation process to improve the dynamic of the active conditioner, as will be shown later on. This way, each filter will develop a specific function using a unified control that treats the load compensation process globally.

7.2.1 Structure of the UPQC

Figure 7.7 shows the power circuit of the LCAC. It includes two full bridge power converters with a common dc link and passive elements for the filtering the high frequencies. One of

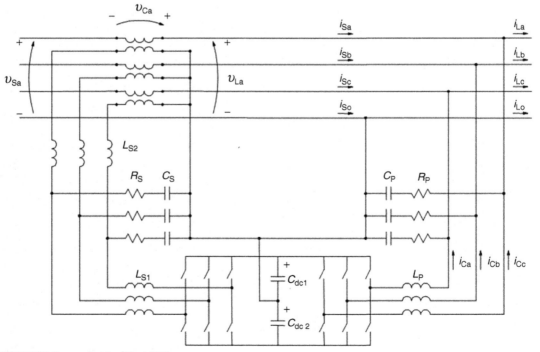

FIGURE 7.7 Power circuit of the LCAC.

the converters is connected is series with the supply, through a coupling transformer, and the other one is connected in parallel with the load. Both are voltage source converters, with pulse width modulation (PWM) control. In the most recent developments, for three-phase four-wire systems, these converters are made with IGBT devices with a three-leg configuration (see Section 4.3.3) and a split dc link to allow the neutral wire connection [8].

The figure also shows the location of the passive elements for the filtering of the higher harmonics. The harmonic traps (L_{S1}, R_S, C_S, L_{S2}) provide filtering of the voltage harmonics generated by the series inverter. The branches R_P, C_P in association with L_{S2} provides a low impedance path for the higher harmonics of i_C or i_L. The same elements attenuate the effects of the noncontrollable harmonics of v_S over the load voltages v_L or the supply currents i_S.

The switching control of the converters is based in the reference signals v_C^*, i_C^* generated in the internal LCAC control, following the compensation strategy exposed previously.

The construction of the voltage waveform v_C for the series converter is made with the scheme shown in Figure 7.8. It is a PWM modulation control by comparison with a triangular wave for a switching frequency of 20 kHz. A sample and hold element is applied to the reference signal v_C^*, each 50 μs, to limit the maximum switching frequency. It also allows to implement this control by DSP as well as to show the robustness of the control strategy explained in this section.

Figure 7.9 shows the control scheme applied to the generation of the current i_C in the shunt converter. The switching control is made with periodic sampling, with a fixed switching frequency. Although this method is not the best for accurate wave generation, the criteria for its selection were the foregoing exposed for the series filter. Actually, both the acquisition of all the measurements as well as the generation of the signal references of the conditioner are sampled every 50 μs.

7.2.2 Control Strategy of the UPQC

The control of the LCAC allows the power circuit to develop a dual function. The series filter works as a controlled voltage source to compensate that part of the supply voltages that differs from a balanced and regulated sinusoidal waveform. That is, the series filter eliminates the voltage harmonics and unbalance to get a direct sequence system. And at the same time it regulates the voltage level applied to the load.

On the other hand, the shunt filter located downstream of the series filter, will compensate the harmonics, reactive and asymmetry components included in the load currents. As

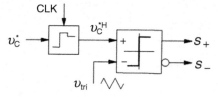

FIGURE 7.8 PWM generation of the series converter voltage v_C.

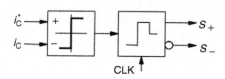

FIGURE 7.9 Generation of the shunt converter current i_C.

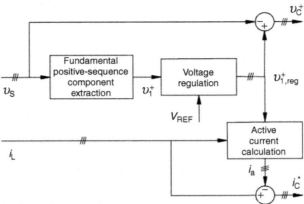

FIGURE 7.10 Overall scheme of the LCAC control.

a consequence, it cancels the neutral current. It will be also responsible for regulating the voltages of the common dc link.

Figure 7.10 shows the overall scheme of the control reference calculations, wherein each filter develops a specialized function through a unified control that treats globally the load compensation process. Thus, the voltage signals that the control circuit uses for determining the compensation currents of the parallel filter (see Figure 7.12) are the reference voltages of the series filter. And, as will be explained in the next section, the crossed correcting signals between converters will be detailed; and their function as stabilizer against transients and possible imperfections in the generation of the reference waveforms will be shown.

The reference voltage of the series compensation filter (v_C^*) is obtained by comparing the supply voltage (v_S) and its fundamental positive-sequence component with controlled amplitude ($v_{1,reg}^+$) as described later. This signal $v_{1,reg}^+$ is also used for the determination of the compensation current (i_C^*) of the shunt converter, based in the calculation of the active current of the load [11] and the internal power balance of the conditioner. The use as voltage signal of $v_{1,reg}^+$ makes the reference active currents (i_a) of each phase to be balanced and sinusoidal.

7.2.2.1 Calculation of the Load Voltage Reference ($v_L^* = v_{1,reg}^+$)
The control of the series filter aims to obtain an ideal voltage supply: balanced voltages without distortion. Figure 7.11 shows the block diagram for this part of the control circuit.

The first stage is to obtain the positive-sequence component (v^+) of the supply voltages. The phase shift of $\pm 1/3$ of a cycle of the voltage signals in phases 2 and 3, and the mean value of the three signals, results in a signal that has as fundamental harmonic the positive-sequence component of the voltages (v_1^+):

$$
\begin{aligned}
v^+(t) &= \frac{1}{3}\left(v_1(t) + v_2\left(t - \frac{T}{3}\right) + v_3\left(t + \frac{T}{3}\right) \right) \\
&= V^+ \cos(\omega t + \varphi^+) + \sum_{h=2}^{\infty} V_h \cos(h\omega t + \varphi_h) + V_{dc}
\end{aligned}
\tag{7.12}
$$

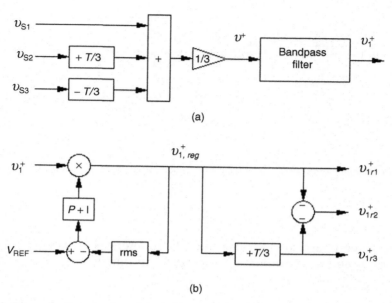

FIGURE 7.11 Block diagram for the extraction of the fundamental positive-sequence amplitude regulated voltage component. (a) Obtaining the fundamental positive-sequence component; (b) Amplitude regulation and three-phase wave building.

There are three terms in equation (7.12): the direct-sequence component of the fundamental harmonic, the rest of harmonic components, and an eventual nonzero average component. Subsequently, through a bandpass filter, the harmonic and dc components are removed and the signal v_1^+ is obtained.

In the second stage, a control loop regulates the rms value of v_1^+ to a preset value V_{REF}. The selection of the bandwidth of the bandpass filter ($\pm 2\%\ f_N$) is oriented to attenuate fast voltage amplitude variations (voltage flicker [7]). And the combination with the amplitude control loop allows a fine voltage regulation. Finally, the three-phase voltage waveforms are built ($v_{1r1}^+(t), v_{1r2}^+(t), v_{1r3}^+(t)$). The use of a double subtraction for phase 2 cancels instantaneously any remaining zero-sequence component and simplifies the scheme.

7.2.2.2 Calculation of the Active Current Reference ($i_S^* = i_a$)

The shunt filter control is based on the calculation of the active load current (i_a), and its block diagram is shown in Figure 7.12. The current reference i_S^* will be this active load current, and it is obtained from the calculation of the active load current using the voltage reference $v_{1,reg}^+$.

The transfer of an active power P from source to load requires the line currents (active currents) to be [11]:

$$i_{a1} = \frac{P}{V^2} v_1; \quad i_{a2} = \frac{P}{V^2} v_2; \quad i_{a3} = \frac{P}{V^2} v_3 \qquad (7.13)$$

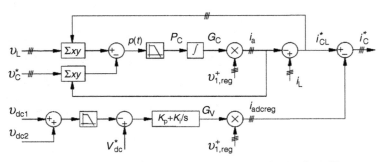

FIGURE 7.12 Block diagram for the calculation of the compensation currents for the shunt filter.

$$V^2 = \frac{1}{T} \int_0^T (v_1^2 + v_2^2 + v_3^2)\,\mathrm{d}t \tag{7.14}$$

where V is the three-phase voltage rms value. The difference between the load currents i_L and the active currents i_a determinate the components that do not transport useful power and are therefore compensable (i_{CL}^*)

$$i_{CL}^* = i_L - i_a \tag{7.15}$$

The determination of the active power P requires a low pass filter that eliminates the oscillatory component of the instantaneous power $p(t)$, with a critical frequency at least the supply frequency. Using a feedback loop and an integrator it is possible to reach, in steady state, some values of i_C^* whose average power P_C is null. Therefore, the output of the integrator remains without change in the previous estimated value of P/V^2. This will be its state until the load changes. The adaptation speed against load variations depends mainly on the dynamic response of the set low pass filter + integrator. The use of $v_{1,\mathrm{reg}}^+$ allows the active currents references i_a to be balanced and sinusoidal, if the equivalent conductance G_C is constant. If there are unbalances and harmonics the instantaneous power is not constant and there must be a tradeoff between the adaptation speed of the control loop and the allowable oscillating components in G_C.

In the second stage, the power flow through the series filter ($v_C^* \cdot i_S^*$) is included in the instantaneous power to enhance the estimation of the power balance, as well as the necessary direct regulation of the dc side voltages of the conditioner (see Section 4.5.2). If the power balance is accurate the variation of the dc side voltages will be reduced and the corresponding control loop will set an equivalent conductance G_V with smaller oscillations.

Finally, the crossed correcting terms of equations (7.10) and (7.11) should be defined. These components have a strong damping effect in the power circuit and can be analyzed together with the dynamic behavior of the passive components around the PWM converters. This joint analysis will allow to enhance the selection of both components.

7.2.3 Parameters Design

In the first stage, the dynamic behavior of the passive components will be analyzed. This will be carried out using the state-space equations, to check if they mitigate adequately the high frequency switching harmonics and no amplifications occur with possible resonances. The damping terms can be included later, with a similar analysis, to study their joint behavior in the different frequency ranges.

Figure 7.7 has shown the location of the passive elements of the active conditioner. These passive components have the function of filtering the higher harmonics (those greater than $f_C/10$, where f_C is the maximum switching frequency of the converters). Among these harmonics are the harmonics generated by the converters themselves. The series converter generates harmonics with frequencies near the switching frequency [8], with an amplitude equal to the maximum assigned compensating voltage (20–30% of the rated voltage V_N). The shunt converter produces harmonics with lower amplitude but also lower frequencies [12,13]. In the same way, the passive components can be assigned the mitigation of those components noncontrollable by the converters (like sharp load current changes, etc.). On the other hand, they should interfere as little as possible with the tasks of the LCAC at low frequencies, mainly to maintain at maximum the energetic efficiency of the conditioner.

Figure 7.13 presents the single-phase equivalent circuit of the conditioner, where the series-coupling transformer has been considered as ideal, and the load as a current source. The series converter, with its PWM generation control, has been considered as a controlled voltage source; and the shunt converter with its output inductance has been considered as a controlled current source.

7.2.3.1 Analysis of the Passive Circuit

The passive circuit of the active conditioner can be considered as a passive *n*-port network [9] that relates the one-port elements among each other, corresponding to the points of supply, load, series and shunt converters of the LCAC. This relation can be expressed with the state-space equations of the *n*-port:

$$
\begin{aligned}
\frac{d}{dt}x &= Ax + B_1u + B_2w \\
y &= Cx + D_1u + D_2w
\end{aligned}
\quad \equiv \quad
\begin{aligned}
\frac{d}{dt}\begin{bmatrix} v_{cap} \\ i_{ind} \end{bmatrix} &= A\begin{bmatrix} v_{cap} \\ i_{ind} \end{bmatrix} + B_1\begin{bmatrix} v_C \\ i_C \end{bmatrix} + B_2\begin{bmatrix} v_S \\ i_L \end{bmatrix} \\
\begin{bmatrix} v_L \\ i_S \end{bmatrix} &= C\begin{bmatrix} v_{cap} \\ i_{ind} \end{bmatrix} + D_1\begin{bmatrix} v_C \\ i_C \end{bmatrix} + D_2\begin{bmatrix} v_S \\ i_L \end{bmatrix}
\end{aligned}
\tag{7.16}
$$

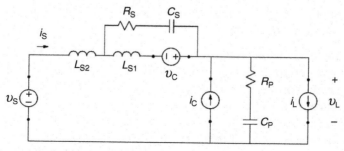

FIGURE 7.13 Single-phase equivalent circuit of the passive components.

where $\mathbf{u} = [v_C, i_C]$ are the variables controlled by the converters of the LCAC; $\mathbf{w} = [v_S, i_L]$ are the variables corresponding to the distortions introduced in the system from the supply or the load sides; $\mathbf{y} = [i_S, v_L]$ are the target variables of the control; and $\mathbf{x}$ are the state variables of the passive network that are usually identified with the energy storing elements (capacitance voltages, v_{cap}, and inductance currents, i_{ind}). For the case of the Figure 7.13 there are four state-space variables.

If the equations are expressed in the Laplace domain, we can calculate the transfer functions to evaluate the effect of the independent variables $\mathbf{u}$ and $\mathbf{w}$ over the target variables $\mathbf{y}$ of the control; through the passive elements of the conditioner:

$$Y(s) = \left[C\,(I_S - A)^{-1} B_1 + D_1 \right] U(s) + \left[C\,(I_S - A)^{-1} B_2 + D_2 \right] W(s) \tag{7.17}$$

If the n-port network is linear, its response can be considered as the superposition to different excitations. Thus, it is possible to split the response to the control sources $\mathbf{u}$ into two parts: the ideal value $\mathbf{u^*}$ that could achieve the desired values of $\mathbf{y}$ (v_L^* and i_S^*); and the part of the control $\mathbf{u}_E$, responsible of some deviations of the response $\mathbf{y}$. The variable $\mathbf{u}_E$ includes the high frequency harmonics generated by the converters, as well as their control limitations. In this way, it is possible to model the behavior of the passive network against perturbations with a formulation similar to (7.16), with the difference that now $\mathbf{u} = \mathbf{u}_E$, and $\mathbf{w} = \mathbf{w}_E$ are the deviations over the ideal control and ideal measurements, and $\mathbf{y} = \mathbf{y}_E$ show the effects of these "noises" over the target values of v_L and i_S.

$$\frac{d}{dt}(x^* + x_E) = A(x^* + x_E) + B_1(u^* + u_E) + B_2(w^* + w_E)$$
$$(y^* + y_E) = C(x^* + x_E) + D_1(u^* + u_E) + D_2(w^* + w_E) \tag{7.18}$$

$$\frac{d}{dt}x_E = Ax_E + B_1 u_E + B_2 w_E$$
$$y_E = Cx_E + D_1 u_E + D_2 w_E \tag{7.19}$$

The analysis made on the passive circuit is based on two criteria:

- The eigenvalues of the matrix $\mathbf{A}$ are the poles of the characteristic equation of the system and show the natural responses of the passive circuit and also its stability. They define the dynamic response of the passive circuit during transients, like the start of the LCAC where the initial conditions of the internal states can be far from the reference values.
- The bode plots of the transfer functions of (7.19), between each element of the $\mathbf{y}_E$ vector, with respect to each element of $\mathbf{u}_E$ and $\mathbf{w}_E$ vectors, will show the attenuation degree obtained at the control targets.

The study has been made on an equivalent single-phase model (Figure 7.13), where the series converter has been modeled as a controlled voltage source and the shunt converter as a controlled current source. The values of the passive components have been chosen through their results with the eigenvalues and transfer functions of this model. As a general target, an attenuation of -40 dB (1%) has been aimed, especially over $f_C/10$. The values of these components are shown in Table 7.1, and the corresponding eigenvalues are shown in Table 7.2. After some initial state-space analysis like the one that is to be explained here-

Table 7.1   Values of the Passive Components for the State-Space Analysis

L_{S1}	25 mH	L_{S2}	5 mH
R_S	800 Ω	R_P	0 Ω
C_S	1 μF	C_P	15 μF

Table 7.2   Eigenvalues of *A* for the Passive Circuit of Figure 7.13

Eigenvalues of A (s^{-1})
$-16.84 \pm 1.476j$ (235 Hz)
-1.283
-190.683

after, the value of the resistance R_P has been finally set to zero, to reduce the influence of the high frequency components of i_C over the load voltage v_L.

Figure 7.14 presents the magnitudes of the transfer functions of $\mathbf{u_E}$ and $\mathbf{w_E}$ over $\mathbf{y_E}$, as a function of frequency. For a better interpretation of the results, the transfer functions have been plotted in per unit values, being 230 V and 1000 VA/230 V the base values for voltages and currents. Some solid lines have been included in Figure 7.14 for a better interpretation and

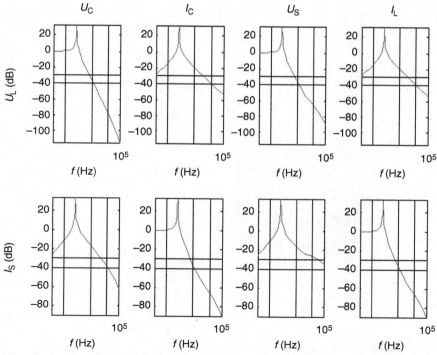

FIGURE 7.14 Transfer functions of the passive network. Reference vertical solid lines at 50 Hz (f_N), 2 kHz (f_c/10), and 20 kHz (f_c). Reference horizontal solid lines at -30 and -40 dB.

resolution. Three vertical solid lines show the frequency references of 50 Hz, 2 kHz ($f_c/10$), and 20 kHz (f_c). There are also two horizontal lines at the attenuation levels of -30 and -40 dB.

A correct attenuation for frequencies greater than 2 kHz can be observed, but a strong resonance in 235 Hz is also appreciated, with a very slow transient response (see first eigenvalue of **A** in Table 7.2). Any inaccuracy in the generation of the compensating voltages v_C of currents i_C will be amplified in a factor greater than 10, and will deteriorate the compensation feature of the conditioner.

7.2.3.2 Control Modification

It is difficult to avoid these types of resonance in the choice of the values of the passive components, since these components should be dissipative the least possible. To enhance the behavior of the system in the range of the controllable frequencies, the basic control is modified with two correcting signals: over v_C, a signal $R(i_S^* - i_S)$ is added, and over i_C the value of $G(v_L^* - v_L)$. That is, correcting signals proportional to the deviation of the targets respect to their control values, and applied in the opposite converter that should control them. These signals will act eliminating the resonances that could occur. Their effect can be analyzed including this control law over $\mathbf{u}_E$ in the state-space equations (7.19); and recalculating the transfer functions and eigenvalues.

The values of R and G can be chosen to make the adjustment of **y** to their control references as fast and stable as possible, as well as to reduce the magnitude of the transfer functions. On the other hand, it would be convenient if these values were not so large, in order to avoid unnecessary effort by the compensator. Figure 7.15 shows the effect of

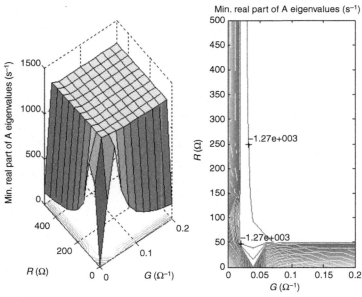

FIGURE 7.15 Minimum negative real part of **A** eigenvalues as a function of the parameters **R** (Ω) and **G** (Ω^{-1}). Three-dimensional and contour graphics.

these parameters on the eigenvalues of the modified matrix **A**. This three-dimensional surface shows the minimum real part of the eigenvalues of **A** for each pair of values of the parameters R and G. The eigenvalue with a lower negative real part can be considered as the dominant pole of the system and it is associated with the slowest dynamic response. This graphic does not show if the eigenvalues are real or complex, but gives an initial selection of the control parameters. In the contour plot at the right side we can see that values for R greater than 50 and for G greater than 0.02 give a value for the minimum real part of -1.270 s^{-1}; which is equivalent to a time constant τ of 0.79 ms for the slowest pole. Larger values of R or G would not obtain much better results in terms of a faster response.

Figure 7.16 shows the new transfer functions obtained with these values, and the effect of the modified control can be observed: The amplification at 235 Hz disappears and the system is better damped against possible perturbations. Table 7.3 presents the corresponding eigenvalues of the modified matrix **A**. The slowest pole has now a value of -1.302 s^{-1}, and the second one is complex but better damped than that of Table 7.2. However, considering the steady state behavior, the attenuation degree of the transfer functions shown

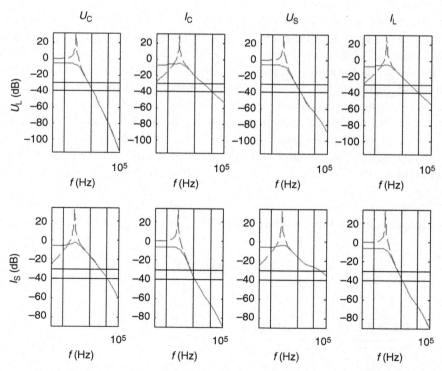

FIGURE 7.16 Transfer functions of the passive network (- -), in comparison with that of the modified control (-), with **R** = 50 and **G** = 0.02. Reference vertical solid lines at 50 Hz (f_N), 2 kHz (f_c/10), and 20 kHz (f_c). Reference horizontal solid lines at -30 and -40 dB.

Table 7.3 Eigenvalues of Matrix **A** With $R = 50$ and $G = 0.02$

Eigenvalues of the modified **A** (s^{-1})
-1.302
$-1.521 \pm 1.421j$ (226 Hz)
-190.683

in Figure 7.16 is quite low for the low frequency range. That means that, for instance, any inaccuracy in the compensation will appear almost in the same proportion in the target variables v_L and i_S.

To enhance this attenuation degree it is possible to increase the parameters R and G, without significantly changing the response speed, as shown in Figure 7.15. Table 7.4 shows the eigenvalues for values of R and G equal to 250 and 0.1, respectively and Figure 7.17 presents the corresponding transfer functions.

The first and last eigenvalues in Table 7.4 are similar to those shown in Table 7.3, and the pair of complex poles has a faster and much better damped dynamic response. And the attenuation degree of the transfer functions shown in Figure 7.17 is also enhanced in the low frequency range.

The analysis done so far has shown the capability of the crossed control terms of equations (7.10) and (7.11) to damp the system dynamic and to reduce the perturbations over the control targets. It is now convenient to include the direct compensation terms of equations (7.10) and (7.11) for a more complete analysis. The distortions of the supply voltage v_S and the load current i_L can be compensated directly and the corresponding transfer functions will be greatly reduced. However, it is usual to compensate only a fraction of these signals (typically 95%) to prevent positive feedbacks due to measurement inaccuracies or control delays. Another possibility is to filter out the high frequency components of these signals, which can be filtered by the passive components and damped with the crossed control terms. This can also reduce the efforts of the converters in the high frequency range.

Considering that the converters can build accurate compensating signals until $f_C/10$, the input signals v_S and i_L will be filtered with a second order Butterworth high-pass filter at 2 kHz:

$$v_C^* = G_1(s)(v_L^* - v_S) + R \cdot (i_S^* - i_S) \tag{7.20}$$

Table 7.4 Eigenvalues of Matrix **A** With $R = 250$ and $G = 0.1$

Eigenvalues of the modified **A** (s^{-1})
-1.251
$-7.767 \pm 774.7j$ (123 Hz)
-181.881

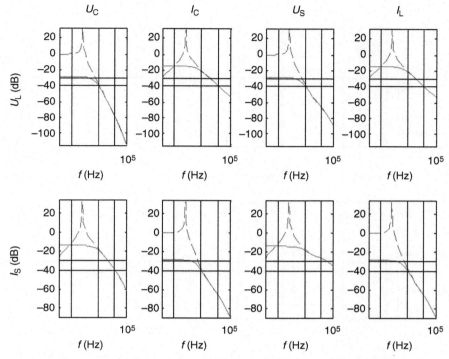

FIGURE 7.17 Transfer functions of the passive network (- -), in comparison with that of the modified control (-), with $R = 250$ and $G = 0.1$. Reference vertical solid lines at 50 Hz (f_N), 2 kHz ($f_c/10$), and 20 kHz (f_c). Reference horizontal solid lines at -30 and -40 dB.

$$i_C^* = G_1(s)(i_L - i_S^*) + G \cdot (v_L^* - v_L) \tag{7.21}$$

where $G_1(s)$ is:

$$G_1(s) = \frac{\sqrt{2}\dfrac{s}{\omega_f} + 1}{\dfrac{s^2}{\omega_f^2} + \sqrt{2}\dfrac{s}{\omega_f} + 1} \tag{7.22}$$

Their effect can also be analyzed including this control law over $\mathbf{u}_E$ in the state-space equations (7.19); and recalculating the transfer functions and eigenvalues. Figure 7.18 presents the corresponding transfer functions, and the eigenvalues are shown in Table 7.5. The first four eigenvalues are equal to those in Table 7.4, and the last four of the double complex pole correspond to those introduced by the Butterworth filters (equation (7.22)). Looking at the transfer functions we can see that the ones for the noise sources in v_C or i_C remain unchanged, but those corresponding to v_S or i_L are greatly reduced, especially in the low frequency range.

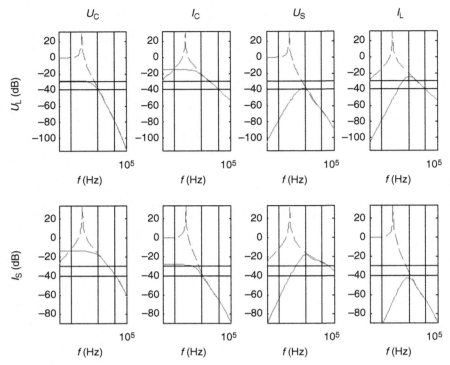

FIGURE 7.18 Transfer functions of the passive network (- -), in comparison with that of the modified control (-), with $R = 250$, $G = 0.1$ and prefiltering of the input signals. Reference vertical solid lines at 50 Hz (f_N), 2 kHz ($f_c/10$), and 20 kHz (f_c). Reference horizontal solid lines at −30 and −40 dB.

Table 7.5   Eigenvalues of Matrix A With $R = 250$, $G = 0.1$ and Prefiltering of the Input Signals

Eigenvalues of the modified **A** (s^{-1})
−1.251
−7.767 ± 774.7j (123 Hz)
−181.881
−8.886 ± 8.886j (1.414 Hz) (double)

EXAMPLE 7.1

The UPQC with the control strategy described in this section will be simulated to check its behavior. The power circuit of the conditioner is shown in Figure 7.7 and the values of the passive components are shown in Table 7.1. The value of the parallel filter inductance L_P, that is not included in the state-space analysis, is 50 mH, 1 Ω. All the dc link capacitors are of 2200 μF, and will be precharged at the reference value at the beginning of the simulation. The UPQC will be applied to an unbalanced nonlinear load composed of three different single-phase loads. In the phase 1 it is a single-phase diode rectifier with a high capacitor and a resistive load in the dc side, and a smoothing reactor in the ac side. For the phase 2, it is a single-phase diode rectifier with a series RL load in the dc side. And ideal resistors for the phase 3, with an ideal switch to produce a load change. Figure 7.19 represents the simulation model.

This load is fed with an unbalanced set of voltages (214, 198, 225 V), and a 5th harmonic component of 10 V with inverse sequence. The equivalent impedance of the supply network has been modeled with a resistance of 1Ω and an inductance of 1 mH in series. This example has been simulated in MATLAB-Simulink using device models of the SymPowerSystem library.

This simulation case will show the behavior of the UPLC, to isolate the load from the harmonics and unbalance of the supply side, and to compensate the harmonics, unbalance, and reactive components of the load currents. The UPQC is activated at 100 ms, and before that the series part of the conditioner is short-circuited and the IGBTs of the shunt conditioner are off. The load changes at 200 ms to show also the dynamic response of the conditioner in those cases.

Figures 7.20 and 7.21 show the phase voltages v_L at the right side of the UPQC and the phase currents i_L of the nonlinear load that has to be compensated by the conditioner, during the whole simulation. Figures 7.22 and 7.23 show the waveforms of the phase voltages v_S at the supply side of the UPQC and the phase currents i_S that flow through the series converter. Before the activation of the UPQC, the supply currents are distorted and unbalanced, mainly due to the distortion produced by the nonlinear load and the harmonics and unbalance of the supply voltages. There is also some level of distortion produced by the resonances among the load, the passive components of the conditioner and the supply line impedance.

Figure 7.24 shows in some detail the effects of the connection of the UPQC, in the supply side of the conditioner. These currents become sinusoidal and balanced, according to the active component of the load current and the power balance of the conditioner. In the first milliseconds of this transition, some high frequency peaks appear, due to the efforts of the converters to control quickly the target magnitudes. After this short transient, the supply voltages v_S remain distorted and unbalanced due to the characteristics of the own voltages sources. Figure 7.25 shows the variation of the load voltages and currents with the start of the UPLC. The resonances, distortion, and unbalance are strongly mitigated and the load voltages become sinusoidal, balanced, and regulated to its nominal value. The load currents seem to be quite robust against the distortion of the applied voltages, mainly because they have a strong distortion by themselves. Only in the resistive load of the third phase a clear enhancement can be observed due to the compensation of the load voltage. And the major visible change after the compensation is observed in the current of the first phase i_{La}. The load of this phase is the most sensible to the voltage variations, especially at the peak values, and the enhancement of this phase voltage produces a visible increase of the corresponding current, with a small transient of around two cycles.

The load change is set at 200 ms, and is concentrated in a change of the resistance of the third phase (see Figure 7.21). This connection produces a fast change of this current that cannot be compensated instantaneously by the shunt converter and produces a small deviation of this phase voltage (see. Figure 7.20) that is compensated in few milliseconds. Afterwards, the induced transient is much smoother, while the control of the UPQC set the new value of the supply currents to deliver the active power of the load, in approximately two and a half cycles (see Figure 7.23).

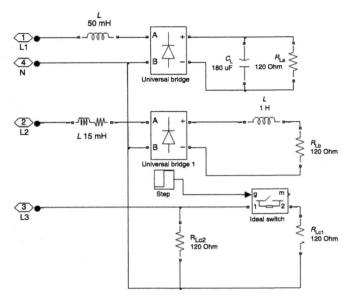

FIGURE 7.19 Nonlinear unbalanced three-phase load for Example 7.1.

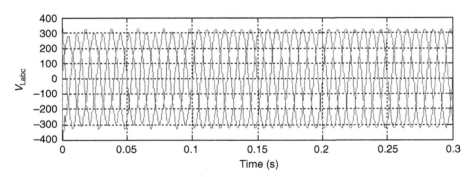

FIGURE 7.20 Phase voltages at the load side of the UPQC.

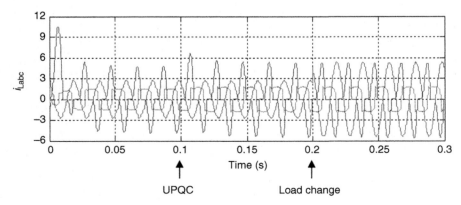

FIGURE 7.21 Phase load currents.

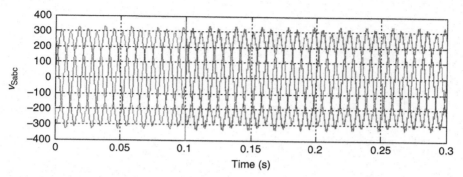

FIGURE 7.22 Phase voltages at the supply side of the UPQC.

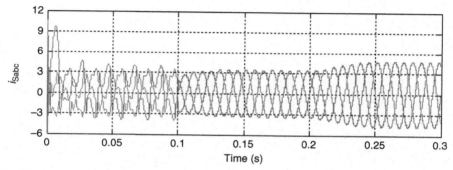

FIGURE 7.23 Phase currents of the supply side.

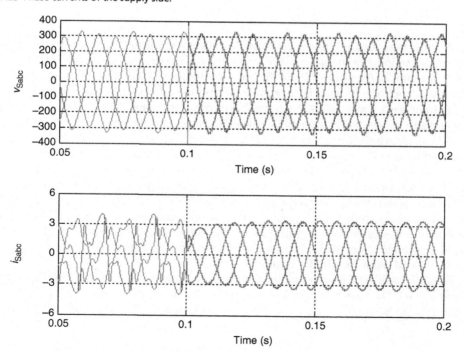

FIGURE 7.24 Detail of the supply voltages and currents before and after the connection of the UPQC.

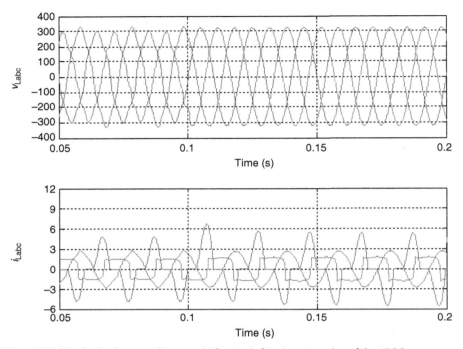

FIGURE 7.25 Detail of the load voltages and currents before and after the connection of the UPQC.

On the other side, Figure 7.26 shows the neutral current of the load i_{Ln} and the supply i_{Sn}. Before compensation, the character of the load produces a considerable amount of zero-sequence currents, of the same order of the phase currents, which flow also through the supply line. The unbalance of the load is clearer when the applied voltages are compensated. Then, the neutral current is even greater; although the shunt converter of the UPQC can compensate this component and the neutral current of the supply line is reduced to a minimum value, mainly due to the voltage regulation of the dc side of the conditioner

Figure 7.27 shows the compensating currents i_C produced by the shunt converter of the UPQC. With the activation of the converters, the conditioner delivers the harmonic, unbalance, and reactive components of the load currents, as well as an active component to regulate the power balance of the conditioner and the voltages on the dc side. The compensating current of the third phase is mainly a fundamental component to compensate the zero- and negative-sequence components of the three-phase load. And with the load change, the differences are mainly in fundamental unbalance components because the currents of the first and second phases of the load remain in similar values.

Figure 7.28 shows the references of the compensating voltages v_C^* produced by the series converter of the UPLC to regulate the load voltage and to isolate the load side from the harmonics and unbalance of the supply side. They are around 25% of the amplitude of the base

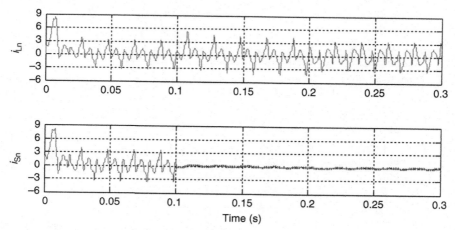

FIGURE 7.26 Neutral currents of the load and the supply sides.

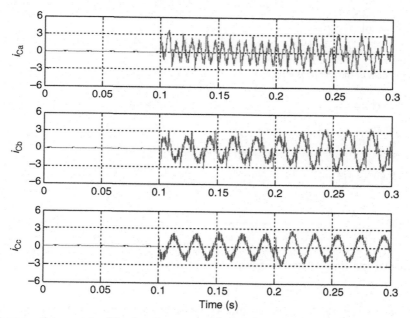

FIGURE 7.27 Compensating currents of the shunt converter of the UPQC.

voltage and differ slightly in each phase, corresponding with the voltage regulation and voltage harmonic present in each one. They also show an increase after the load change at 200 ms to compensate the voltage drop of the increasing supply currents.

Figure 7.29 presents the instantaneous active and reactive powers required by the load during the simulation. They show during the whole period a clear shape of an unbalanced and distorted behavior. Even after the compensation of the load voltages, they indicate a

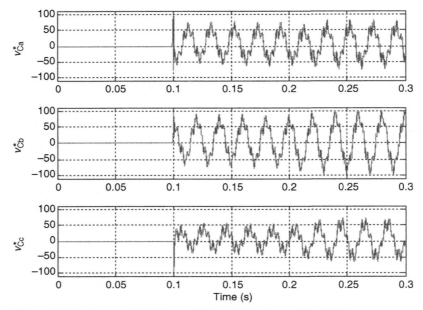

FIGURE 7.28 References of the compensating voltages of the series converter of the UPQC.

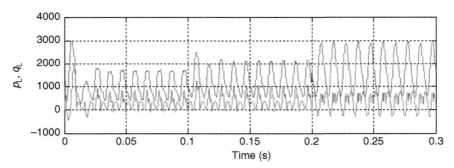

FIGURE 7.29 Instantaneous active and reactive powers at the nonlinear load.

stronger unbalance component that becomes greater with the load change. Figure 7.30 shows the instantaneous active and reactive powers in the point between the converters where both magnitudes are controlled: the load voltages v_L and the supply currents i_S. After the compensation both variables become balanced and sinusoidal, and therefore the instantaneous powers behave as constant values. The instantaneous reactive power is almost null because these voltages and currents are in phase. And the average value of the instantaneous active power increases after the beginning of the compensation due to a better voltage regulation of the load side.

Figure 7.31 presents the instantaneous active and reactive powers delivered by the supply. After the compensation of the UPQC, the oscillating components of the power are clearly

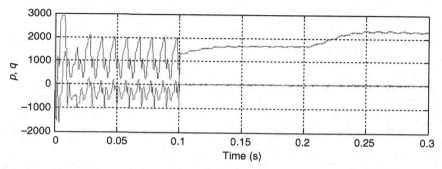

FIGURE 7.30 Instantaneous active and reactive powers between the converters of the UPQC.

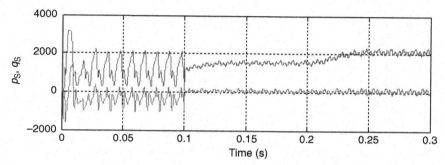

FIGURE 7.31 Instantaneous active and reactive powers provided by supply side.

reduced due to the enhancement of the supply currents. However they do not disappear completely because the supply voltages remain unbalanced and distorted.

Figures 7.32 and 7.33 show the instantaneous active and reactive powers that flow through the series and shunt converters of the UPQC, respectively. The powers through the series converter are significantly lower, to compensate the relatively low values of voltage distortion, unbalance and lack of regulation of the supply side, and the supplementary efforts to control the supply currents. They are increased with the change of the load and the corresponding supply current. One significant part of the average value of the active power delivered by the series converter corresponds to the voltage drop of the supply voltages in its fundamental balanced component. The oscillating powers through the shunt converter reach significantly higher values. The strong unbalance of the load produces a visible 100 Hz component that must be delivered with the shunt converter along with the rest of high frequency components. With the change of the load, the active power shows a clear first order transient that lasts around two and a half cycles. This is produced by the power balance control of the conditioner. While the supply current reference does not reach the active current of the new load condition, the UPQC must provide this power to the load. This will reduce the voltage of the dc side capacitors, which will have to be recharged slowly along the new steady stage.

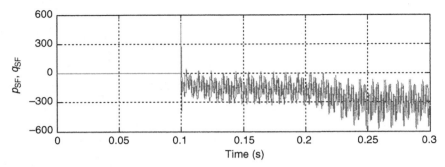

FIGURE 7.32 Instantaneous active and reactive powers of the series converter of the UPQC.

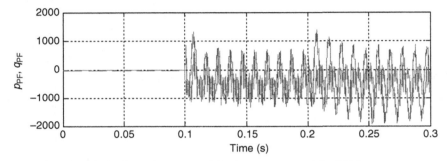

FIGURE 7.33 Instantaneous active and reactive powers of the shunt converter of the UPQC.

Figure 7.34 presents the evolution of the voltages of the dc side capacitors during the present case. At the beginning of the compensation they have a relatively slow decrease of their average value. The difference between them reaches a visible amount while the conditioner compensates the transient components of the currents and voltages during the first two cycles; until this values become balanced. In this stage an oscillating and alternating component remains due to the zero-sequence components of voltages and currents that are being compensated. After the load change, the average value has a greater variation. At the beginning of the compensation, the internal power balance was set to the previous values and the change of the load voltages in this step was not so great as to change significantly the active power of the load. In comparison, the load change implies a bigger variation of the active power of the load. In the first two cycles, the reference of the supply current is not high enough and the capacitors are being discharged. When the power balance is set again, this decrement stops and the dc voltages begin to recover to their reference values through the additional active current of the shunt converter that controls these parameters.

Finally, a harmonic and unbalance analysis of the resultant voltage and current waveforms has been performed. Figure 7.35 shows the harmonic spectrum of the phase voltages and currents at the supply and load side after compensation, in a time window from 160 ms to 180 ms. The data is presented with the harmonic distortion index, HD, of each individual harmonic with respect to the fundamental value, and the harmonic order h has been used

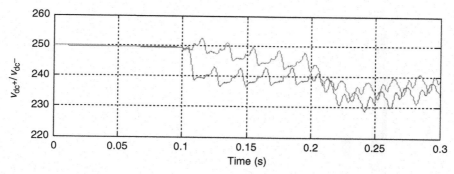

FIGURE 7.34 Voltages at the dc link of the UPQC.

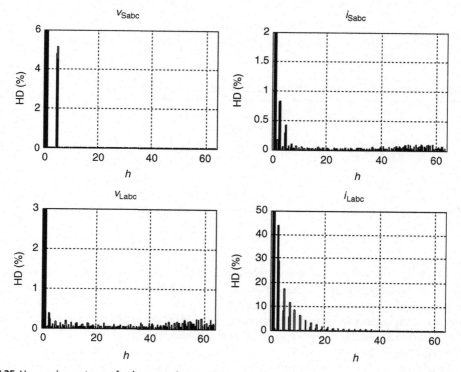

FIGURE 7.35 Harmonic spectrum of voltages and currents.

instead of the frequency. The values of the three phases of each magnitude are presented together to appreciate the global envelope of the harmonic spectrum of the three-phase set. Figure 7.36 shows a magnification of the harmonic spectrum of the supply currents for the first 20 harmonics, where the harmonic content of each phase can be distinguished.

Table 7.6 shows the rms and total harmonic distortion (THD) values of these 12 variables obtained in this analysis. We can see that, after compensation, the load voltages v_{Labc} are

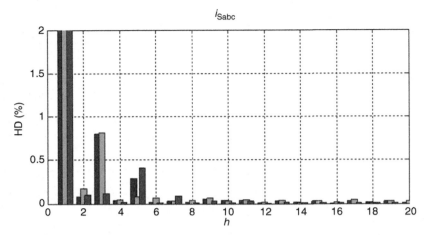

FIGURE 7.36 Detail of the harmonic spectrum of the supply currents for the first 20 harmonics (50 Hz–1 kHz).

Table 7.6   Individual rms and THD Values After Compensation

	rms (V)	THD (%)		rms (A)	THD (%)
v_{Sa}	211.8	4.730	i_{Sa}	2.385	0.916
v_{Sb}	195.8	5.117	i_{Sb}	2.443	0.896
v_{Sc}	222.8	4.499	i_{Sc}	2.425	0.500
v_{La}	227.6	0.840	i_{La}	2.963	45.03
v_{Lb}	227.8	0.809	i_{Lb}	1.643	37.78
v_{Lc}	228.2	0.529	i_{Lc}	1.904	0.529

much more balanced and regulated in amplitude than the supply voltages v_{Sabc}. The average voltage deviation respect to the nominal value is around 0.9%. The harmonic content of the load voltages is also greatly reduced. Looking at the harmonic spectrum, the fifth harmonic component of the supply voltages is clearly compensated by the series converter. The harmonic content of the load voltages is more spread, with small components around 0.2%. On the other side, the high THD values of the load currents of the first two phases indicate the nonlinear character of these two loads. After compensation, the supply currents become almost balanced and sinusoidal, with a low THD around 0.9%, and almost in phase with the supply voltages.

Table 7.7 shows some relevant three-phase indices of this analysis, as defined in Std. 1459, [14]: Three-phase rms effective voltage V_e; voltage total harmonic distortion, VTHD; voltage unbalance fundamental, VUF; three-phase rms effective current I_e; current total harmonic distortion, ITHD; current unbalance fundamental, IUF; active, reactive, and apparent powers of the fundamental components, P_1, Q_1, S_1; and total effective apparent power, S_e, of each three-phase set. These values have been calculated in the supply and load sides, as well as between the series and shunt converters; where the voltages are v_{Labc} and the compensated currents i_{Sabc}.

Table 7.7   Three-Phase Indices of the Harmonics and Unbalance Calculations

	Supply Side (v_s, i_s)	Between Conv. (v_L, i_s)	Load Side (v_L, i_L)
V_e	210.3 V	227.9 V	227.9 V
VTHD	4.763%	0.663%	0.663%
VUF	4.577%	0.113%	0.113%
I_e	2.418 A	2.418 A	2.425 A
ITHD	1.187%	1.187%	48.99%
IUF	2.208%	2.208%	44.43%
P_1	1521 W	1652 W	1306 W
Q_1	48 VAr	12 VAr	382 VAr
S_1	1522 VA	1653 VA	1361 VA
S_e	1526 VA	1653 VA	1658 VA

In Table 7.7, VUF is the voltage unbalance fundamental index and IUF is the current unbalance fundamental index [15], which are defined as:

$$\text{VUF} = \frac{V_{u1}}{V_{b1}}, \text{IUF} = \frac{I_{u1}}{I_{b1}} \tag{7.23}$$

These three-phase indices show that the unbalance, harmonics, and deregulation of the supply voltages are compensated in the load side, where the voltages have an unbalance fundamental index of only 0.11%, a total THD index of 0.66%, and an effective rms voltage of 227.9 V. On the other hand, the conditioner compensates the load currents and therefore the supply currents are much more balanced, with an unbalance fundamental index of 2.21% and with a low distortion, 1.19%. The effective value is slightly lower than at the load side due to the compensation of the unbalance and harmonic components. Although, the losses of the conditioner must be considered, as well as the low fundamental voltage at the supply side, that requires higher currents to deliver the same active power.

Regarding the power components during the active conditioning, we can see that the supply delivers an active power at the fundamental component equivalent to that requested by the load and the losses of the conditioner. The active power between the converters is higher than at the supply side because the series converter must inject a direct-sequence voltage to regulate the load voltages. It draws this power from the common dc link and the shunt converter equilibrates the power balance increasing the reference of the active current. On the other hand, the fundamental reactive power of the load is delivered mainly through the shunt converter. Between the converters and in the supply side the fundamental apparent power is almost equal to the active power and the displacement power factor is almost one. If we compare the fundamental apparent power, S_1, with the total effective apparent power, S_e, we can see that they are almost the same between the converters, where both voltages and currents are balanced and sinusoidal. At the load side the nonlinear currents imply relevant harmonic components in the apparent power and the total power factor is 0.787. At the supply side the distorted voltage sources increase slightly the effective apparent power, reducing the power factor to 0.997.

7.3 Experimental Prototype of UPQC

An experimental platform has been set to perform a validation of the proposed strategies and to verify the behavior of the conditioner. A series– parallel compensation has been applied to an unbalanced and nonlinear load supplied by an unbalanced set of voltages. Figure 7.37 shows the equivalent power circuit per phase of the series–parallel conditioner with their corresponding matching transformers and the passive elements for the filtering of the high frequency components. The series and shunt three-phase IGBT converters (Semikron SKM50GB123D) are connected back to back, with a common dc link composed by two electrolytic capacitors, C_{dc+} and C_{dc-} of 2200 μF and 400 V each one [16,17]. The middle point of the dc link is connected with the neutral wire of the three-phase line. The turn ratio of the matching transformers T_P and T_S are 1:2 and 1:1, respectively, to achieve reasonable voltage values in the dc link. The filtering inductances L_P and L_S are 50 mH and 25 mH each one, and the values of the parallel filtering capacitor C_P and resistance R_P are 15 μF and 2.5 Ω, respectively. A panel selector switch allows to bypass the series converter to make tests with only parallel active compensation.

The switching control of the active converters and the calculation of the compensation references are made with a data acquisition and control system (dSPACE-DS1103). The sampling time of the main processor was set to 80 μs for the series–parallel compensation. The target references for the compensation are the fundamental positive-sequence components of the load currents i_L and supply voltages v_S, in order to achieve the lowest unbalance and harmonic components in the compensated system. When the series–parallel compensation is performed, the references for the series compensating voltage v_C and the shunt compensating current i_C are:

$$v_C^* = G_1(s)(v_L^* - v_S) + R(i_S^* - i_S) \tag{7.24}$$

$$i_C^* = G_1(s)(i_L - i_S^*) + G(v_L^* - v_L) \tag{7.25}$$

where v_L^* is reference for the load voltage: the fundamental direct sequence component of the supply voltage, regulated in amplitude to its nominal value, and i_S^* is a set of balanced currents in phase with the fundamental direct sequence component of the supply voltage,

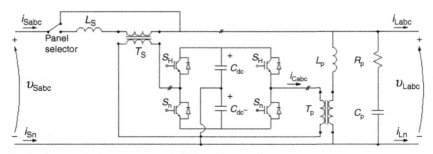

FIGURE 7.37 Single-phase equivalent power circuit of the active conditioner.

that transports the average active power of the load and the losses of the conditioner. This way, the unbalanced and harmonic components of the load current are compensated. The analysis made in Section 7.2.3 gives the prefiltering $G_1(s)$ of the first terms of (7.24) and (7.25); and the second terms of these equations are the crossed correcting signals proportional to the deviation of the target magnitudes to the reference values, which damp the systems behavior in strong transients and with different kind of loads. With this kind of compensation, the load voltage v_L and the supply current i_S should have a very reduced content of unbalance and harmonic components.

The compensating voltage of the series converter v_C is built with a 20 kHz PWM generator (DS1103 slave), using its calculated reference v_C^*. The compensating current of the shunt converters is built with a periodic sampling control at 20 kHz, in an external circuit, using the deviation of the compensating current i_C respect to its reference i_C^*.

On the other side, an independent measurement system to calculate the indices is implemented [15] in another data acquisition card (dSPACE-DS1005), with a signal conditioning system formed by six voltage sensors (LEM-LV25-P) and eight current sensors (LEM LA35-NP). These signals are taken simultaneously to avoid phase differences between the measurements, to enhance the accuracy of the calculated results. The configuration of the virtual instrument was made following the IEC Std 61000-4-7 and IEC Std 61000-4-30 recommendations, using a window equal to five cycles of the fundamental component and a sampling frequency of 6400 Hz to avoid problems of aliasing and leakage errors.

Finally, the load is composed of three different single-phase loads. In the phase 1 it is a single-phase diode rectifier with a high capacitor and a resistive load in the dc side, and a smoothing reactor in the ac side. For the phase 2, it is a single-phase diode rectifier with a series RL load in the dc side; and a resistor for the phase 3. This load is fed through a variable autotransformer with an unbalanced set of voltages (214, 198, 225 V), considering a nominal phase to neutral voltage of 230 V_{rms}. These applied voltages have a small distortion similar to the existing in the supply network of the laboratory. Figure 7.38 shows the voltage and current waveforms of the load fed with this power supply.

7.3.1 Results of Practical Case

Figures 7.39 and 7.40 show the resulting voltage and current waveforms with series–parallel compensation in the load and supply side, respectively. The supply currents i_{Sabcn} are practically balanced, sinusoidal and in phase with the fundamental positive-sequence component of the supply voltage. The load voltage is also enhanced, compensating the distortion and unbalance present in the supply voltage.

Figure 7.41 shows the corresponding harmonic spectrum of the phase voltages and currents at the supply and load side after compensation, where the data is presented with the harmonic distortion index, HD, of each individual harmonic respect to the fundamental value, and the harmonic order h has been used instead of the frequency. The values of the three phases of each magnitude are presented together to appreciate the global envelope of the harmonic spectrum of the three-phase set.

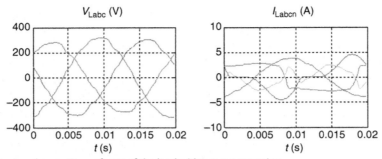

FIGURE 7.38 Voltage and current waveforms of the load without compensation.

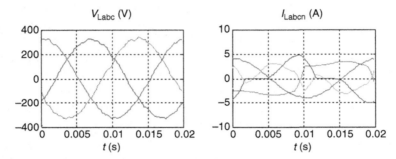

FIGURE 7.39 Voltage and current waveforms of the load side with active series–parallel compensation.

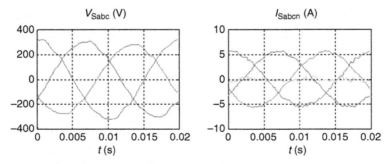

FIGURE 7.40 Voltage and current waveforms of the supply side with active series–parallel compensation.

Table 7.8 presents the rms value and the total harmonic distortion, THD, of these 12 variables. We can see that, after compensation, the load voltages v_{Labc} are much more balanced and regulated in amplitude that the supply voltages v_{Sabc}. The harmonic content of the load voltages is not so reduced, mainly is the phases a and b. Looking at the harmonic spectrum, the various harmonics present in the supply voltages are compensated by the series converter and the harmonic content of the load voltages is more reduced but more spread, with small components around 1%. On the other side, the high THD values

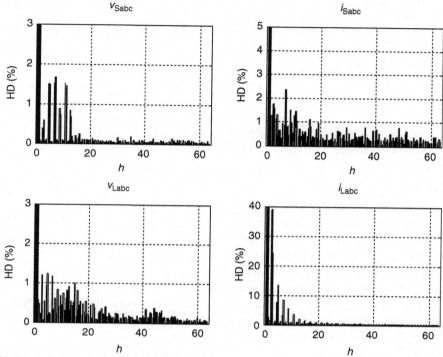

FIGURE 7.41 Harmonic spectrum of voltages and currents.

of the load currents of the first two phases indicate the nonlinear character of these two loads. After compensation, the supply currents become almost balanced and sinusoidal, with a reduced THD around 4%, and almost in phase with the supply voltages.

Table 7.9 shows some relevant three-phase indices of this analysis, as defined in Std. 1459, [14]: Three-phase rms effective voltage V_e, voltage total harmonic distortion, VTHD, voltage unbalance fundamental, VUF, three-phase rms effective current I_e, current total harmonic distortion, ITHD, current unbalance fundamental, IUF, active and apparent powers of the fundamental components, P_1, S_1; and total effective apparent power, S_e, of each three-phase set. These three-phase indices show that the unbalance, harmonics, and

Table 7.8   Individual rms and THD Values After Compensation

	rms (V)	THD (%)		rms (A)	THD (%)
v_{Sa}	212.6	2.804	i_{Sa}	3.861	4.124
v_{Sb}	196.7	2.640	i_{Sb}	3.896	3.623
v_{Sc}	224.7	2.808	i_{Sc}	3.887	4.762
v_{La}	229.2	2.294	i_{La}	2.712	39.70
v_{Lb}	231.9	3.371	i_{Lb}	2.849	29.41
v_{Lc}	230.1	1.887	i_{Lc}	2.821	1.900

Table 7.9   Three-Phase Indices of the Harmonics and Unbalance Calculations

	Supply Side (v_S, i_S)	Load Side (v_L, i_L)
V_e	211.6 V	230.4 V
VTHD	2.681%	2.296%
VUF	4.707%	0.619%
I_e	3.885 A	3.000 A
ITHD	6.066%	34.89%
IUF	0.947%	41.13%
P_1	2448 W	1732 W
S_1	2461 VA	1957 VA
S_e	2466 VA	2077 VA

deregulation of the supply voltages are compensated in the load side, where the voltages have an unbalance fundamental index of only 0.62%, a total THD index of 2.30%, and an effective rms voltage of 230.4 V. On the other hand, the conditioner compensates the load currents and therefore the supply currents are much more balanced, with an unbalance fundamental index of 0.95% and with a reduced distortion, 6.07% mainly due to the regulation of the dc link voltages. The effective value is higher than at the load side due to the losses of the conditioner but also due to the low fundamental voltage at the supply side, which requires higher currents to deliver the same active power.

Regarding the power components during the active conditioning, we can see that the supply delivers an active power at the fundamental component equivalent to that requested by the load and the losses of the conditioner. The reactive, unbalance, and harmonics powers of the load are delivered by the conditioner and the power factor at the supply side reaches a value of 0.993. At the load side the nonlinear currents imply relevant harmonic and unbalance components in the apparent power and the total power factor is 0.834, while the displacement power factor is 0.885.

7.4 Universal Active Power Line Conditioner

The concept of an active power line conditioner, APLC, emerges from the development of active power filters made with PWM force-commutated converters [2,8]. In the first proposals they were focused on the compensation of the currents of unbalanced nonlinear loads, including the reactive component. The capability of these PWM converters to produce controlled current or voltage waveforms in an important frequency range ($\sim$1 kHz) with high fidelity and dynamic response, made them to be applied also in the field of flexible ac transmission systems (FACTS); for power flow control and dynamic and static stability enhancement. The UPFC is one of the most promising devices in the FACTS concept [18–22]. It is based on the combination of series and shunt power converters with a shared dc link, and has the potential of active and reactive power flow control and/or voltage stability in power transmission systems; inserting balanced fundamental voltages and currents in the line circuits.

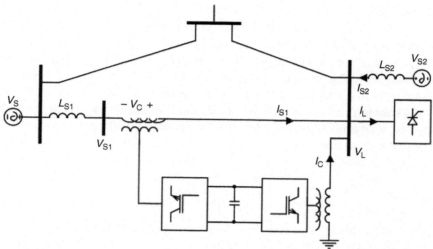

FIGURE 7.42 Structure of a UPLC.

When the combined series–shunt power conditioner includes also the tasks of a UPQC to compensate supply and load side harmonics and unbalances, it receives the name of universal active power line conditioner, UPLC [8,23].

Figure 7.42 shows an UPLC inserted in a line of the power system, near a nonlinear load. Acting as UPQC, it can compensate the distortion and unbalance of this nearby load, as well as to provide harmonic isolation and/or dumping for distortions and unbalances at both ends of the power line. When it works as UPFC, it can control the power flow of the line where the series converter is connected, and regulate the voltage V_L injecting fundamental reactive power with the shunt converter. The combination of both modes gives the UPLC the capability to compensate and to regulate the working conditions of this part of the network.

7.4.1 Unified Power Flow Controller

The UPFC device consists of a combination of series and shunt power converters with a common dc link, like that shown in Figure 7.42, where the series converter acts as a controllable voltage source V_C, whereas the shunt converter acts as a controllable current source I_C. The main purpose of the series converter is to regulate the active and reactive power flow of the controlled line L_{S1}, changing the receiving end voltage V_{S1} with the series voltage V_C and therefore the current I_{S1} of the line. The shunt converter can help to regulate the voltage at the bus where it is connected, injecting a current I_C that provides the necessary reactive power. The active power delivered by the shunt converter is not free, unless it had significant energy storage components because it must compensate the active power provided by the series converter and the internal losses of the conditioner.

The phasor diagram shown in Figure 7.43 illustrates the principle of voltage regulation with the shunt converter. When the current I_C leads 90° with respect the voltage V_L (Figure 7.43a), it produces a voltage V_{LS2} in the equivalent impedance of the nearby network, thus increasing the voltage V_L. When the current I_C lags 90° (Figure 7.43b) it produces the opposite effect and reduces the voltage V_L. The amount of reactive power to be injected will be similar to other classical techniques to control the Q–V relation in electrical power systems.

Figure 7.44 shows a single-phase equivalent circuit of the series part of the UPFC to explain the principle of power flow control. The inductance L and the resistance R represent the series impedance of the transmission line, where it is usual to remove the resistance R because $\omega_0 L \gg R$ in this kind of line. Thus, the line current phasor I_{S1} is given by:

$$I_S = \frac{V_S - V_L + V_C}{j\omega_0 L} \tag{7.26}$$

For the sake of simplicity, the assumption of $V_S = V_L$ leads to the phasor diagram shown in Figure 7.45. When the series voltage V_C is in quadrature with the voltage V_L, it promotes the circulation of a current I_{S1} in phase with this voltage, as shown in Figure 7.45a, and

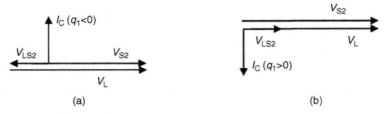

FIGURE 7.43 Voltage regulation with reactive power.

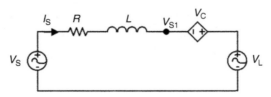

FIGURE 7.44 Single-phase equivalent circuit.

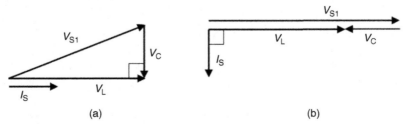

FIGURE 7.45 Phasor diagrams for series power control. (a) Active power control; (b) Reactive power control.

changes the flow of active power. Controlling V_C to be in phase with V_L, Figure 7.45b, makes the current I_{S1} lag by 90° with respect to V_L, thus resulting in reactive power flow.

This power flow principle is most interesting feature of the UPFC. As we can derive from equation (7.26), the amount of power flow that the voltage source V_C can control is inversely proportional to the line fundamental impedance. Furthermore, the series converter will be working with balanced fundamental voltage waveforms, easier to achieve with low switching frequencies and reduces losses.

7.4.2 Power Flow Control, Voltage Regulation, and Power Quality Improvement

In this section, the control implementation of the UPFC principles will be discussed and detailed, as well as the integration of these control blocks in the general UPLC.

Considering the model exposed in Figures 7.44 and 7.45, and equation (7.26), the control can be made using the line current as target, with the corresponding active and reactive powers. An open loop determination of the control references for V_C can be made with equation (7.26), although a robust feedback scheme must be used because the voltages V_S at the other end of the line and its own impedance (or the Thevenin equivalent of the whole network at this side of the UPFC) are usually unknown in real time.

The simplest feedback scheme is the proportional control. On the d–q frame coordinates of the voltage and current vectors, the d-axis current i_d corresponds to the active power, and so it can be controlled by the q-axis voltage v_{Cq}. Therefore, the reference for the v_{Cq} component is:

$$v_{Cq}^* = K_p(i_d^* - i_d) \tag{7.27}$$

$$i_d^* = \frac{P_{REF}}{v_d} \tag{7.28}$$

where P_{REF} is the preset value for the active power and v_d is the d-axis component of the voltage (The d–q frame is supposed to be in phase with the fundamental direct-sequence component of the voltage and therefore the q-axis component is null).

The q-axis current i_q corresponds to the reactive power, and so it can be controlled by v_{Cd}. Therefore, the reference for the v_{Cd} component is:

$$v_{Cd}^* = K_q(i_q^* - i_q) \tag{7.29}$$

$$i_q^* = \frac{Q_{REF}}{v_d} \tag{7.30}$$

where Q_{REF} is the preset value for the reactive power, considered positive for leading currents.

These two control laws provide the base for an independent control of P and Q, but they are obtained from steady state phasor relations. In fact, this control scheme presents interferences between the d- and q-axis that induce poor damped transient behaviors [18]. The control scheme proposed in [18] includes a proportional control between the magnitudes of the same axes:

$$\begin{bmatrix} v_{Cd}^* \\ v_{Cq}^* \end{bmatrix} = \begin{bmatrix} -K_r & -K_q \\ K_p & -K_r \end{bmatrix} \cdot \begin{bmatrix} i_d^* - i_d \\ i_q^* - i_q \end{bmatrix} \tag{7.31}$$

The new terms defined by K_r act as a damping resistor that enhances strongly the dynamic behavior. The transient analysis exposed in [18] deduces a second-order system for the series control of the UPFC and relates its damping factor with the line impedance and the proportional gains of the control:

$$\varsigma = \frac{K_r + R}{\sqrt{(\omega_0 L + K_p)(\omega_0 L + K_q) + (K_r + R)^2}} \tag{7.32}$$

Using equations (7.32) and (7.26) the values of K_p, K_q, and K_r can be preset. The values of K_p and K_q should be of the same order as the line impedance, for an open loop control estimation based on equation (7.26). And for a given value of ζ, K_p, and K_q, the K_r gain should be:

$$K_r \approx \frac{\varsigma}{\sqrt{1-\varsigma^2}} \sqrt{(X_L + K_{pq})^2} - R \approx \frac{\varsigma}{\sqrt{1-\varsigma^2}} (X_L + K_{pq}) \tag{7.33}$$

For instance, for a damping factor $\zeta = 0.8$ and $K_p = K_q = X_L$, the corresponding value for K_r is 2.67 times the line reactance X_L.

Nonetheless, the results presented in [18] with the proportional control still have cross couplings between the P and Q controls in steady state. The addition of slow integral gains allows the system to set the final values in the preset references.

Similar considerations can be made for the implementation of the voltage regulation with the reactive power of the shunt converter. Taking references in the magnitudes presented in Figure 7.42, the equation that relates the phasor voltage V_L and the voltage of the right side subsystem V_{S2} is:

$$V_L = V_{S2} - j\omega_0 L_{S2} \cdot I_{S2} \tag{7.34}$$

If all the injected current I_{Cq} flows through the line S_2, the leading values of I_{Cq} will produce and increment of V_L and the lagging values of I_{Cq} will reduce the voltage (see Figure 7.43), and a similar control strategy could be applied:

$$i_{Cq}^* = K_{qv}(v_d^* - v_d) \tag{7.35}$$

However, there are important differences in the shunt converter control. The first one is that the control of the active power is not free because it is focused in the power balance of the conditioner. The second one, and maybe the most important, is that according to equation (7.34) the effect of the reactive current injected on the voltage to be regulated, is "reduced" by a factor equal to the line impedance, in per unit values; and there are easier and classical methods to provide the same amount of reactive power. Furthermore, the implementation of a control scheme similar to equation (7.31) would produce important oscillating values of the instantaneous power that will interfere in the power balance of the conditioner.

7.4.2.1 Integration in the UPLC Control

When the UPFC control is applied in a system with distortions and unbalances, additional requirements must be considered. First of all, the d–q components of the space vectors no longer behave as constant values. The dc components of these d–q coordinates must be obtained, in order to avoid the introduction of additional distortions through the control signals of the UPFC. And this filtering introduces a delay that changes the response of the damping factor K_r, deteriorating the transient response in the low frequency range. Finally, the existing unbalances and distortions mean that, even with an ideal compensation, the instantaneous power that flows through the converters could have important oscillatory components that must be considered in the dimensioning of the dc link of the conditioner.

When the features of an UPFC are to be included in a UPQC to set a UPLC, the last consideration is already included in the efforts of the harmonic and unbalance compensation. Furthermore, the harmonic damping term $R \cdot i_{sh}$ (equation (7.10)) of the UPQC can be extended to give the response of the factor K_r because they have essentially the same role.

On the other hand, the control of the UPQC explained in Section 7.2 is formulated with the phase variables, instead of d–q frame variables. The adaptation of the control scheme to this approach implies to reorder the control calculations. The UPQC control has an internal variable, $v_{1,reg}^+$, that is in phase with the fundamental direct-sequence component of the load voltage. These internal signals have unity amplitude and can be used as a phase reference. And also, an instantaneous quadrature voltage vector can be defined with these signals [24]:

$$e_{d,\text{abc}} = v_{1,\text{reg,abc}}^+$$

$$e_q = \begin{bmatrix} e_{qa} \\ e_{qb} \\ e_{qc} \end{bmatrix} = \frac{1}{\sqrt{3}} \begin{bmatrix} 0 & -1 & 1 \\ 1 & 0 & -1 \\ -1 & 1 & 0 \end{bmatrix} \cdot \begin{bmatrix} e_{da} \\ e_{db} \\ e_{dc} \end{bmatrix} \tag{7.36}$$

The vector $\mathbf{e_q}$ leads 90° the signal vector $\mathbf{e_d}$ and has also unity amplitude. The instantaneous reference of the currents in the d–q axis can be calculated as:

$$i_d^* = \frac{P_{\text{REF}}}{\bar{v}_d} \tag{7.37}$$

$$i_q^* = \frac{Q_{\text{REF}}}{\bar{v}_d} \tag{7.38}$$

where v_d is the dc component of the d-axis component of the voltage vector:

$$\bar{v}_d = \frac{1}{T}\int_0^T (v_a e_{da} + v_b e_{db} + v_c e_{dc})\,dt \tag{7.39}$$

The average d–q components of the actual currents can be calculated as:

$$\bar{i}_d = \frac{1}{T}\int_0^T (i_a e_{da} + i_b e_{db} + i_c e_{dc})\,dt \tag{7.40}$$

$$\bar{i}_q = \frac{1}{T}\int_0^T (i_a e_{qa} + i_b e_{qb} + i_c e_{qc})\,dt \tag{7.41}$$

These components are related to the average values of the active and reactive instantaneous powers, produced by those currents that are in phase with the corresponding voltage waveforms. Thus, the voltage v_C^* references are:

$$v_{Cq}^* = K_p(i_d^* - \bar{i}_d)\cdot\begin{bmatrix} e_{qa} \\ e_{qb} \\ e_{qc} \end{bmatrix}, v_{Cd}^* = K_q(i_d^* - \bar{i}_q)\cdot\begin{bmatrix} e_{da} \\ e_{db} \\ e_{dc} \end{bmatrix} \tag{7.42}$$

However, for the damping component K_r, the voltages references have been calculated directly by comparison of the phase current and their references:

$$i_{Sabc}^* = i_d^*\cdot\begin{bmatrix} e_{da} \\ e_{db} \\ e_{dc} \end{bmatrix} + i_q^*\cdot\begin{bmatrix} e_{qa} \\ e_{qb} \\ e_{qc} \end{bmatrix} = \frac{P_{REF}}{\bar{v}_d}e_d + \frac{Q_{REF}}{\bar{v}_d}e_q \tag{7.43}$$

$$v_{Cr}^* = -K_r\begin{bmatrix} i_{Sa}^* - i_{Sa} \\ i_{Sb}^* - i_{Sb} \\ i_{Sc}^* - i_{Sc} \end{bmatrix} \tag{7.44}$$

The direct use of the instantaneous values of the currents avoids the delay related with the calculation of average d–q components that deteriorate the dynamic response of the control. Furthermore, it includes the harmonic damping term that was incorporated in the series converter of the UPQC; and it has similar gain values.

Finally, the voltage references for the series converter of the UPLC will be:

$$v_C^* = (v_S - v_L^*) + v_{Cq}^* + v_{Cd}^* + v_{Cr}^* \tag{7.45}$$

The control of the shunt converter of the UPLC has to be modified only to include the term i_{Cq}, and no substantial modifications in the power balance should be expected:

$$i_C^* = (i_L - i_S^*) + G\cdot(v_L^* - v_L) + i_{Cq} \tag{7.46}$$

EXAMPLE 7.2

The UPLC with the control strategy described in this section will be simulated to check its behavior. The power circuit of the conditioner is the same as in Example 7.1, with the connection of a second power supply in the load side that plays the role of the right side subsystem. Figure 7.46 shows the general power circuit for this case. The UPLC will be applied to the same load of Example 7.1, with the same unbalanced and distorted supply system in the left side of the conditioner. Only the line impedance has been changed to produce a 4% voltage drop with nominal load, through a resistance of 1Ω and an inductance of 7 mH. The power supply of the right side will provide balanced nominal voltages, but the line impedance will be greater, with a resistance of 1Ω and an inductance of 14 mH to produce an 8% voltage drop with nominal load.

The resultant waveforms of this simulation case will be presented in per unit values: the voltages and currents will be divided by the amplitude of their nominal references (400/230 V, 50 Hz, 3 kVA) and the instantaneous power will be referred to the nominal apparent power of the system. The gains K_p, K_q, K_r for the proportional control of the S1 line have been set to 0.2, 0.2, and 5 pu, respectively. The gain K_{qv} for the voltage regulation has been set to 1 pu.

This simulation case of 400 ms will show the behavior of the UPLC through the following events:

- Activation of the UPLC at 100 ms. Before that moment, the shunt converter is off and the voltage reference of the series converter is set to 0 V. When it starts, the reference for the instantaneous active and reactive powers of the controlled line are set to 0.5 and 0 pu, respectively.
- Change in the reference of the instantaneous active power P_{REF} from 0.6 pu to −0.6 pu, at 200 ms.
- Change in the reference of the instantaneous reactive power Q_{REF} from 0 pu to −0.3 pu, at 300 ms.

Figures 7.47 and 7.48 show the phase voltages v_L at the right side of the UPLC and the phase currents i_L of the nonlinear load that has to be compensated by the conditioner, during the whole simulation. Figures 7.49 and 7.50 show the waveforms of the phase voltages v_{S1} at the left side of the UPLC and the phase currents i_{S1} that flow through the series converter and the controlled line of the subsystem S1. Before the activation of the UPLC, the supply currents are strongly distorted and unbalanced, mainly due to the distortion and unbalance of the voltage sources of the S1 subsystem. There is also some level of distortion produced by the nonlinear load, as well as some resonances among the load and the subsystems S1 and S2.

Figure 7.51 shows in some detail the effects of the activation of the UPLC in the left side of the conditioner. These currents become sinusoidal and balanced, according to the references of the active and reactive power of the controlled line. The v_{S1} voltages remain distorted and unbalanced due to the characteristics of the own voltages sources of this subsystem. Figure 7.52 shows the variation of the load voltages and currents with the start of the UPLC. The resonances, distortion, and unbalance are strongly mitigated and the load voltages become sinusoidal and balanced. It shows how the UPLC isolates the load side from the distortion and unbalance of the S1 supply side. The load currents are quite robust against the distortion of the

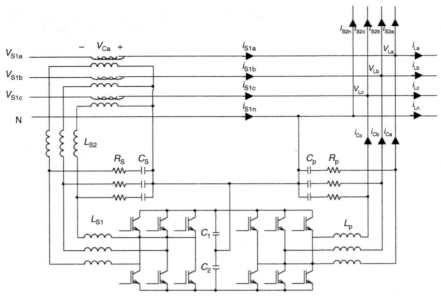

FIGURE 7.46 Power circuit for the simulation case of Example 7.2.

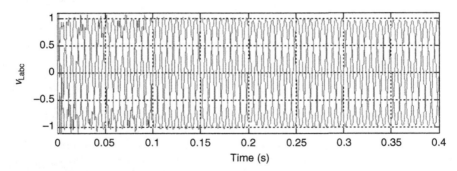

FIGURE 7.47 Phase voltages at the right side of the UPLC (load voltages).

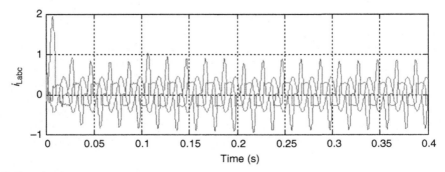

FIGURE 7.48 Phase load currents.

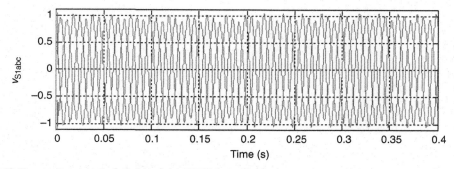

FIGURE 7.49 Phase voltages at the left side of the UPLC (controlled line).

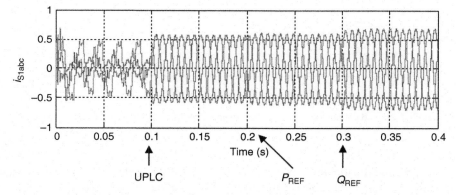

FIGURE 7.50 Phase currents of the left side subsystem S1 (controlled line).

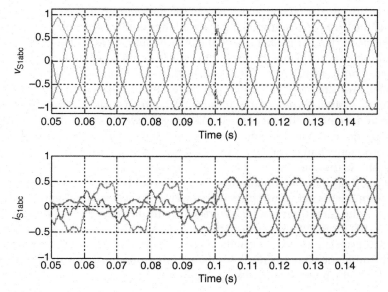

FIGURE 7.51 Detail of the S1 supply voltages and currents before and after the activation of the UPLC.

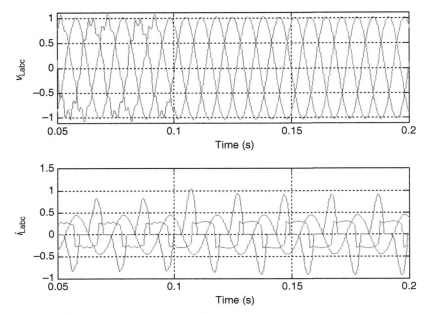

FIGURE 7.52 Detail of the load voltages and currents before and after the activation of the UPLC.

applied voltages, and the major change with the compensation is observed in the current of the first phase i_{La}. The load of this phase is the most sensible to the voltage variations, especially in the peak values, and the enhancement of this phase voltage produces a visible increase of the corresponding current, with a small transient.

On the other side, the UPLC compensates also the harmonic, unbalance, and reactive components of the load current, improving the working conditions of subsystem S2. Figure 7.53 shows the neutral current of the load i_{Ln} and the lines of both the supply subsystems, i_{S1n} and i_{S2n}. Before compensation, the unbalance in S1 produces a strong circulation of zero-sequence currents between both subsystems. With the activation of the conditioner, the neutral current of the controlled line is reduced to a minimum value. The compensation of the shunt converter enhances also the neutral current of the subsystem S2, after a small transient in the first cycle. The evolution of the neutral current of the load during the whole simulation indicates also a steady behavior after compensation that reflects the stability of the voltages applied to the load during the subsequent changes of the power references of the controlled line.

Figure 7.54 shows the compensating currents i_C produced by the shunt converter of the UPLC. In the first stage of the compensation, the conditioner delivers the harmonic, unbalance, and reactive components of the load currents, as well as an active component to regulate the power balance of the conditioner and the voltages on the dc side. This is quite accurate in this stage, from 100 ms to 200 ms because the references of the active and reactive powers of the controlled line are similar to those of the load with only an UPQC compensation. In the next stages, the differences are mainly in fundamental components, due to the changes in the controlled line and the power balance of the conditioner.

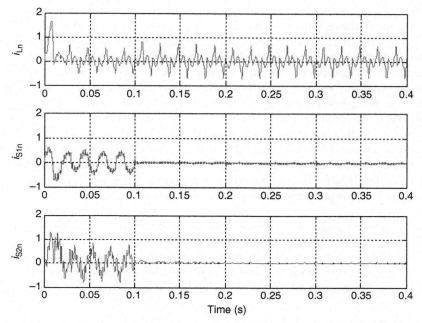

FIGURE 7.53 Neutral currents of the load, the supply S1 (controlled line) and the supply S2 (right side subsystem).

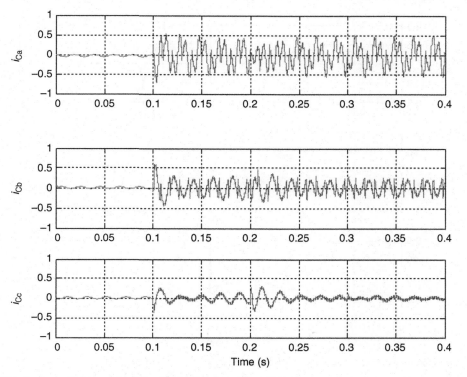

FIGURE 7.54 Compensating currents of the shunt converter of the UPLC.

Figure 7.55 shows the references of the compensating voltages v_C^* produced by the series converter of the UPLC to regulate the power flow in the controlled line and to isolate the load side from the harmonics and unbalance of the S1 subsystem. They are around 25% of the amplitude of the base voltage and differ slightly in each stage corresponding with the changes in the active and reactive power references. They show steady values among each step change, reflecting a very fast transient behavior of the power control. Figure 7.56 presents the instantaneous active and reactive powers in the controlled line. From the very beginning of the compensation they are set in the reference values with a very fast transient response. When the active power reference is changed from 0.6 pu to −0.6 pu at 200 ms, the transient is a little bit longer due to the fluctuations in the S2 subsystem, induced by the great change in the current of the controlled line. In the real systems the changes in the power references should be limited in a range and/or rate change to avoid these unnecessary short transients and provide a smoother transition.

Figure 7.57 shows in detail the evolution of the voltages and currents of the right side subsystem S1 during the step change of the active power reference, from 0.6 pu to −0.6 pu. The transition of the supply current is very fast, in few milliseconds, to produce a complete reversion of the power. During this transition, small disturbances appear in the supply voltage, related to the larger changes produced in the currents, mainly in the second and third phases.

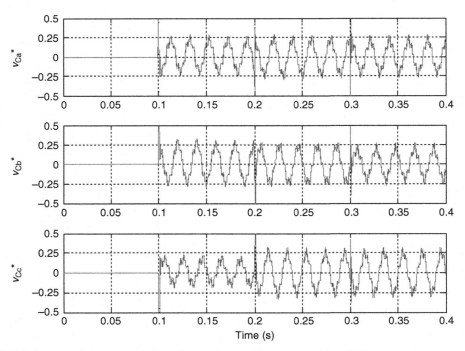

FIGURE 7.55 References of the compensating voltages of the series converter of the UPLC.

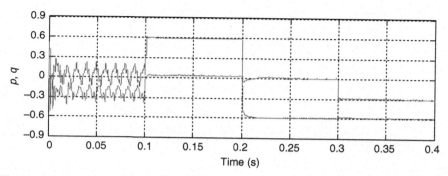

FIGURE 7.56 Instantaneous active and reactive powers at the controlled line.

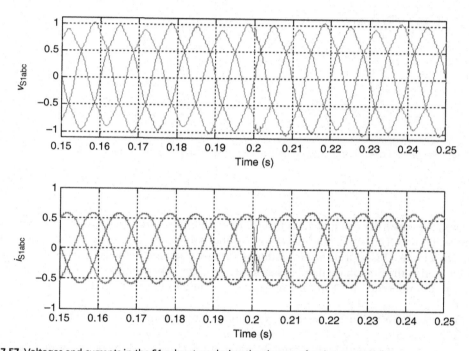

FIGURE 7.57 Voltages and currents in the S1 subsystem during the change of active power reference.

Figure 7.58 shows in detail the evolution of the voltages and currents of the right side subsystem S1, this time during the step change of the reactive power reference, from 0 pu to −0.3 pu. In this case, the controlled line demands active and reactive power to the left side subsystem; and the line currents increase their amplitude and phase. Again, the control transient is very fast; however the changes in the controlled currents are smaller and therefore the supply voltages of the left side exhibit smaller changes.

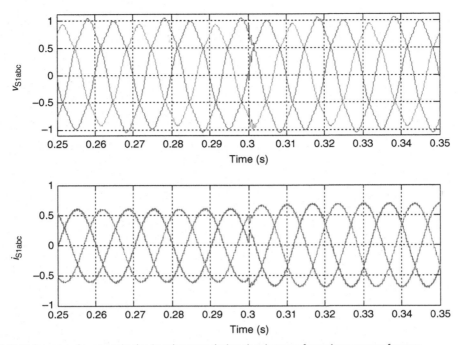

FIGURE 7.58 Voltages and currents in the S1 subsystem during the change of reactive power reference.

Figure 7.59 shows the voltages and currents of the nonlinear load in the stages of power reference changes. The active power reversion induced at 200 ms produces a voltage drop in the first half cycle. This short transient induces a longer transient in the current of the voltage sensitive load of the first phase that lasts around five cycles. The changes induced by the reactive power reference at 300 ms are not so sharp, and the corresponding transients are much smaller.

Figure 7.60 shows the instantaneous active and reactive powers demanded by the nonlinear load during the whole simulation. At a first glance, they show a very stable behavior, before and after the activation of the UPLC, and during the step changes of the power references. The average active power increases slightly with the start of the UPLC, from 0.42 pu to 0.44 pu, remaining quite constant afterwards. Meanwhile, the average reactive power has a similar behavior, with 0.13 pu before compensation and around 0.14 pu for the rest of the time.

Figure 7.61 shows the instantaneous active and reactive powers provided by the left side subsystem S1. Before compensation, the system is strongly distorted and unbalanced, supplying an average active power of 0.06 pu and receiving a reactive power of 0.27 pu. After compensation, the instantaneous powers become very constant with a small ripple due to the distortion and unbalance of the supply voltage. Their average values follow very near the control references, with small differences due to the voltages that the series converter must apply to control the active and reactive power of the line.

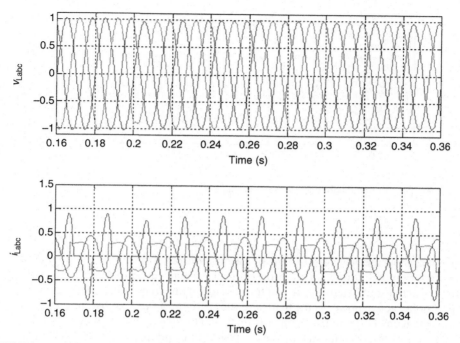

FIGURE 7.59 Voltages and currents in the load during the changes of the power references.

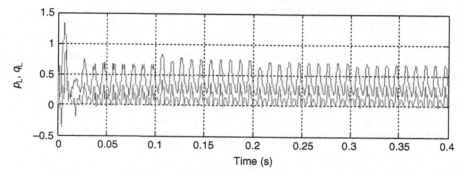

FIGURE 7.60 Instantaneous active and reactive powers at the nonlinear load.

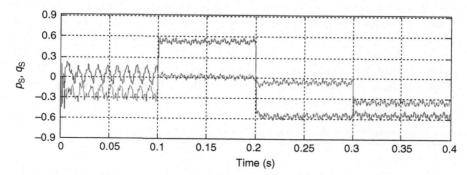

FIGURE 7.61 Instantaneous active and reactive powers provided by the left side subsystem.

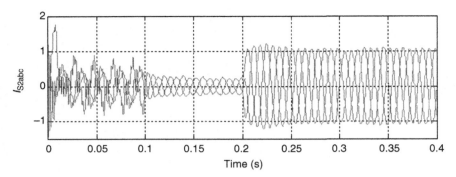

FIGURE 7.62 Phase currents of the right side subsystem (supply S2).

Figure 7.62 show the waveforms of the phase currents of subsystem S2. Before the start of the compensation they supply an important part of the load currents. After compensation, the references of active and reactive power of the controlled line make subsystem S1 to provide the major part of the active power of the load, and the currents of the S2 subsystem are reduced. When the active power reference changes from 0.6 pu to −0.6 pu at 200 ms, the S2 line must provide the total amount requested by the controlled line and the nonlinear load. The fast control of the current i_{S1} (see Figure 7.57) induces a transient in the load side of the conditioner with the uncontrolled transmission line. The strong response of the currents of S2 due to its relatively small line impedance, is partially damped with the UPLC and the subsystem reach steady values after three cycles. The reaction produced by the change in the reactive power reference is smaller because the variations of the controlled currents i_{S1} are also lower.

Figure 7.63 shows the square root of the instantaneous aggregate values of the load voltages. After compensation the oscillating component is strongly reduced and the average value is 1.02 pu. The transient induced by the active power reference of the controlled line produces a short voltage drop of about 12% during one cycle but it reaches quickly a steady value of 0.99 pu. And when the controlled line requires a reactive power of 0.3 pu the new steady value is 0.97 pu.

Figure 7.64 shows the voltages of the capacitors of the dc link of the UPLC and the total amount. It shows that the highest dc voltage variation occurs at the activation of the conditioner, where the instantaneous active power flow is bigger. The change in the references of the active and reactive powers of the controlled line does not imply great oscillations of the active power and the control of power balance of the conditioner manages it with lower variations. Figure 7.65 shows the instantaneous active and reactive powers that flow through the series converter of the conditioner. They present reduced average values along all the stages of the compensation and illustrate the effectiveness of the UPLC series converter to control the flow of the power line with a small effort. Figure 7.66 shows the instantaneous active and reactive powers that flow through the shunt converter of the conditioner. They have higher oscillating components due to the characteristics of the compensated load.

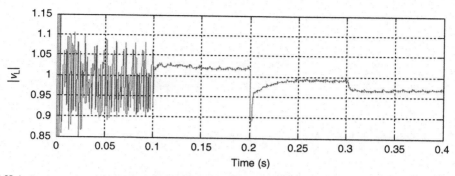

FIGURE 7.63 Instantaneous aggregate values of the load voltages v_L.

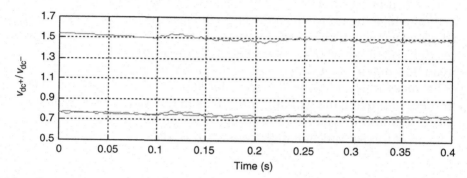

FIGURE 7.64 Voltages at the dc link of the UPLC.

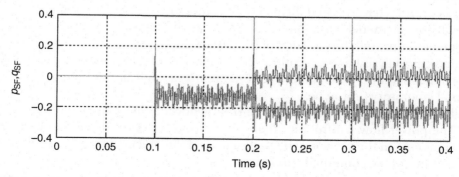

FIGURE 7.65 Instantaneous active and reactive powers of the series converter of the UPLC.

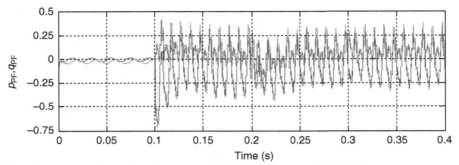

FIGURE 7.66 Instantaneous active and reactive powers of the shunt converter of the UPLC.

The results of Example 7.2 illustrate the extended capabilities of the UPLC. It can provide a fast and accurate control of the power flow at the controlled line, which should be even slowed to reduce the transients induced in the nearby parts of the network. It regulates also the load voltages with reactive power from the shunt converter. At the same time it enhances the EPQ of the whole system. Before the activation of the UPLC, there was a strong spread flow of harmonics, resonances and unbalance. With the UPQC compensation the unbalances and distortions of the left side subsystem are isolated and the current is compensated, improving the working conditions in this side. In the right side subsystem, the distortion and unbalance introduced by the nonlinear load are compensated, and the voltages at this side are enhanced for nearby loads.

7.5 Summary

This chapter has illustrated various of the capabilities of the set made with the combination of series and shunt active power filters. It started with the considerations for adequate compensation strategies for a complete load compensation, with appropriate voltages and compensated currents. These considerations have led to the definition of the compensating control references and the power structure to achieve these targets.

The state-space model has allowed us to analyze the transient behavior of the conditioner, and has improved the selection of the passive components and control parameters. The theoretical analysis has been contrasted with the results obtained in simulation cases, developed in a platform based on MATLAB-Simulink; as well as with the results of an experimental laboratory prototype.

Finally, the extended capabilities of the UPFC concept have been included in a UPLC design. The power flow control and the reactive-voltage regulation have been adapted and implemented in the previous UPQC design, and the practical case has shown the enhanced features of the combined conditioner.

References

[1] Akagi H. Trends in active power line conditioners. IEEE Trans Power Electr 1994;9(3):263–8.

[2] Akagi H. New trends in active filters for power conditioning. IEEE Trans Ind Appl 1996;32(6):1312–22.

[3] Akagi H, Fujita H, Wada K. A shunt active filter based on voltage detection for harmonic termination of a radial power distribution line. IEEE Trans Ind Appl 1999;35(3):638–45.

[4] Wang Z, Wang Q, Yao W, Liu J. A series active power filter adopting hybrid control approach. IEEE Trans Power Electr 2001;16(3):301–10.

[5] Campos A, Joos G, Ziogas P, Lindsay J. Analysis and design of a series voltage compensator for three-phase unbalanced sources. IEEE Trans Ind Electr 1992;39(2):159–67.

[6] Akagi H, Fujita H. A new power line conditioner for harmonic compensation in power systems. IEEE Trans Power Deliv 1995;10(3):1570–5.

[7] Akagi H, Fujita H. The unified power quality conditioner: the integration of series and shunt active filters. IEEE Trans Power Electr 1998;13(2):315–22.

[8] Aredes M, Heumann K, Watanabe EH. An universal active power line conditioner. IEEE Trans Power Deliv 1998;13(2):545–51.

[9] Kamran F, Habetler TG. Combined deadbeat control of a series–parallel converter combination used as a universal power filter. IEEE Trans Power Electr 1998;13(1):160–8.

[10] Moran S. A line voltage regulator/conditioner for harmonic-sensitive load isolation. In: Proc. 1989 IEEE/IAS Annual Meeting; 1989. p. 947–51.

[11] Montaño JC, Salmerón P. Instantaneous and full compensation in three-phase systems. IEEE Trans Power Deliv 1998;13:1342–47.

[12] Dixon JW, Tepper SM, Moran LT.Analysis and evaluation of different modulation techniques for active power filters. In: Applied Power Electronics Conference and Exposition, APEC'94, Proc. vol. 2; 1994. p. 894–900.

[13] Superti - Furga G, Tironi E, Ubezio G. Shunt active filter for four wire low-voltage systems: theoretical operating limits and measures for performance improvement. ETEP 1997;7(1):41–7.

[14] IEEE Std 1459-2010, IEEE Power and Energy Society, IEEE Standard definitions for the measurement of electric power quantities under sinusoidal, nonsinusoidal, balanced or unbalanced conditions; 2010.

[15] Salmerón P, Pérez A, Litrán SP. New approach to assess unbalance and harmonic distortion in power systems. In: Proc. of International Conference on Renewable Energies and Power Quality (ICREPQ'13); March, 2013, Bilbao, Spain. p. 1–4.

[16] Prieto J, Salmerón P, Vázquez JR, Alcántara FJ. A series–parallel configuration of active power filters for VAR and harmonic compensation. In: Proc. of IECON'02; 2002Sevilla. p. 1–6. Paper FILTQ.4.

[17] Prieto J, Salmerón P, Herrera RS. A unified power quality conditioner for wide load range: practical design and experimental results. In: Proceedings of IEEE St. Petersburg PowerTech Conference; 2005Russia. p. 429.1–7.

[18] Fujita H, Watanabe Y, Akagi H. Control and analysis of a unified power flow controller. IEEE Trans Power Electr 1999;14(6):1021–7.

[19] Gyugyi L. Unified power-flow control concept for flexible ac transmission systems. Proc Inst Electr Eng 1992;139(pt. C):323–31.

[20] Yu Q, Round SD, Norum LE, Undeland TM. Dynamic control of a unified power flow controller. In: IEEE PESC'96. pp. 508–514.

[21] Edris A, Zelinger S, Gyugyi L, Kovalsky LJ. Squeezing more power from the grid. IEEE Power Eng Rev 2002;22:4–6.

[22] Gyugyi L, Shauder CD, Williams SL, Rietman TR, Torgerson DR, Edris A. The unified power flow controller: a new approach to power transmission control. IEEE Trans Power Deliv 1995;10(2):1085–93.

[23] Akagi H, Watanabe EH, Aredes M. Instantaneous power theory and applications to power conditioning. IEEE Press; 2007.

[24] Montaño JC, Salmerón P. Identification of instantaneous current components in three-phase systems. IEE Proc Sci Meas Technol 1999;146(5):227–33.

8

Distributed Generation

CHAPTER OUTLINE

This chapter begins with a description of different distributed generation (DG) technologies. Thermal solar plants, wind power plants, cogeneration, photovoltaic (PV) plants, small hydrogenation, fuel cell, flywheel, superconducting coils, and supercapacitors are some of technologies currently most widely used, or it is expected that they may be commonly connected to electrical networks in the near future. These are presented from the point of view of their working principles and how they connect to the network.

In addition, an introduction is made to the load flow problem in DG systems. The different types of nodes are defined and the equations of the problem are formulated.

Connecting DG systems to the network of involves a change in the control of the system and the quality of the electrical power. In some cases, this connection leads to a deterioration of the quality indices of electric power. In other cases, it may serve to improve these

P. Salmerón Revuelta, S.P. Litrán, and J.P. Thomas: Active Power Line Conditioners. http://dx.doi.org/10.1016/B978-0-12-803216-9.00008-0

indices. Therefore, this chapter includes a section, which analyzes the impact of DG on the quality of electric power.

The last section is devoted to presenting and analyzing different control strategies of the inverters that are used as interface for the connection between DG and network. The main target of these strategies is to deliver the power generated by the source to the network. However, they are also designed to allow these interface devices to improve the electrical power quality.

8.1 Introduction

Until some years ago electrical systems were based on a centralized model. It was characterized by a system with large power plants, with high power generators connected to the transmission network. And the transmission system is used to transport the energy generated from these large power plants to the consumers (Figure 8.1). Most of the times this transmission is made over long distances, therefore with high voltage values. In Europe, 400 kV is usual but in other places, as North America or China, the transmission voltage can reach 750 kV.

Once the energy has been transported to nearby places of the consumption centers, it is necessary to distribute it through a greater number of lines, with shorter length and lower rated power. From a technical and economical point of view it is desirable to reduce the voltage levels. All these networks and substations together form what is called the distribution system. Traditionally, this distribution network did not have generators units connected, so it was considered as a passive network. In this condition the power flow is one-way, from the transmission to the distribution network.

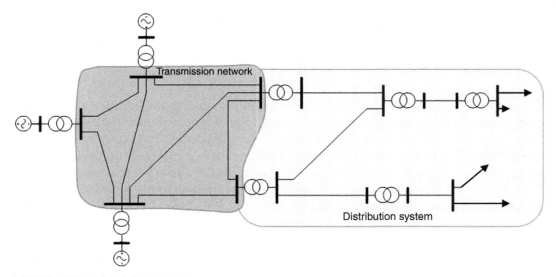

FIGURE 8.1 Conventional electric system.

Approximately in the 1990s, this conception of the electrical system begins to change. There was an increased interest in harnessing energy resources that require small power plants, usually located dispersedly. These resources are often related to renewable energies. Thus, the concept of DG, distributed energy, or dispersed generation is introduced (Figure 8.2). While it was called initially in one way or another, it seems that in recent years the term, distributed generation is the most used in the technical literature, falling in disuse the remaining two. The concept of DG is not defined in a formal and unique way [1]. In some countries, this definition is based on the rating of the power plant, or the voltage level to which it is connected, as it is usually associated with the technical documents used to specify the connection conditions and the operation of this type of generation.

Despite there is no complete agreement on the definition of DG, considering the different proposals, the main features of this kind of generation can be established as follows:

- It is connected to the distribution network.
- It is usual that part of this generation is consumed by the same facility and the rest is exported to the distribution network (e.g., cogeneration).
- There is no unified planning of this generation and is not usually dispatched centrally.
- The power rating of these units used to be lower than 50 MW.

The DG units can be connected to the system at different voltage levels, from low to high voltage. Traditional systems were designed to feed the loads with power flows from the higher voltage levels to the lower ones, hence the distribution network was considered as a passive system. With the integration of DG, the distribution system is no longer passive and becomes active, where the power flow can be in the opposite direction to conventional systems [2]. These changes bring as a result, the overload of networks and transformers, an increased risk of overvoltages, and a deterioration of the power quality.

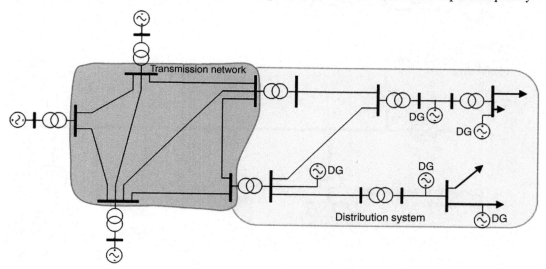

FIGURE 8.2 Electric system with DG.

On the other hand, the installation of the backup units close to the demand center avoids the cost of transmitting the power and any associated transmission losses. DG is foreseen to fit well in the new restructured energy market with all stockholders enjoying major benefits:

- For electricity distribution companies, DG will increase transmission and distribution capacity and, therefore, limit market influence on the energy cost increase.
- For large or small electricity consumers, DG uses a locally available mix of prime fuel sources, thus decreasing the dependency on importing. It can also be used for emergency backup and, moreover, it can be considered as an income-generating vehicle if properly interconnected with a hosting grid.

DG is closely related to the use of renewable energies. Among them, energy from the Sun and the wind are the ones that are having a greater development. The introduction of this type of energy in the electricity system marks another change in the safely and efficient operation and control of the network. This change can be partially solved by microgrids, which are entities that coordinate distributed energy resources in a consistently more decentralized way, thereby reducing the control burden on the grid and allowing them to provide their full benefits [3].

Microgrids can be ac or dc depending on the type of load and generation type that is installed, as shown in Figure 8.3 [4]. Each microgrid can be efficiently connected to the

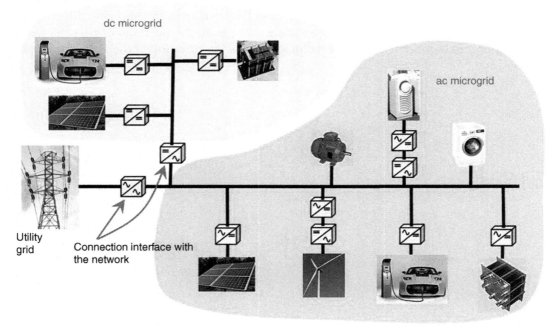

FIGURE 8.3 Microgrids.

public network through an interface that controls the power flow between them, in optimal quality conditions.

Recently, the concept of "smart grid" has appeared [5]. It defines a type of self-managing network with dynamic optimization techniques that uses real-time measurements to minimize losses, to maintain the voltage levels, to increase security, and to improve the system management. The data collected by the smart grid and its subsystems will enable system operators to quickly identify the best strategy to ensure proper system operation.

8.2 Different Technologies for Distributed Generation

In the last years, a wide variety of small power plants that are connected to the distribution network have proliferated [6]. Some of them use technologies already developed to generate electricity as wind turbines, PV solar, small hydro, or cogeneration units. Other techniques are more recent as fuel cells, solar thermal, microturbines, biomass, and marine renewable technologies. On the other hand, new energy storage technologies are also being developed such as flywheel, batteries, and supercapacitors. This section is intended to give an overview of those technologies that are presently taking more impact.

8.2.1 Thermal Solar Plant

Energy from the Sun is the primary energy source for life in our planet. Solar energy is really the main renewable energy available. This energy reaches the earth in the form of electromagnetic waves. The amount of this radiation that reaches a given area depends on many factors, such as the season, humidity, air mass, time of the day, etc. Most of the energy incident on the surface is used for the warming cycle, the weather cycle, wind, and waves. A small fraction is used for the photosynthesis in plants and the rest is emitted into space. The energy provided by the Sun can be approximately equal to 10,000 times the energy needs of the world [7]. Mankind has always tried to take advantage of solar energy and today this approach may be the solution to the energy problem of our societies.

One way of using this type of energy is through thermoelectric plants. Such plants use solar energy to heat a boiler, in order to produce the steam needed to run a turbine generator. A range of different technologies are used depending on the way that the solar energy is concentrated onto a receiver in which a working fluid is housed.

The solar concentrating systems (concentrating solar power (CSP) system) consist of a set of lenses designed to focus the energy on a receptor that acts as a boiler to generate steam. The system usually includes a tracking control system to maximize its efficiency. There are different types of reflectors, but the most widely used are parabolic mirrors, as shown in Figure 8.4.

Another approach concentrates the solar energy on a tower in which the receiver is placed, Figure 8.5. The use of a large number of reflectors, called heliostats, is needed, whose amount will depend on the system capacity. Water/steam is used as the heat-transfer fluid. Most recent designs use molten nitrate salt.

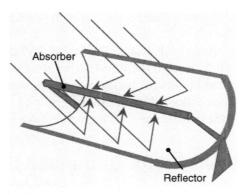

FIGURE 8.4 CSP system with parabolic mirror.

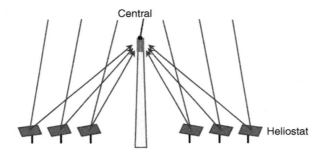

FIGURE 8.5 CSP system of tower.

Another technology used is based on the use of a Fresnel lens system. This type of lens design allows the construction of large aperture lens and a short focal length without the weight and volume of material that is needed in a lens of conventional design. The compact linear Fresnel reflector (CLFR) uses Fresnel lenses located along a single axis to concentrate solar energy and generate steam, Figure 8.6. It has the great advantage that the mirrors are cheaper to manufacture than the parabolic ones.

All the technologies described use solar energy to heat a fluid and produce steam to power a steam turbine. The technology to generate electricity from steam is the same as that used in conventional thermal power plants.

Finally, another technology that uses the heat of the Sun, based on the Stirling engine is presented (Figure 8.7). This thermal machine operates by cyclic compression and expansion of air or other gas, called working fluid, at different temperatures such that a net conversion of heat energy to mechanical energy occurs. The alternator is coupled to the machine shaft, to transform the mechanical energy into electric energy.

8.2.2 Photovoltaic Plant

PV plants convert solar radiation into electrical energy directly. Today it is a sufficiently mature technology; since its beginnings, it has gone from being a way to feed small loads

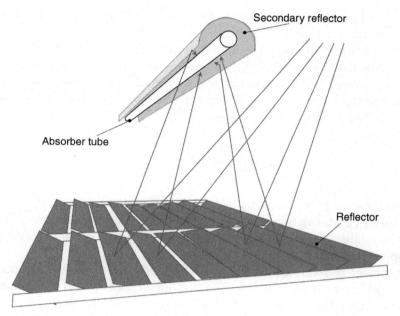

FIGURE 8.6 Compact Linear Fresnel Reflector, CLFR.

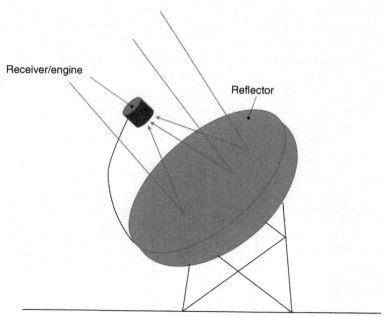

FIGURE 8.7 Motor stirling.

that operate in isolation in remote areas to PV plants of certain power that connect to the electricity distribution network.

The amount of energy produced by a PV module is directly proportional to the area of the module (the terms PV module and PV panel are used interchangeably).

There are many models of "PV cells or bank" for calculating the output power due to the numerous technologies currently available. One alternative equation to approximate the output power of PV panels is

$$P_{mp} = \frac{G}{G_{ref}} P_{mp,ref} [1 + \gamma(T - T_{ref})] \tag{8.1}$$

where G is the incident irradiance, P_{mp} is the maximum power output, $P_{mp,ref}$ is the maximum power output under standard testing conditions, T is the temperature, T_{ref} is the temperature for standard testing conditions reference (25°C), $G_{ref} = 1000 \text{ W/m}^2$, and γ is the maximum power correction for temperature.

The rated power generated by a module is limited. The maximum output voltage is corresponding to open circuit, which is usually reduced with respect to the mains voltage. Therefore, to increase the voltage of the plant, several PV modules are connected in series, composing strings. Furthermore, when an increase in power plant is required, several strings are connected in parallel to form an array.

Figure 8.8 shows a general wiring diagram of a PV plant to be connected to a distribution network. The output of a panel string can be connected to the dc bus through a dc/dc converter in order to adjust to the proper voltage levels. A dc/ac inverter converts the dc voltage to the appropriate ac voltage level of the distribution network.

In this basic scheme, a circuit of maximum power point tracking (MPPT) can be added to make the PV modules work in the maximum power point, with energy storage elements placed at the MPPT circuit output that is connected to the dc/dc converter. At the output of the inverter it usually has an isolating transformer and a filter to remove the harmonic components. Figure 8.9 shows the hardware structure to connect a PV system to the network.

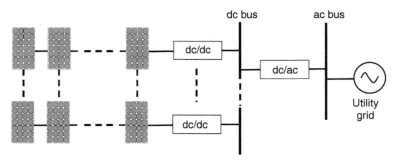

FIGURE 8.8 PV plant.

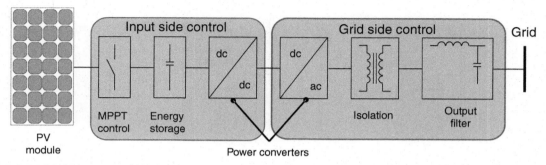

FIGURE 8.9 Grid connection scheme of a PV plant.

PV inverters normally operate with unity power factor, not generating reactive power, so they do not take part in the voltage control of the system.

8.2.3 Wind Power Plants

Wind energy is one of the most abundant resources and is the renewable technology that is growing most rapidly worldwide. The improvement in the turbines and the design of power converters has led to a significant drop in the cost of this power generation technology, which places it as the most competitive renewable energy from the economical point of view. This has caused it to become the second largest energy renewable resource behind the hydroelectrics.

A wind turbine operates by drawing kinetic energy from the wind passing through the rotor. The power output of a wind turbine is given by

$$P = \frac{1}{2}C_{\mathrm{p}}\rho V^3 A \tag{8.2}$$

where C_{P} is the power coefficient, a measure of the effectivity of the rotor aerodynamics; P is the power in W; V is the wind speed in m/s; A is the swept area of rotor disk in m²; and ρ is the air density (1.25 kg/m³).

The power depends on the wind speed cubed, so it is important to install such facilities in places with high annual average wind speeds. This requires a detailed study to establish the most suitable location.

The generator is coupled to the turbine, usually through a reduction gear. There are several configurations of these, depending on system speed [8].

- Constant speed wind turbine. This type of turbine is coupled via a multiplier to a squirrel cage induction generator (Figure 8.10a). The generator is connected directly to the network or through a soft starter. A capacitor bank is necessary in addition, to compensate the reactive power of the machine.
- Variable speed wind turbine with variable rotor resistance. These use a wound rotor induction generator (Figure 8.10b). As in the previous configuration a soft starter and the capacitor bank are included.

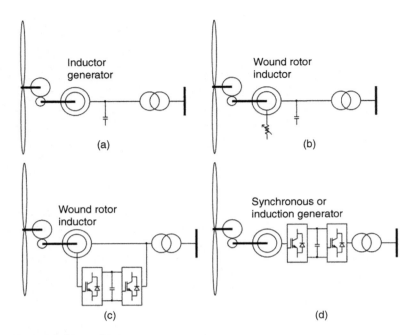

FIGURE 8.10 (a–d) Wind turbine architectures.

- Variable speed wind turbine with partial-scale frequency converter. A double-fed induction generator (Figure 8.10c) is used to generate the energy. The machine control is performed with power electronics. In this way, the turbine is operated at the maximum power point according to the wind conditions to control the active and reactive power of the machine.
- Variable speed wind turbine with full-scale frequency converter. It may include a synchronous or induction generator (Figure 8.10d). The network connection is made with ac/dc and dc/ac converters. The power control is performed with power electronics.

Variable speed turbines with permanent magnet synchronous generators are often used in small wind turbines. They do not usually include reduction gear, and their speed varies widely. The output is ac of variable frequency, so it includes a ac/dc converter and an inverter as coupling interface to the voltage and frequency of the distribution network.

Nowadays there is a growing interest in offshore wind farm projects. They have advantages over those projected on land: reduced visual impact, better wind conditions for speed and reduced turbulence, and low wind shear leading to lower towers. The main disadvantage is their high cost.

8.2.4 Cogeneration Plant

Cogeneration is the simultaneous production of electricity and heat for industrial or residential use. The electricity produced is often consumed by the same industrial plant,

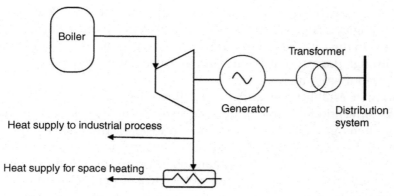

FIGURE 8.11 Cogeneration scheme with steam turbine.

although the surplus energy can be exchanged with the public distribution system. On the other hand, the heat generated is used in industrial processes or heating in a local area or district. This makes possible to achieve an overall system performance up to 67%. This means a 35% less fuel consumption than at a power plant in which the steam generated is only used to move the turbine.

Although a number of configurations can be given, Figure 8.11 shows a simple topology for a steam turbine. The generator is generally synchronous, connected directly to the network.

8.2.5 Fuel Cell

Hydrogen is an important source of renewable and clean energy, and there is plenty of it in the universe. Among its highlight qualities are that it is nontoxic, colorless, and odorless. Hydrogen can be used as a transportable energy; it is storable and can be developed in places where it is needed. When used there are no carbon emissions since it only produces water.

To utilize the energy of hydrogen, a fuel cell can be taken (Figure 8.12). Its operation is as follows: the hydrogen that arrives at the anode side is dissociated in protons and

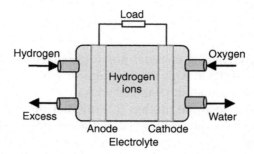

FIGURE 8.12 Fuel cell scheme.

electrons. Protons are driven to the cathode through the membrane, but the electrons are forced to go through an external circuit (producing energy) because the membrane is electrically isolated. In the catalyst of the cathode, the oxygen molecules react with electrons (conducted through the external circuit) and protons to form water. In this case, the only waste is water vapor or liquid water.

It is important to establish the fundamental differences between conventional batteries and fuel cells. Conventional batteries are energy storage devices, that is, the fuel is therein and it produces energy until it is consumed. However, in the fuel cell, the reagents are supplied as a continuous flow from the outside, allowing uninterrupted power generation.

The output voltage of a fuel cell is constant, so its connection to an ac network requires a dc/ac converter. Thus, for connecting the fuel cell to a power grid, a similar circuit to that used in PV plants is often used (Figure 8.9).

8.2.6 Small Hydro Generation

The generation of electrical energy from the potential energy of water is a relatively mature technology that has been developed in Europe and North America. Most of the large water resources have already been developed. So, in recent years the trend is to give value to the small power availabilities. From the point of view of technology, research continues on more efficient turbines that are also capable to exploit variable flow rates and heads.

The output power of a hydraulic turbine is given by

$$P = QH\eta\rho g \qquad (8.3)$$

where P is the output power (W), Q is the flow rate (m^3/s), H is the effective head (m), η is the overall efficiency, ρ is the density of water (1000 kg/m^3), and g is the acceleration due to gravity (m/s^2).

The feasibility study of a small hydroelectric plant requires knowledge of the mean annual flow that can be used and the height of the waterfall. For the first point, the most useful is to express the resource as a flow duration curve, which shows the percentage time that a given flow is equaled or exceeded.

The small hydroelectric plant can use induction or synchronous generators although it tends to use squirrel cage induction generators for its robustness and simplicity. The network connection of such machines is performed similarly to that shown in Figure 8.10.

8.2.7 Energy Storage

A major problem is that renewable energies are not manageable, that is, electricity production does not always coincide with the moment that it is needed, so it is practically impossible to achieve generation–load balance with this type of energy source. Although today there are sufficiently accurate models that predict the environmental conditions and the amount of energy of some sort that can be generated within a day in advance, is necessary to have additional plants that are manageable. As an example, solar energy is available only during daylight and is reduced significantly on a cloudy day.

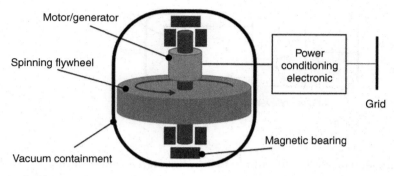

FIGURE 8.13 Flywheel scheme.

To mitigate this problem, energy storage is used. There is no energy storage process that has an acceptable efficiency and that can store energy in large enough quantities. In many cases, projects to achieve these aims are in the experimental stage. In the case of solar thermal energy, to increase production in times when there is no Sun, molten nitrate salts have been used. This allows the heat generated during the day to be stored, so that the generator can keep injecting energy during periods of time when the energy from the Sun is not available.

Another method of energy storage uses a flywheel. The flywheel is a metal disc, which starts spinning when a torque is applied, see Figure 8.13. Once it is rotating, it decelerates when subjected to a resisting torque. The equation of the stored kinetic energy is as follows

$$E = \frac{1}{2} I \omega^2 \tag{8.4}$$

where E is the energy stored; I, the moment of inertia that is a function of the rotating mass and its distance to the axis of rotation; and ω is the angular speed.

A motor/generator absorbs energy from the grid and uses it to push/brake the rotor mass to reach a certain angular velocity. Under ideal conditions, once the flywheel is rotating, its speed will remain constant until a resisting torque is applied. In practice, we look for the friction losses to be minimal. This is achieved by installing the wheel in vacuum-sealed containers and using magnetic bearings where there is no mechanical friction.

There are facilities, which use flywheels of various megawatts. They can provide an amount of energy in a relatively short time, so they can play an important role in the primary regulation of the power–frequency control.

Another method of energy storage is the use of superconducting coils. The energy stored in a coil is given by the expression

$$E = \frac{1}{2} L I^2 \tag{8.5}$$

where L is the inductance (H) of the coil and I is the circulating current (A).

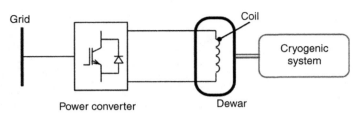

FIGURE 8.14 Superconducting coil.

A constant electrical current is made to flow through a superconducting coil. If the conductor has no energy losses, that energy is stored permanently until the circuit changes. The coil will behave as a current source. An electronic conditioning system (inverter/rectifier) will be needed to allow the power flow in both directions and to prevent overvoltages in the system (Figure 8.14).

Another way of storing energy is the use of supercapacitors. These are devices that have an operation principle similar to a traditional condenser. However, their capacity and discharge current is much higher and therefore they can be used as storage devices in power systems. The energy stored is given by

$$E = \frac{1}{2} C V^2 \tag{8.6}$$

where C is the capacity (F) and V is the terminal voltage (V).

It should be noted that the capacity is not constant and depends on the voltage across its terminals. Supercapacitors have the ability to be loaded and unloaded in very short periods of time, in the order of seconds or less, which makes them particularly suitable for responding to supply interruptions of short duration. With superconducting coils, a power converter for the network connection is also necessary.

Batteries are other devices that are used to store energy. Batteries are electrochemical devices that convert electrical energy (in the form of direct or constant current) into chemical energy as the battery charges, and convert chemical energy back into electrical energy when it is discharged. In energy storage systems, only rechargeable batteries can be used. The cost, duration, efficiency, battery life, and energy that can provide per unit volume are some of the most important features to consider before selecting a type of battery.

Pumped-storage hydroelectricity facilities are another efficient way of storing energy. Here, water is pumped to a reservoir located at a certain height to store potential energy; which can be exploited later using the gravitational field of the Earth on the water to drive a turbine that is coupled to a generator to produce electricity. The main problem with this approach is that two reservoirs located at different elevations are usually required, which often implies high investment costs.

Finally, the energy storage facilities of compressed air (CAES) could be mentioned. They function like big batteries. Powerful electric compressors compress the air into an underground geological formation (abandoned mines, cavities filled with mineral solutions or aquifers) during the periods of time when the use of electricity is lower, for example at night. And when the maximum power is needed during periods of high demand, the precompressed air is used in modified combustion turbines to generate electricity. Natural gas or other fossil fuels are still needed to run the turbines, but the process is more efficient. This method uses up to 50% less natural gas than a normal system power production.

8.3 Power Flow Control of a Distributed Generation

The integration of renewable energy sources as a part of the grid must also be examined through the power flow analysis, to ensure acceptable values of voltages and currents for a certain generation and distribution of the electrical energy throughout the network. The load flow problem calculates the voltage in each of the nodes in steady state and balanced conditions, as well as the active and reactive power flows through the system elements [9], for a given set of generation and loads. Knowledge of these parameters is necessary for the safe and effective operation and planning of the electric system.

The classical methods of analysis, by mesh currents or nodal voltages, do not apply directly when the load data are given in terms of power. The problem statement will lead to a system of nonlinear equations for whose solution is necessary to apply iterative techniques

Let "i" be a generic node with a power source that supplies to the node a current $\mathbf{I}_i$ and a complex power $\mathbf{S}_i$, with $\mathbf{U}_i$ as node voltage, Figure 8.15 (hereinafter the complex quantities will be represented in bold). The complex power can be expressed as

$$\mathbf{S}_i = \mathbf{U}_i\,\mathbf{I}_i^* \tag{8.7}$$

In the load flow analysis only balanced three-phase systems are considered. Thus, any system can be represented by their single-phase equivalent. Applying node analysis to this equivalent circuit, for node "i" the equation can be expressed as

$$\mathbf{I}_i = \sum_{k=1}^{n}\mathbf{Y}_{ik}\,\mathbf{U}_k \tag{8.8}$$

where n is the number of nodes of the system, $\mathbf{Y}_{ik}$ for $k{\neq}i$ is the mutual admittance between nodes i and k with the sign changed, $\mathbf{Y}_{ii}$ is the self admittance of the node i, and $\mathbf{U}_k$ is the voltage at node k.

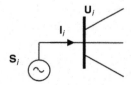

FIGURE 8.15 Generic bus.

When (8.8) is replaced in (8.7), it leads to

$$S_i = U_i \left(\sum_{k=1}^{n} Y_{ik} U_k \right)^*$$

(8.9)

In this expression, phasors and complex numbers can be represented in exponential form, that is

$$U_i = U_i \, e^{j\delta_i}; \quad Y_{ik} = Y_{ik} \, e^{j\gamma_{ik}}; \quad U_k = U_k \, e^{j\delta_k}$$

(8.10)

where δ is the phase angle of the voltages and γ is the argument of the admittances.

Thus equation (8.9) can be rewritten in the form

$$S_i = U_i \left(\sum_{k=1}^{n} Y_{ik} U_k \, e^{j(\delta_{ik} - \gamma_{ik})} \right)$$

(8.11)

where δ_{ik} is the phase difference of the node voltages i and k, that is, $\delta_{ik} = \delta_i - \delta_k$.

Equation (8.11) can be decomposed into the following two real equations

$$P_i = U_i \sum_{k=1}^{n} Y_{ik} U_k \cos(\delta_{ik} - \gamma_{ik})$$

(8.12)

$$Q_i = U_i \sum_{k=1}^{n} Y_{ik} U_k \sin(\delta_{ik} - \gamma_{ik})$$

(8.13)

In equation (8.12), P_i is the real part of the apparent power S_i, that is, the incoming active power in node i. In equation (8.13), Q_i is the imaginary part of S_i, incoming reactive power in node i.

Each node has two equations and four variables associated with it: the magnitude of the voltage U_i, the phase angle of the voltage δ_i, the injected active power P_i, and the injected reactive power Q_i. To solve the resulting system of equations, it is necessary to use iterative methods such as Newton–Raphson or Gauss–Seidel.

On the other hand, the nodes of the system can be classified according to three basic categories:

- Generator nodes or PV nodes. In this node, the active power P and the voltage magnitude U can be taken as system data as these two variables are easily adjusted at the power plant (Figure 8.16a).
- Load nodes or PQ nodes. They are the nodes where the consumption of active and reactive power are connected and where either no generation exists or it is fixed and is taken as given (Figure 8.16b).
- Swing node or slack node (Figure 8.16c). This is a generation node (usually designated as node 1) in which instead of the active power the argument of the voltage is specified and fixed at $\delta_1 = 0$. This node is used as reference for the arguments of the

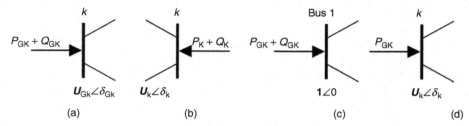

FIGURE 8.16 Bus type: (a) P–V bus; (b) PQ or load bus; (c) swing or slack bus; and (d) P constant active power.

voltages for the rest of nodes. It uses to be the node of an important generator of the system. The function of a slack bus is to balance power consumption and power loss with the net-injected generated power.

PVs, wind, or other types of generation coupled to the network through inverters [10], can be considered as a new node category, as these kind of power sources used to work with unity power factor, therefore producing only active power. So, when the unit is connected to a local power network, it can be specified according to Figure 8.16d.

EXAMPLE 8.1

The Figure 8.17 shows the single line diagram of a four-node system with a rated voltage of 66 kV. The generator connected to node 1 is set to a voltage of 1.04 per unit. A PV generator is connected at node 3, whose inverter operates at unity power factor. Two loads are connected at nodes 2 and 4, whose complex apparent powers are those shown in Figure 8.17. The resistances and inductances of the distribution lines are indicated in Table 8.1. Determine the solution of the power flow when the PV generator injects 1.5 MW.

To solve the load flow problem a tool included in MATLAB–Simulink has been used, specifically in the SimPowerSystems library. It is necessary to build the Simulink model of the network shown in Figure 8.17 to use this tool. Each of the nodes must be defined by type. Thus node 1 will be the swing node, nodes 2 and 4 are PQ type and node 3 is PV type.

The iterative method of Newton–Raphson is used, where a tolerance of 10-4 has been set. Table 8.2 presents the results obtained with the node voltages and the net active and reactive powers of each one. The base voltage is 66 kV. To solve the power flow, Simulink allows to the load nodes to choose between having constant active and reactive powers or to maintain a constant impedance. In this case, the constant power option has been selected.

As most remarkable results, the active power losses of the system are 0.93 MW and 1.93 Mvar of reactive power. The powers defined in the statement remain constant. In node 1 the active and reactive powers are 99.43 MW and 31.93 Mvar.

8.4 Distributed Generation Impact in Power Quality

Inclusion of DG on the electrical system may result in a worsening of the electric power quality [11,12]. The connection of a large number of generators may have effects on the distribution network locally or globally, may even affect the transmission system [7].

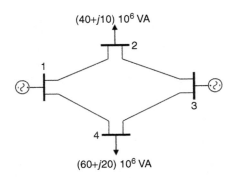

FIGURE 8.17 One-line diagram (Example 8.1).

Table 8.1 Impedances Line, Example 8.1

	Line 1-2	Line 2-3	Line 3-4	Line 4-1	Line 2-4
Resistance (Ω)	0.8	0.5	0.6	0.8	0.4
Inductance (mH)	5	4	3.5	5.5	3

Table 8.2 Load Flow Results, Example 8.1

Bus ID	Type	U (pu)	U_angle (deg)	P (MW)	Q (Mvar)
Bus_1	Swing	1.04	0	99.43	31.93
Bus_2	PQ	1.0261	−0.818	40	10
Bus_3	PV	1.0259	−0.846	1.5	0
Bus_4	PQ	1.0249	−0.884	60	20

The grid connection of DG is performed basically with three technologies [13]: synchronous generators, asynchronous generators, and power electronic inverters. The impact on the network depends on the technology used [14], although in general, DG can impact positively or negatively on the power quality in the following aspects:

- Voltage fluctuations. Such disturbances can be produced due to the variability of the power fed into the grid of some energy sources. Wind power generation may be a clear example, as the wind speed is usually not constant and rapid changes may appear even in windy conditions, so that the voltage at the connection point undergoes random variations, which may result in the phenomenon of flicker. This phenomenon is more evident in the case of weak networks.
- Harmonics. Some connected interfaces emit high frequency harmonics. However, the problem of harmonics is not referenced in the literature, perhaps because it has not yet had sufficient penetration of DG in the system.
- Voltage unbalances. Many small generators are usually single-phase and cannot be connected to a three-phase distribution network. That can lead to unbalances in the system voltage.

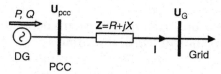

FIGURE 8.18 Equivalent circuit with DG connected.

- Voltage dips. In general, it can be said that DG improves the behavior bus voltage where it is connected. Voltage dips are less severe, although this depends on the size of DG, improving when the power is greater.

The presence of DG does not always have a negative impact on the system [15]. It is in fact possible to use an intentionally DG to improve the power quality. On the other hand, it is clear that there is also an impact from the point of view of control and operation of the electric system that is not covered in this section.

The following subsections will discuss how DG affects each of these aspects.

8.4.1 Voltage Fluctuations

Voltage fluctuations can be produced by sources whose available power varies strongly with time [16]. Figure 8.18 shows the equivalent circuit when a DG is connected to the utility grid. According to this model the current can be obtained as

$$I = \frac{P - jQ}{\mathbf{U}_{PCC}}$$

(8.14)

where **I** is the line current, $\mathbf{U}_{PCC}$ is the voltage at the point of common coupling (PCC), P is the output active power of DG, and Q is the output reactive power of DG.

Voltage between the connection points of the DG the utility grid is the voltage in the **Z** impedance, which is given by

$$\Delta\mathbf{U} = \mathbf{U}_g - \mathbf{U}_{PCC} = \mathbf{I}\,\mathbf{Z} = \mathbf{I}\,(R + jX)$$

(8.15)

where $\mathbf{U}_g$ is the voltage at the grid; R is the line resistance; and X, the line reactance. When (8.14) is replaced in (8.15), the result is

$$\Delta\mathbf{U} = \frac{RP + XQ}{\mathbf{U}_{PCC}} + j\frac{PX + QR}{\mathbf{U}_{PCC}}$$

(8.16)

As can be deduced, variations in the active power or reactive power injected into the network cause voltage fluctuations in the grid. DG produced by solar energy and wind are two clear examples of systems with output powers that vary randomly.

In the case of wind power, wind is unstable in nature, which causes fluctuations in the output power of wind turbines. When it comes to large turbines in a wind farm operating at the same time, significant impacts may occur in the system. The impact on the voltage level of the wind turbines depends on the generator type. For induction generator systems connected directly to the network, fixed speed systems depend primarily on the X/R relation of the network [17]. A relation of two or three times generates a low impact, provided that the system is equipped with capacitor banks to compensate the reactive power. When the wind turbine system includes power converters connected to the grid, the inverter control usually includes a loop to control the voltage level and reactive power. Moreover, it usually includes a smooth control to avoid switching transients that occur when capacitor banks are included.

Wind turbulence together with the turbine dynamic itself creates power variations in the region of 0.01–10 Hz. The use of converters provides the possibility of reducing the dynamic voltage fluctuations. This is because incoming power fluctuations can be offset by slight changes in rotor speed.

For PVs, solar radiation varies randomly, causing a change in the system power flow [18] and hence variations in the output voltage. These variations also depend on the functioning of the whole PV system, PV modules, inverter, filters, control mechanisms etc. Some studies have shown that short fluctuations in irradiance and cloud over play an important role in distribution networks of low voltage with high penetration of PV.

8.4.2 Harmonics

When PV systems, fuel cells, micro-turbines connect with the grid or supply power for ac load, they need an inverter device to carry out the dc-ac conversion. If a constant speed, constant frequency, asynchronous wind generator connects to grid directly, its soft start-up phase needs power electronic devices such as thyristor, and variable speed constant frequency wind generator systems also need a rectifier and an inverter device to connect with grid. Therefore, from a harmonic modeling and simulation standpoint, a distributed generator is usually a converter-inverter type unit and can therefore be treated as a nonlinear load injecting harmonics into the distribution feeder. Under the present framework of IEEE 519-2014, the supplier of electricity is responsible for the quality of the voltage supplied. The end-user is responsible for limiting harmonic current injections based on the size of the end-use load relative to the capacity of the system. Distributed resources are small relative to system capacity, but the smaller sizes are much more likely to achieve significant penetration levels for economic reasons.

The harmonic emission from DG is smaller than that emitted by modern loads, whereby the voltage distortion is rarely a problem [7]. By contrast, DG may cause increases in harmonic components that traditionally in a power system were small as the even harmonics, interharmonics and high frequency harmonics.

The injected current in the grid should not have a total harmonic distortion larger than 5%. A detailed image of the harmonic distortion with regard to each harmonic is given in Table 8.3 for six-pulse converters, according to IEEE Std 519 [19].

On the other hand, the IEC 61000-3-15 standard [20] establishes emission limits and immunity requirements for DG systems in low voltage.

Table 8.3 Distortion Limits as Recommended in IEEE Std 519 for Six-Pulse Converters

Odd Harmonics	Distortion Limit (%)
3rd–9th	<4.0
11th–15th	<2.0
17th–21st	<1.5
23rd–33rd	<0.6

8.4.3 Voltage Unbalances

In an electrical system the main sources of voltage unbalance are mainly due to transmission networks nontransposed or poorly transposed and connection of large single-phase loads. For the transmission networks, the system unbalance depends on the transport power. In the second case, the impact will depend heavily on how large installations with power unbalances are connected. In this sense, the IEC 61000-3-13 [21] standard establishes recommendations for connecting unbalanced network facilities.

DG can generate voltage imbalances in different ways. Single-phase units can produce an increase of these imbalances, although these units are limited to low power systems and connections in low voltage, so that the incidence is very low. In the case of many single-phase generator connected to the system, they have resulted in increased levels of voltage unbalance. Balance is very complicated when the system has a large number of generators, so there is a high probability that the system voltage is unbalanced.

In the case of three-phase generators, they limit the voltage imbalance provided that currents injected to the system are balanced. Generally the presence of these generators connected to the distribution system reduces the negative sequence components of the voltage.

8.4.4 Voltage Dips

DG does not affect the increase of voltage sags to the system directly, but there may be indirect effects as follows [22]:

- The replacement of large power plants by DG systems leads to a weakening of the network, resulting in increased voltage dips.
- The high penetration of DG requires new lines and strengthening others. This is most obvious in the case of large wind farms. This increase in the line number causes an increase in number of sags to the consumers connected to nearby lines.
- The weakening of the transmission system leads to an increase in the timing to remove faults. This affects less to the distance and differential protection but it does affect overcurrent protection that is sometimes used in these lines.
- After clearing the fault, rotating machines absorb a large reactive power. In weak networks, this causes decreasing voltage, which generates a voltage dip.
- When the network connection of DG is done through power converters, these can help maintain the load voltage locally.

More directly, the impact on voltage sags may be caused by wind generation due to its high penetration in the present electrical system. When a wind turbine is connected to the grid, high currents occur that may cause voltage dips [23]. This may be aggravated in the case of turbines that use a capacitor as compensation device, since the value of reactive power is proportional to the square of the output voltage. If at the moment of connection the value of the voltage is low, the current consumption is greater and thus the dip depth.

In the past, [16] wind generators were not allowed to stay connected when the voltage at the common connection point was below 85% of rated voltage, forcing disconnection [2]. This caused stability problems in the system. The present wind energy penetration into the electrical network has forced system operators to adapt their grid codes to this new generation, preventing an unacceptable effect on the system safety and reliability.

8.5 Distribution Line Compensation

Much of distributed generation uses converters as the interface for connection to the grid. The inverter control determines the operation and synchronization with the network. For injection of a given active power can use different control strategies, although all generally include two control loops [24–27]. A current loop that regulates the current injected to the network and a voltage control loop that controls the voltage at the dc-link.

The current control loop is responsible for the quality of the power and dc-link voltage controller is designed to control the power flow in the system.

Under normal operating conditions, DG operates at unity power factor. Mainly there are two control strategies that enable this objective:

- Instantaneous unity power factor control strategy. The aim of this strategy is to maintain constant active power and reactive power zero. By contrast, the network currents may be distorted.
- Positive sequence control strategy. In calculating the reference current, the positive sequence component of the fundamental harmonic of the voltage at the PCC is used. With this strategy injected currents are sinusoidal and balanced.

In addition to these strategies, some authors propose other aims to improve the above when a fault occurs in the system. In this network situation, oscillations appear in the active power and reactive power of double frequency of the mains frequency. These oscillations cause voltage oscillations at the dc side of the inverter, making it more difficult to control network in such conditions [28], leading to imbalance in the network current.

8.5.1 Instantaneous Unity Power Factor Control Strategy

Figure 8.19 shows a DG system connected to the network through three-phase dc/ac converter, which injects a power P at the PCC. In a three-phase system as shown, an instantaneous equivalent conductance can be defined by

$$g = \frac{P}{|\mathbf{v}|^2}$$

(8.17)

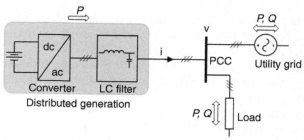

FIGURE 8.19 System with DG and utility grid.

where $|\mathbf{v}|^2$ is instant norm module of the phase voltage vector at the inverter output.

The instantaneous current vector, $\mathbf{i}^*$ can be obtained from the product of the equivalent conductance g by voltage vector $\mathbf{v}$,

$$\mathbf{i}^* = g\,\mathbf{v} = \frac{P}{|\mathbf{v}|^2}\,\mathbf{v} \tag{8.18}$$

This vector current allows injecting a P power of DG at the PCC, to the grid voltage whose phase values are the vector components, $\mathbf{v}$.

Current vector components, $\mathbf{i}^*$ form a balanced and sinusoidal system when the network voltage vector is balanced and sinusoidal too. If the voltage vector is unbalanced, the current vector is unbalanced. If any of the components of $\mathbf{v}$ has any harmonic, the current system will also be distorted.

Moreover, the current vector, $\mathbf{i}^*$ is proportional to the voltage vector; therefore, it has no orthogonal component with respect to the mains voltage; hence, there is no reactive power transfer between the network and DG.

EXAMPLE 8.2

A DG system is connected to a distribution network via a dc/ac converter, with a configuration as shown in Figure 8.19. The system injects an active power of 15 kW with unity power factor. The supply voltage is 400 V and 50 Hz frequency. At the PCC, a load has been connected. It transfers an active power of 1 MW and a reactive power of 0.75 Mvar to the rated voltage.

The instantaneous unity power factor control strategy is applied to the converter. Its performance will be verified in two different situations voltage:

- Sinusoidal, balanced voltage system.
- Unbalanced network voltage system with a negative sequence component of 20% and nonsinusoidal with a 5th order harmonic of 15%.

To verify the operation of this strategy, the system is simulated in MATLAB–Simulink. The DG system is connected to the network by means of an LC filter formed by an inductance of 50 mH and a capacity of 0.7 μF.

The distribution network is modeled by means of a programmable three-phase source of infinite power. This source allows include imbalances and harmonics in the voltage.

The dc-link controller has been omitted in order to not influence the creation of the current references. As a consequence, dc power sources are used to supply the necessary voltage in dc-link. The dc-link voltage is set to 1500 V.

The control strategy is implemented according to the block diagram shown in Figure 8.20, which is the development of equation (8.18). The input signals to the control scheme are the phase voltage vector, **v**, and active power of DG, *P*.

In the first case, the voltage is sinusoidal with an rms value of 400 V and 50 Hz. Figure 8.21a shows the waveform of the system voltage in the PCC. The rms value of the phase voltage is 230.9 V. The currents injected by the DG system have the waveform shown in Figure 8.21b. As noted, it is a balanced and sinusoidal system. The rms value of the current is 21.5 A.

With respect to the power balance, DG system provides 14,884 W and distribution network 985,116 W, thus the active power consumed by the load is 1 MW.

In the second case, the voltage of the distribution network has a negative sequence component of 20% and a 5th order harmonic of 15%. The three-phase source that model the distribution network is programmed with these values. The voltages waveforms are shown in Figure 8.22a. The voltage rms values of each phase are 279, 214, and 214 V.

Current waveforms injected into the system by the DG system are shown in Figure 8.22b. It is noted that these currents are not balanced and sinusoidal. Current rms values of each phase are: 23, 20, and 21 A.

Respect to powers, in this situation DG supplies 14,657 W, distribution network supplies 1,027,676 W, and load consumes 1,042,333 W.

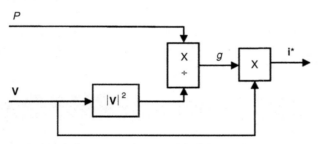

FIGURE 8.20 Instantaneous unity power factor control strategy, block diagram.

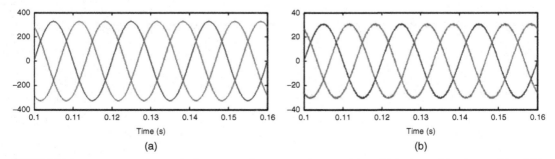

FIGURE 8.21 Instantaneous unity power factor control strategy. Balance and sinusoidal voltage: (a) voltage grid and (b) grid current.

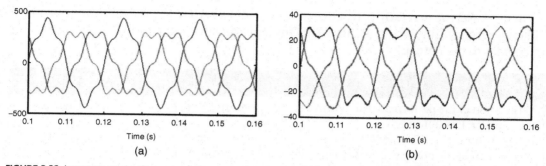

FIGURE 8.22 Instantaneous unity power factor control strategy. Unbalance and nonsinusoidal voltage: (a) voltage grid and (b) grid current.

8.5.2 Positive Sequence Control Strategy

In situations where the voltage vector is in balance and/or presents harmonics, the above strategy, instantaneous unity power factor, can cause a deterioration in the waveform of the current and thereby, loss of power quality. In this situation, for the circuit of Figure 8.19, an instantaneous equivalent conductance can be defined by means of

$$g^+ = \frac{P}{\left|\mathbf{v}_f^+\right|^2}$$

(8.19)

In this equation, $\left|\mathbf{v}_f^+\right|^2$ represents the magnitude of the instantaneous norm of the positive sequence voltage vector at the fundamental frequency [29]. The reference current vector is given by

$$\mathbf{i}^* = g^+ \mathbf{v}_f^+ = \frac{P}{\left|\mathbf{v}_f^+\right|^2} \mathbf{v}_f^+$$

(8.20)

This strategy requires an algorithm for detecting positive sequence components of the voltage.

When the voltage is balanced and sinusoidal, instantaneous active power coincides with the active power, P, and reactive power is zero. When the voltage is unbalanced the instantaneous active power is different from the power P.

When this strategy is applied, the instantaneous active power can be expressed in terms of the positive and negative sequence components by means of

$$p = \mathbf{v} \cdot \mathbf{i}^* = \mathbf{v}^+ \cdot \mathbf{i}^* + \mathbf{v}^- \cdot \mathbf{i}^*$$

(8.21)

The first term in equation (8.21) is the power P and the second is $\tilde{p}$, which represents a power oscillation of double frequency of the fundamental.

Similarly, the instantaneous reactive power can be calculated as

$$p = \left|\mathbf{v} \times \mathbf{i}^*\right| = \left|\mathbf{v}^+ \times \mathbf{i}^*\right| + \left|\mathbf{v}^- \times \mathbf{i}^*\right|$$

(8.22)

In this case, the first term is reactive power, Q with a zero value and the second is an oscillatory term, $\tilde{q}$, double frequency of the fundamental that appears in the reactive power injected into the network.

EXAMPLE 8.3

Verification of the performance of positive sequence control strategy for the DG system is described in Example 8.2.

Figure 8.23 shows the block diagram of the control circuit designed. The positive sequence component of the phase voltage vector, $\mathbf{v}$ is obtained using the block v_a^+. This component is calculated by means of

$$v_a^+ = \frac{1}{3}\left(v_a + a\,v_b + a^2 v_c\right)$$

The operator a can be obtained from two all pass filters connected in cascade and a^2 inverting input signal in the all pass filter. For the design of this filter, it is used as a Simulink block that generates the transfer function

$$F(s) = \frac{s - 181.4}{s + 181.4}$$

Figure 8.24 shows the block diagram that calculates the positive sequence component.

The direct sequence component of the fundamental harmonic is obtained with the block v_{fa}^+. This block is similar to that used in Chapter 5 when the fundamental component is determined. After obtaining the fundamental harmonic of the positive sequence component of the phase a, the b and c phase components are determined. In Figure 8.23, this calculation is performed by the block v_f^+. Here, the inverse transformation of Fortescue is applied, this is

$$\mathbf{v}_f^+ = \begin{bmatrix} v_{fa}^+ & v_{fb}^+ & v_{fc}^+ \end{bmatrix}^T = \begin{bmatrix} v_{fa}^+ & a^2 v_{fa}^+ & a v_{fa}^+ \end{bmatrix}^T$$

Figure 8.25 shows block diagram of $\mathbf{v}_f^+$.

The rest of the diagram in Figure 8.23 is the same as that of Example 8.2.

When the voltage system is balanced and sinusoidal, positive sequence control strategy and instantaneous unity power factor control strategy show a similar behavior.

In the case of nonsinusoidal and unbalanced voltage system, network currents in DG have a sinusoidal waveform, as shown in Figure 8.26. For each of the phases, the current rms values are 21.15, 21.2, and 21.2 A. Thus, these form a balanced three-phase system.

Finally, DG supplies to the grid an active power of 14,647 W, the distribution network supplies 1,027,686 W, and the load consumes 1,042,333 W.

8.6 Power Quality Improvement in Distributed Environment

The increased penetration of DG can generate problems in power quality if appropriate measures are not taken. However, if the appropriate control strategies are taken, DG can help improve that quality. In this section, different strategies to achieve those improvements are analyzed.

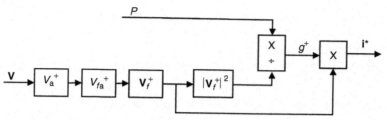

FIGURE 8.23 Positive sequence control strategy, block diagram.

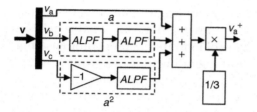

FIGURE 8.24 Block diagram to determine the positive sequence component.

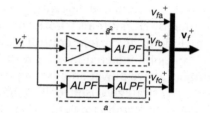

FIGURE 8.25 Fortescue inverse transformation.

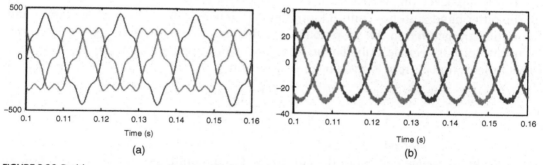

FIGURE 8.26 Positive sequence control strategy. Unbalance and nonsinusoidal voltage: (a) grid voltage and (b) grid current.

In the distribution network, the voltage at different points differs from its rated value. The voltage profile of the network can be improved by installing DG in the appropriate places [30]. This takes system planning, so that the optimal locations and proper size of DG are chosen.

Figure 8.27a shows a network that feeds a set of loads distributed throughout the network. Usual voltages profile in an actual facility is also shown. Voltage in the feeding point nearest to the loads is greater than that of the most distant loads. When a DG system is connected to the network modifies the voltage profile with an increase in the tension of the farthest loads (Figure 8.27b). Connecting these units at different points in the network

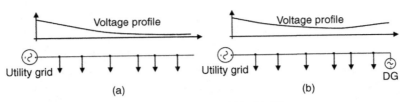

FIGURE 8.27 Voltage profile of an electric line. (a) Without DG and (b) with DG.

can generate a flat profile, which would be ideal from the point of view of system operation. The voltages profile obtained depends on the situation of DG and its size.

The presence of current harmonics in the network due to nonlinear loads can be mitigated by DG systems whose connection interface is an inverter. For this, it is necessary to include in its current control strategy a new control loop that allows compensating harmonics be injected [31,32]. In this case, the DG system inverter has a dual function, to inject active power and to act as an active filter to compensate harmonics.

The strategies used in this new control loop are varied, thus, in [33] the inverter operates as an active inductor to absorb harmonic currents. Another method is that the inverter acts as an active conductance to dampen the harmonics of the distribution network [34]. In Ref. [35] a control strategy is presented for renewable energy interface based on p–q theory.

In this work, a strategy that integrates two control loops is described. On the one hand, a control loop based on positive sequence control strategy to inject the active power, and on the other hand, another control loop is added to compensate the reactive power and harmonics transferred to the set of loads. For this, instantaneous active power vectorial theory, which has been described in previous chapters is applied. According to this theory, the compensation current is given by the expression

$$\mathbf{i}_C = \mathbf{i}_L - \frac{P_L}{\left|\mathbf{v}_f^+\right|^2}\mathbf{v}_f^+ \tag{8.23}$$

where P_L is the transferred power to the load, $\mathbf{i}_L$ is the load current vector. The current vector, $\mathbf{i}_C$, is the reference current of the control loop that compensates reactive power and current harmonics.

DG transfers the active power, P_{DG}, when the inverter generates an instantaneous current, $\mathbf{i}_P$, defined according to (8.20). The sum of $\mathbf{i}_P$ and $\mathbf{i}_C$ is the reference current for the inverter control, this is shown in Figure 8.28.

$$\mathbf{i}_{DG}{}^* = \mathbf{i}_P + \mathbf{i}_C = \frac{P_{DG}}{\left|\mathbf{v}_f^+\right|^2}\mathbf{v}_f^+ + \mathbf{i}_L - \frac{P_L}{\left|\mathbf{v}_f^+\right|^2}\mathbf{v}_f^+ \tag{8.24}$$

The above equation can be rewritten in the form

$$\mathbf{i}_{DG}{}^* = \mathbf{i}_L + \frac{P_{DG} - P_L}{\left|\mathbf{v}_f^+\right|^2}\mathbf{v}_f^+ \tag{8.25}$$

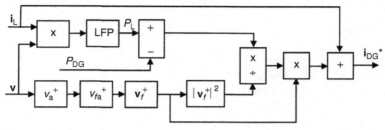

FIGURE 8.28 Block diagram of control strategy applied in Example 8.4.

Figure 8.28 shows the block diagram that implements the equation (8.25). The blocks are the same as described in Example 8.3 with the exception of block "LFP" consisting of a low-pass filter to obtain the average value of the output voltage by load current, resulting in active power transferred to the load, P_L.

EXAMPLE 8.4

Figure 8.29 shows a power system with DG connected to the distribution network of 400 V and 50 Hz. In a system bus, a linear load of 30 kW active power and reactive power of 20 kvar and a nonlinear load are connected. The nonlinear load is formed by uncontrolled rectifier with a resistor of 15 Ω connected in series with an inductance of 100 mH at the dc side.

DG should inject an active power of 10 kW, compensate reactive power of all loads and mitigate current harmonics of the distribution network.

For the rated voltage, the set of linear and nonlinear loads transfer an active power of 46,812 W and reactive power of 22,413 var. The load power factor is 0.74 inductive. The current presents a waveform clearly nonsinusoidal, as shown in Figure 8.30a. Its rms value is 75.2 A and has a THD of 7.89%. The most significant harmonics are of order 5 and 7. When the DG system is not connected, this will be the waveform and power transferred from the distribution network.

The DG system inverter is controlled with a strategy based on positive sequence strategy and instantaneous active power vectorial theory. For this, the block diagram shown in Figure 8.28 is implemented in MATLAB–Simulink. When the DG system is connected, the waveform of the current supplied by the distribution network is shown in Figure 8.30b. The rms value of this current is 54.0 A and has a THD of 2.26%. From the viewpoint of the quality of power is a significant improvement in the current waveform as there is a large decrease THD. Furthermore, regarding the power balance, the active power transferred by the distribution network is 37,377 W with a power factor of 1.00.

Respect to the power provided by the DG, it is 9435 W. This value is somewhat lower than the expected power of 10,000 W. This is due to losses occurring in the inverter output LC filter, since the resistive effect of the inductance has been considered. Regarding the current waveform generated by the DG, it is shown in Figure 8.31, the rms value of the current is 35.8 A.

On the other hand, microgrids are proposed to isolate the problems associated with DG. Different solutions have been proposed to mitigate these problems of lack of quality

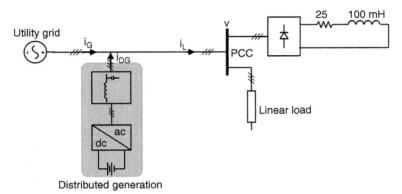

FIGURE 8.29 Power system for Example 8.4.

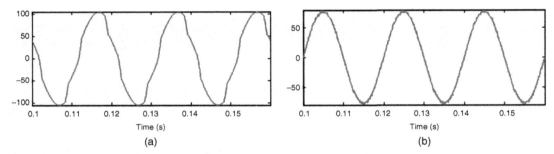

FIGURE 8.30 Example 8.4. (a) load current waveform and (b) grid current waveform with DG.

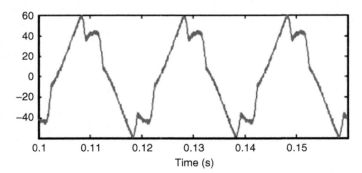

FIGURE 8.31 DG current waveform (Example 8.4).

in microgrids. Some of these are based on improving the functioning of the converters that connect equipment of DG to the network [36]. Other proposals are based on the use of a power quality compensator [37,38]. When the microgrid is connected to the public network, it should not cause disturbances to the rest of the system, so that a connection interface between the microgrid and the main grid is necessary to avoid these issues. These interfaces are based on a configuration of series-parallel active power filter [39].

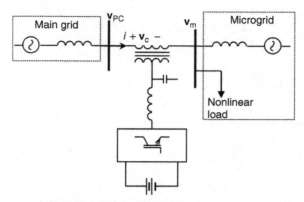

FIGURE 8.32 Scheme of interface between utility grid and microgrid.

The interface consists of an inverter connected in series with the system between main grid and microgrid. Figure 8.32 shows the general scheme, where the main network is represented by its Thevenin equivalent and the microgrid by a set of source and nonlinear load.

The series APF consists of an inverter bridge of IGBTs. At the ac side a ripple filter is connected, which allow the high frequency components to be eliminated. Connection to the system is done with coupling transformers.

The objective of the proposed control strategy is to avoid transferring voltage unbalances and current harmonics from the main grid to microgrid or from microgrid to main grid [40]. Thus, the inverter must generate a compensation voltage such that at the network side, voltage unbalances are not transmitted to the microgrid and vice versa. Furthermore, the generated harmonic currents should not be transferred between two networks. Therefore, two control loops will be designed: one to compensate voltage unbalances and another to mitigate current harmonics.

Respect to loop control to compensate voltage unbalances, the direct sequence component of voltage vector of the "a" phase can be calculated by means of the following expression

$$v_a^+ = \frac{1}{\sqrt{3}}(v_a + a\,v_b + a^2 v_c) \tag{8.26}$$

where v_a, v_b, and v_c are the components of the voltage vector and a, operator is defined as $a = e^{j2\pi/3}$, which supposes a 120° phase shift. This operator can be implemented by an all pass filter [41].

Once (8.26) is applied, the Fortescue inverse transformation allows voltage vector of direct sequence component to be obtained. It is calculated by means of the expression

$$\vec{v}^+ = \begin{bmatrix} v_a^+ & v_b^+ & v_c^+ \end{bmatrix}^T = \frac{1}{\sqrt{3}}\begin{bmatrix} v_a^+ & a^2 v_a^+ & a v_a^+ \end{bmatrix}^T \tag{8.27}$$

where, v_a^+, v_b^+, and v_c^+ are the components of the direct sequence voltage vector for each phase.

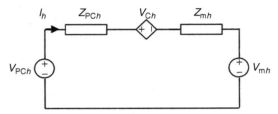

FIGURE 8.33 Single-phase equivalent circuit of Figure 8.32 for a harmonic of h order.

These components are obtained at the main network side ($\mathbf{v}_{PC}$) and at microgrid side ($\mathbf{v}_m$). So, the compensation voltage vector ($\mathbf{v}_c$) will be calculated by

$$\mathbf{v}_c = \mathbf{v}_{PC} - \mathbf{v}_{PC}^+ - \mathbf{v}_m + \mathbf{v}_m^+ \tag{8.28}$$

To mitigate the current harmonics, the inverter is controlled to present zero impedance at the fundamental frequency and high impedance at the frequencies of the load harmonics. For it, the compensation voltage will be proportional to the current harmonics.

$$v_C = k\, i_h \tag{8.29}$$

where k is the proportionality constant.

Figure 8.33 shows the single-phase equivalent circuit of Figure 8.32 for a harmonic of h order. Here, the main grid and the microgrid are represented by its Thevenin equivalent and the inverter by a controlled voltage source of v_C value. By applying Kirchhoff's voltage law and the equation (8.29), the current is given by

$$I_h = \frac{v_{PCh} - v_{mh}}{Z_{PCh} + Z_{mh} + k} \tag{8.30}$$

According (8.30), the h order harmonic of the current can be mitigated when k has a high value, ideally k should be infinite, however, in the practice it is enough when $k >> Z_{PCh} + Z_{mh}$ to the most significant harmonics.

EXAMPLE 8.5

Figure 8.34 shows a serial interface connection between a microgrid and DG system. The values of the passive components are shown in Table 8.4. The main grid impedance is considered less than microgrid impedance. This is intended to consider the microgrid weaker than the main grid.

The proposed control strategy will be applied and the system will be subjected to disturbances at the network side and microgrid side.

To verify the performance the system will be subjected to two different situations. In one, the power flow is from the main grid to the microgrid, and in the other, the microgrid is what gives power to the main grid. So, it will be subjected two different network situations:

1. The voltages at the main network side have an unbalance of 25% and further have a 5th order harmonic of 12% of the fundamental harmonic. In this case, the power flow is from the main grid to microgrid.
2. The voltages at the microgrid side have an unbalance of 25% and contain a 5th order harmonic of 12% of the fundamental harmonic. In this test, power flow is considered from the microgrid to the main grid.

In the first test, the main grid voltage contains a 5th order harmonic of 12% and a negative sequence component of 25%. Figure 8.35 shows the voltage waveforms at the PCC and the current before connecting the inverter.

When the interface is connected, the voltage waveform at the main grid side and at the microgrid side shown in Figure 8.36 is obtained. The proposed control strategy was applied to the inverter. The proportionality constant defined in (8.29) was fixed in 50. The unbalance factor was calculated to be 5.2%. This represents a significant reduction of voltage unbalance but does not eliminate it completely. The remainder is due to the current system being unbalanced, resulting in different voltage drops across the equivalent impedances in each phase. These current unbalances cannot be compensated with a series topology, so this slight unbalance voltage due to the current unbalance will always remain.

On the other hand, taking into account the "a" phase, the current THD is reduced from 24.6% to 4.9%. This reduction is even higher in the other phases. Therefore, it is possible to state that the interface "isolates" the microgrid from the main grid harmonics.

A similar analysis can be performed with respect to the harmonics of the voltage. The results shown in Figure 8.36a and b illustrate how the interface "isolates" the microgrid of the voltage harmonics from the main grid.

In the second test, the voltages at the microgrid side have an unbalance of 25% and contain a 5th order harmonic of 12% of the fundamental harmonic. At the main grid side, the voltages are balanced and sinusoidal.

Figure 8.37 shows the waveform before connection to the interface. Voltages and currents present a waveform that is not sinusoidal, with the 5th harmonic being the most significant.

After, the interface is connected to the system with the control strategy proposed. The proportionality constant was fixed at 50. The voltage and current waveforms are shown in Figure 8.38. Voltage at the main grid and current are almost sinusoidal. However, the voltages at the microgrid side are strongly distorted. Therefore, the interface isolates the main grid of harmonics generated in the microgrid side.

With respect to the voltage unbalance, with the proposed control is possible to reduce it. The calculated value of unbalance factor is of 3.8%. It is an important reduction however, as it occurred in test 1, its value is not null. This is justified by the current unbalance that causes an unbalance in the voltages due to drops in system impedances. These cannot be avoided with an interface topology of series connection.

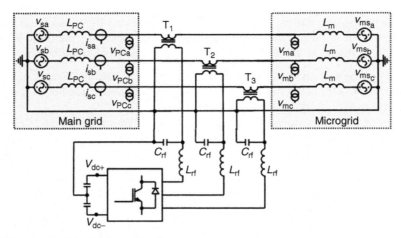

FIGURE 8.34 Simulation circuit for the series interface.

Table 8.4   Values Passive Elements

Main grid	$R_{PC} = 1\ \Omega$	$L_{PC} = 1$ mH
Microgrid	$R_m = 3\ \Omega$	$L_m = 3$ mH
Ripple filter	$C_{rf} = 50\ \mu$F	$L_{rf} = 10$ mH

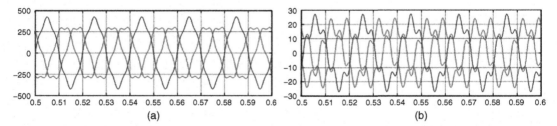

FIGURE 8.35 Test 1, waveform voltage and current, before the interface is connected. (a) Voltage v_{PC} and (b) current.

In the technical literature, other control strategies and different circuit topologies have been proposed for the improvement of electrical power in environments with high penetration of DG [42]. It was not possible, given the scope of this chapter, to discuss them all, so the authors have limited the discussion to those proposals that they consider clearer for its simplicity.

8.7 Summary

Nowadays DG is a reality in the power system. The increased penetration of these technologies means that solutions must be found to the problems that this produces. In this chapter different technologies have been introduced, highlighting how they couple to the

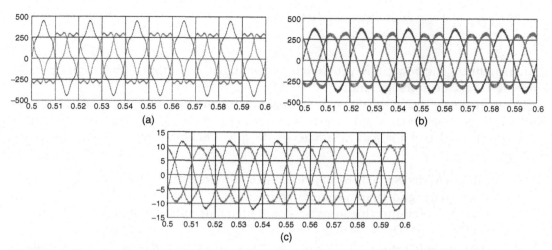

FIGURE 8.36 Test 1, waveform voltage and current, after the interface is connected. (a) Voltage v_{PC}, (b) voltage v_m, and (c) current.

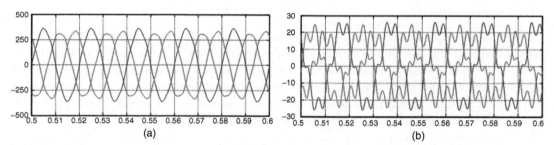

FIGURE 8.37 Test 2, waveform voltage and current, before the interface is connected. (a) Voltage v_{PC} and (b) current.

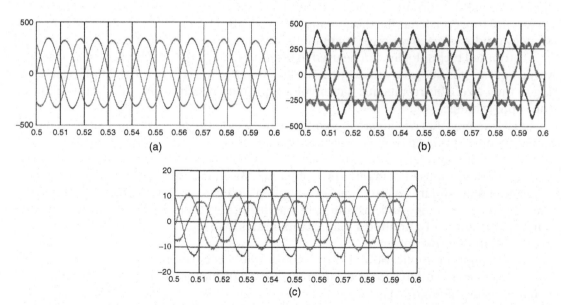

FIGURE 8.38 Test 2, waveform voltage and current, after the interface is connected. (a) Voltage v_{PC}, (b) voltage v_m, and (c) current.

grid. In addition, the power flow problem is reformulated, defining a new category node to characterize this technology kind.

The inclusion of DG in the power system generates an impact on the distribution network. In some cases, it can lead to a worsening of power quality and in others, it can be used to improve power quality. The impact on the grid is analyzed according to the following topics:

- Voltage fluctuations. Such disturbances can be produced due to the variability of the power fed into the grid by some energy sources.
- Harmonics. Some connecting interfaces emit high harmonics of the main frequency.
- Voltage unbalances. Many small generators are usually single-phase and are connected to a three-phase distribution network. This can lead to unbalances in the system voltage.
- Voltage dips. In general, it can be said that DG improves the behavior bus voltage where it is connected.

Much of DG uses a power converter (dc/ac) as an interface for connection to the network. When the aim is to inject active power to the grid, there are two basic control strategies: instantaneous unity power factor and positive sequence. In the first strategy, generation system works with unity power factor. When the voltage system is balanced and sinusoidal, the injected current is balanced and sinusoidal. However, when the voltage system is unbalanced and/or nonsinusoidal, current is unbalanced and/or nonsinusoidal, which may contribute to imbalance system and/or current harmonic circulation to the mains. In the second strategy, positive sequence control strategy, when the supply voltage is unbalanced and/or nonsinusoidal, the currents are sinusoidal and balanced, which is an improvement from this viewpoint. However, it has the disadvantage that in fault situations in the network, the latter strategy produces reactive power fluctuations and active power fluctuations in the system.

Distributed technologies can be used to improve the voltage profile of an electrical network. For this it is necessary to connect these systems to the grid in the right place according to their size. This results in a flatter voltage profiles, which means an improvement in the supply voltage.

In DG systems that use inverter as coupling devices to the network, the electrical power quality can be improved with the right strategy. For this application a new loop is provided to the inverter control, so as to determine the waveform of the current compensation according to the proposed compensation objective. As an example, generation system has been simulated and its inverter has been applied a control strategy based on positive sequence control strategy and instantaneous active power vectorial theory. This mitigates current harmonics of a nonlinear load, compensates load reactive power and feed into the grid a specified active power.

Finally, another application of active power filters is based on its use as an interface device between the utility grid and a microgrid. With proper control strategy, this interface can isolate the microgrid of generated disturbances in the utility grid and isolate the utility grid of those generated disturbances in the microgrid.

References

[1] Ackermann T, Andersson G, Soder L. Distributed generation: a definition. Elec Power Syst Res 2001;57(3):195–204.

[2] Guerrero JM, Blaabjerg F, Zhelev T, Hemmes K, Monmasson E, Jemei S, et al. Distributed generation: toward a new energy paradigm. IEEE Ind Electr Mag 2010;4(1):52–64.

[3] Hatziargyriou N, Asano H, Iravani R, Marnay C. Microgrids. IEEE Power Energy Mag 2007;5(4):78–94.

[4] Justo JJ, Mwasilu F, Lee J, Jung JW. AC microgrids versus DC-microgrids with distributed energy resources: a review. Renew Sustain Energy Rev 2013;24:387–405.

[5] Farhangi H. The path of the smart grid. IEEE Power Energy Mag 2010;8(1):18–28.

[6] Jenkins N, Ekanayake JB, Strbac G. Distributed generation. London, UK: The Institution of Engineering and Technology, IET; 2010.

[7] Bollen M, Hassan F. Integration of distributed generation in the power system. New Jersey, USA: Wiley and Sons; 2011.

[8] Blaabjerg F, Teodorescu R, Liserre M, Timbus AV. Overview of control and grid synchronization for distributed power generation systems. IEEE Trans Ind Electr 2006;53(5):1398–409.

[9] Saadat H. Power system analysis. New York, USA: PSA Publishing; 2010.

[10] Keyhani A. Design of smart power grid renewable energy systems. New Jersey, USA: Wiley and Sons; 2011.

[11] Hariri A, Faruque MO. Impacts of distributed generation on power quality. North American Power Symposium (NAPS); 2014. pp. 1–6.

[12] Coster EJ, Myrzik JMA, Kruimer B, Kling WL. Integration issues of distributed generation in distribution grids. Proc IEEE 2011;99(1):28–39.

[13] Jenkins N. Embedded generation. Power Eng J 1995;9(3):145–50.

[14] Freitas W, Vieira JCM, Morelato A, da Silva LCP, da Costa VF, Lemos FAB, et al. Comparative analysis between synchronous and induction machines for distributed generation applications. IEEE Trans Power Syst 2006;21(1):301–11.

[15] IEEE guide for conducting distribution impact studies for distributed resource interconnection. IEEE Standard 1547.7, 2013.

[16] Jauch C, Sørensen P, Norhem I, Rasmussen C. Simulation of the impact of wind power on the transient fault behaviour of the Nordic power system. Elec Power Syst Res 2007;77:135–44.

[17] Thiringer T, Petru T, Lundberg S. Flicker contribution from wind turbine installations. IEEE Trans Energy Conv 2004;19(1):157–63.

[18] Deng W, Pei W, Qi Z. Impact and improvement of distributed generation on voltage quality in microgrid. Third International Conference on Electric Utility Deregulation and Restructuring and Power Technologies; 2008. pp. 1737–1741.

[19] IEEE Std 519-2014. IEEE recommended practice and requirements for harmonic control in electric power systems.

[20] Standard IEC 61000-3-15. Electromagnetic compatibility (EMC) – Part 3-15: Limits – assessment of low frequency electromagnetic immunity and emission requirements for dispersed generation systems in LV network.

[21] Standard IEC 61000-3-13. Electromagnetic compatibility (EMC) – Part 3-13: Limits – assessment of emission limits for the connection of unbalanced installations to MV, HV and EHV power systems.

[22] Martínez-Velasco JA, Martín-Arnedo J. Distributed generation impact on voltage sags in distributions networks. 9th International Conference Electrical Power Quality and Utilization; October 2007, Barcelona.

[23] Ramos ACL, Batista AJ, Leborgne RC, Emiliano P. Distributed generation impact on voltage sags. Power Electronics Conference, 2009. COBEP '09; 2009, Brazil. p. 446–50.

[24] Saccomando G, Svensson J. Transient operation of grid-connected voltage source converter under unbalanced voltage conditions. In: Proc. IAS; vol. 4, 2001, Chicago, IL. p. 2419–24.

[25] Agirman I, Blasko V. A novel control method of a VSC without ac line voltage sensors. IEEE Trans Ind Appl 2003;39(2):519–24.

[26] Teodorescu R, Blaabjerg F. Flexible control of small wind turbines with grid failure detection operating in stand-alone or grid-connected mode. IEEE Trans Power Electr 2004;19(5):1323–32.

[27] Song S-H, Kang S-I, Hahm N-K. Implementation and control of grid connected ac–dc–ac power converter for variable speed wind energy conversion system. Proc IEEE APEC 2003;1:154–8.

[28] Timbus AV, Liserre M, Teodorescu R, Rodriguez P, Blaabjerg F. Evaluation of current controllers for distributed power generation systems. IEEE Trans Power Electr 2009;24(3):654–64.

[29] Timbus AV, Rodriguez P, Teodorescu R, Liserre M, Blaabjerg F. Control strategies for distributed power generation systems operating on faulty grid. IEEE International Symposium on Industrial Electronics, vol. 2; 2006. pp. 1601–1607.

[30] Acharya N, Mahat P, Mithulananthan N. An analytical approach for DG allocation in primary distribution network. Int J Elec Power Energy Syst 2006;28(10):669–78.

[31] Bojoi R, Limongi LR, oiu D, Tenconi A. Enhanced power quality control strategy for single-phase inverters in distributed generation systems. IEEE Trans Power Electr 2011;26(3):798–806.

[32] Singh M, Khadkikar V, Chandra A, Varma RK. Grid interconnection of renewable energy sources at the distribution level with power quality improvement features. IEEE Trans Power Deliv 2011;26(1):307–15.

[33] Borup U, Blaabjerg F, Enjeti PN. Sharing of nonlinear load in parallel-connected three-phase converters. IEEE Trans Ind Appl 2001;37(6):1817–23.

[34] Jintakosonwit P, Fujita H, Akagi H, Ogasawara S. Implementation and performance of cooperative control of shunt active filters for harmonic damping throughout a power distribution system. IEEE Trans Ind Appl 2003;39(2):556–64.

[35] Pinto JP, Pregitzer R, Monteiro LFC, Afonso JL. 3-phase 4-wire shunt active power filter with renewable energy interface. presented at the Conf. IEEE Rnewable Energy & Power Quality; 2007, Seville, Spain.

[36] Teorodescu R, Blaabjerg F, Liserre M, Loh PC. Proportional-resonant controllers and filters for grid-connected voltage-source converters. IEE Proc Elec Power Appl 2006;153(5):750–62.

[37] Tang X, Tsang KM, Chan WL. A power quality compensator with DG interface capability using repetitive control. IEEE Trans Energy Conv 2012;27(2):213–9.

[38] He J, Li YW, Munir MS. A flexible harmonic control approach through voltage-controlled DG-grid interfacing converters. IEEE Trans Ind Electr 2012;59(1):444–55.

[39] Li YW, Vilathgamuwa M, Loh PC. A grid-interfacing power quality compensator for three-phase three-wire microgrid applications. IEEE Trans Power Electr 2006;21(4):1021–31.

[40] Litrán SP, Salmerón P, Prieto J, Pérez A. Control strategy for an interface to improve the power quality at the connection of AC microgrids. In: Proc. International Conference on Renewable Energies and Power Quality (ICREPQ'14), April 2014.

[41] Salmerón P, Litrán SP. A control strategy for hybrid power filter to compensate four-wires three-phase systems. IEEE Trans Power Electr 2010;25(7):1923–31.

[42] Chen Z, Blaabjerg F, Pedersen JK. Hybrid compensation arrangement in dispersed generation systems. IEEE Trans Power Deliv 2005;20(2 Part 2):1719–27.

Appendix
Simulink Schemes

This appendix provides examples of simulation in Matlab–Simulink environment that appear in the book. Examples are divided up chapterwise. The simulation model for each scheme is described. In addition, the reader can access the Elsevier website to access the file for each example.

Chapter 1
Example 1.1

The purpose of Example 1.1 is to obtain the Norton model of a nonlinear load. This consists of a three-phase noncontrolled rectifier with a resistor and an inductance connected in series at the dc side. This load is connected to a three-phase sinusoidal source with a source impedance modeled by an inductance and a resistor. In order to modify the network conditions an RL branch is connected to the system by means of the switch, k. Figure A.1 shows the power system scheme in Simulink. Block modeling nonlinear three-phase load and measuring block to obtain the harmonic rms values of voltage and current are also shown. In the latter, a SimPowerSystem block is used, which allows us to obtain the Fourier analysis of the input signal over a running window of one cycle of the fundamental frequency.

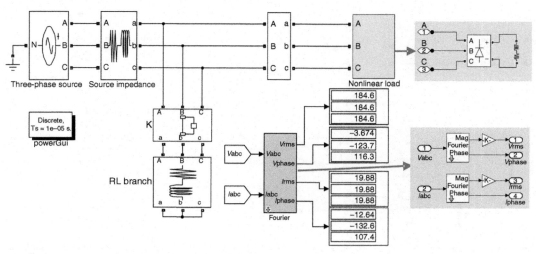

FIGURE A.1 Power circuit scheme, Example 1.1.

P. Salmerón Revuelta, S.P. Litrán, and J.P. Thomas: Active Power Line Conditioners. http://dx.doi.org/10.1016/B978-0-12-803216-9.00009-2

Chapter 2

Example 2.2

In this example we have tried to obtain the characteristic values of the voltages and currents, as well as the different power terms exclusively in the Simulink environment. It has been used as both SimPowerSystems and Signal Processing blockset; blocks of the latter have been used to determine the Fast Fourier Transform (FFT) and subsequent algebraic manipulation of output.

Figure A.2 shows the display of three-phase power system consisting of an unbalanced source and two loads in star: balanced nonlinear load and unbalanced linear load.

Figure A.3 shows the balanced nonlinear load. It consists of three bridge rectifiers connected in a star with an RC branch in the dc side (for it has been used as a Universal Bridge block of SimPowerSystems). Figure A.4 shows unbalanced linear load consisting of three star RL branches.

Figures A.5 and A.6 show block diagrams for computing the sum of the squares of the rms values of the harmonics of the phase-to-neutral voltages, line-to-line voltages, line currents and neutral current. We used buffer blocks (to store the signal samples), magnitude FFT (to compute the absolute value of the DFT), submatrix (to extract the values of interest in the results matrix), and matrix sum (to sum the elements of row/column/all elements of the input array) of the signal processing blockset.

From the results of the block diagrams of Figures A.5 and A.6 effective values of voltage and current are determined by the Std. IEEE 1459, Figure A.7.

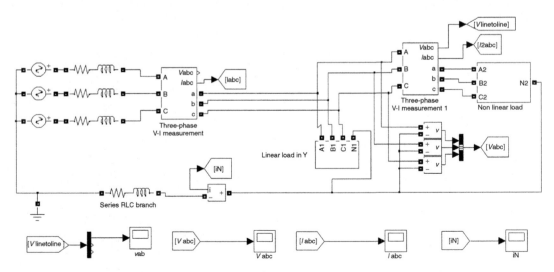

FIGURE A.2 Three-phase power circuit consists of an unbalanced source and two loads in star: balanced nonlinear load and unbalanced linear load.

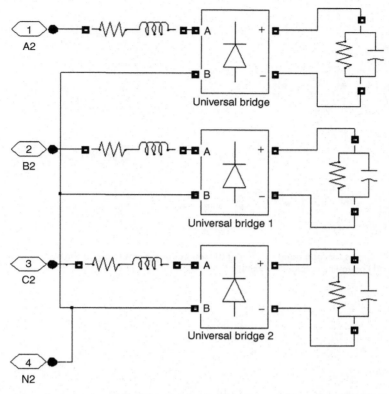

FIGURE A.3 Balanced nonlinear load consists of three bridge rectifiers connected in a star with RC branch in his dc side used in the Example 2.2.

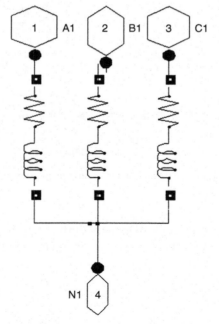

FIGURE A.4 Unbalanced linear load consists of three star RL branches for Example 2.2.

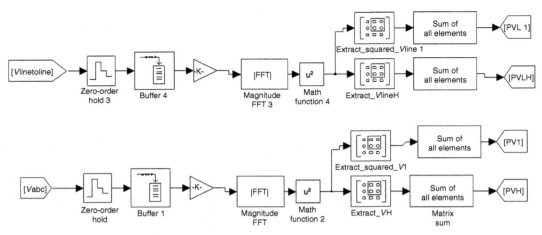

FIGURE A.5 Block diagrams for computing the sum of the squares of the rms values of the harmonics of the phase-to-neutral voltages, and line-to-line voltages in Example 2.2.

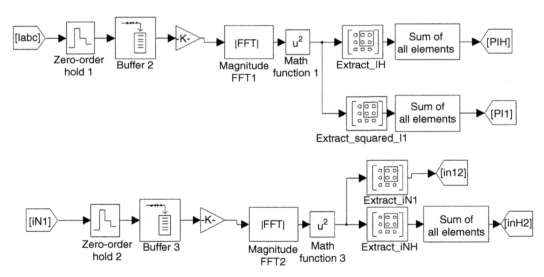

FIGURE A.6 Block diagrams for computing the sum of the squares of the rms values of the harmonics of line currents and neutral current in Example 2.2.

Finally, Figure A.8 shows the block diagrams that determine the power terms; blocks 3-phase sequence analyzer of SimPowerSystem for the positive sequence components of the voltage and current are used.

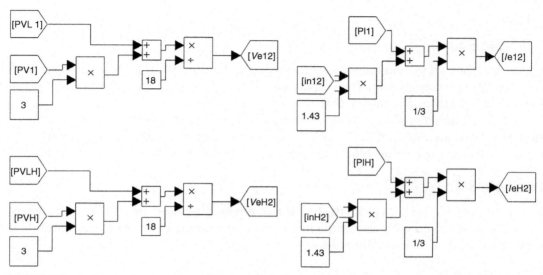

FIGURE A.7 Block diagrams for determining effective values of voltage and current by the Std. IEEE 1459.

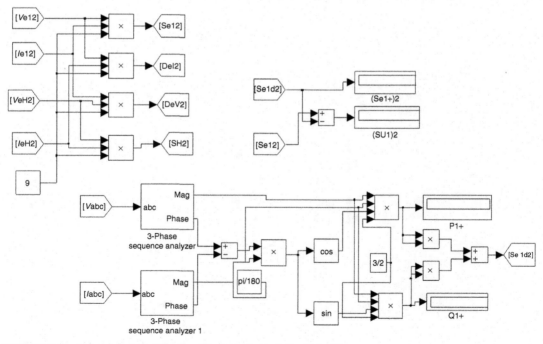

FIGURE A.8 Block diagrams that determine the power terms in Example 2.2.

Chapter 3

Example 3.4a

Figure A.9. shows a system consisting of a three phase power source and a nonlinear load for simulating Example 3.4a. The display of the current source after compensation has been simulated through a switch that switches at 80 ms.

The simulation model of an ac–ac converter, and block diagrams for obtaining compensation currents are included in Chapter 3, Figures 3.11 and 3.12. The block diagram of Figure 3.12 uses the dot product block of Simulink to determine the different power variables.

Example 3.4b

Figure 3.16 of Chapter 3 includes the block diagram for obtaining the currents of constant power compensation with zero-sequence current null. Transfer Fcn block of Simulink is used to simulate a low-pass filter of second order.

Example 3.5

The power system of Example 3.5 is the same as that used in Example 3.4. Figure 3.21 of Chapter 3 shows the block diagram for generating compensation currents.

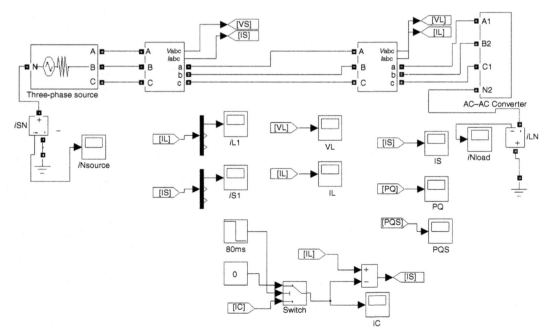

FIGURE A.9 System consisting of a three-phase power source and a nonlinear load for simulating the Example 3.4a. The display of the current source after compensation has been simulated through a switch that switches 80 ms.

Example 3.6

Figure 3.23 of Chapter 3 shows the block diagram for obtaining the compensation currents. Transfer Fcn block of Simulink is used to simulate a low-pass filter of second order.

Example 3.8

The power circuit of Example 3.8 uses blocks three-phase source and universal bridge of SimPowerSystems, Figure A.10. The voltages at the PCC before and after compensation are simulated using a switch of Simulink that switches at $t = 0.08$ s.

Figure A.11 shows block diagrams of the subsystems that perform transformation of phases coordinates to $0\alpha\beta$ coordinates (left figure) and $0\alpha\beta$ coordinates to phases coordinates (right figure). The Simulink blocks are used: matrix concatenate input signals to create a contiguous output signals, and product to perform the matrix multiplications.

Figure A.12 presents the calculation of the instantaneous power variables indicated in Figure 3.34, Chapter 3; The Simulink block and transfer Fcn were reused to simulate a low-pass filter of second order to determine the oscillatory component of the instantaneous power.

Simulink blocks, permute dimensions, and matrix concatenate, provide different current components as shown in Figure 3.34 of Chapter 3 to get the compensation voltage, Figure A.13.

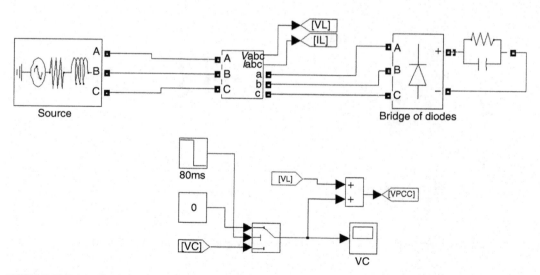

FIGURE A.10 Power circuit for the simulation of compensation through a series active filter of load-type rectifier with capacitive branch on the dc side.

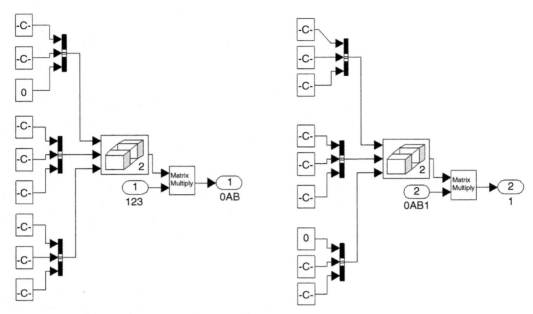

FIGURE A.11 Block diagrams of the subsystems that perform transformation of phases coordinates to $0\alpha\beta$ coordinates (to the left) and $0\alpha\beta$ coordinates to phases coordinates (to the right).

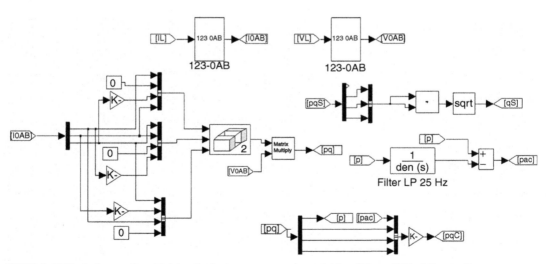

FIGURE A.12 Block diagram for obtaining the instantaneous power variables of Figure 3.34 of Chapter 3.

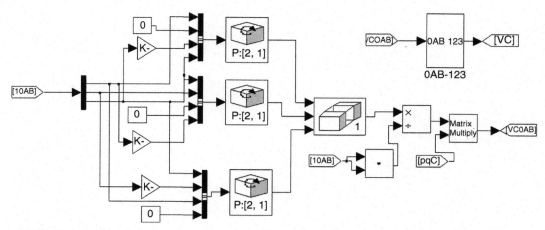

FIGURE A.13 Block diagram for obtaining the current components as shown in Figure 3.34 of Chapter 3 to determine the compensation voltage.

Chapter 4

Example 4.1

Figure A.14 shows the simulation model of a system with compensation through an ideal shunt active power filter consisting of three controlled current source of SimPowerSystems applied at $t = 0.18$ s. The nonlinear load consists of three single-phase rectifiers star-connected as shown in Figure 4.2 of Chapter 4.

Figure A.15 shows how to obtain reference currents; the real power is determined from a Butterworth low-pass filter of fourth order from analog filter design block of signal processing blockset.

Example 4.2

Figures 4.7 and 4.8 of Chapter 4 present the simulation models of Example 4.2.

Example 4.3

Figure A.16 shows the compensated power system of Example 4.3. Thyristor and the universal bridge blocks of SimPowerSystems are used to model the load, and inverter as static compensator, respectively.

Data-type conversion, logical operator, and relay blocks of Simulink, and analog filter design of signal processing blockset as in Example 4.1, are used to model the control circuit of static compensator, Figure A.17.

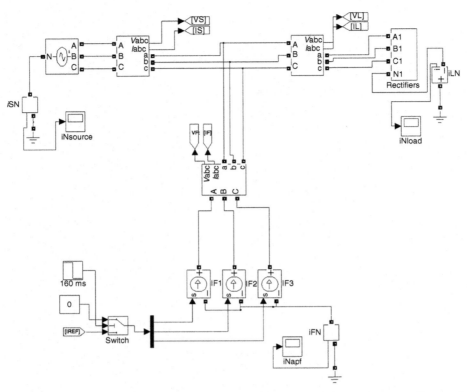

FIGURE A.14 Simulation model of a system with compensation through an ideal shunt active power filter consists of three current source applied at $t = 0.18$ s.

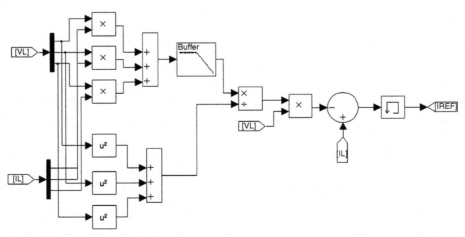

FIGURE A.15 Block diagram for obtaining the reference currents; the real power is determined from a fourth order Butterworth low-pass filter.

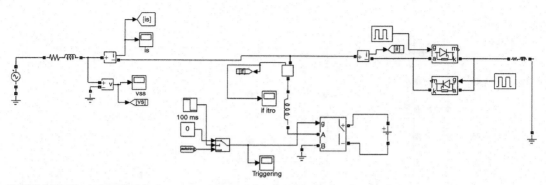

FIGURE A.16 Simulation model of compensated power system of Example 4.3.

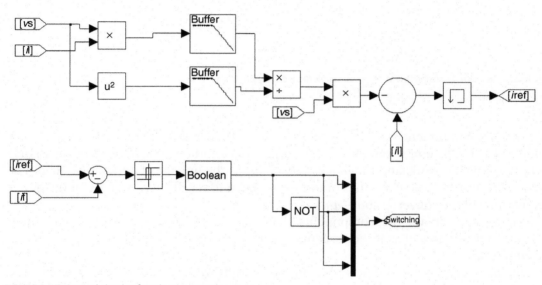

FIGURE A.17 Control circuit of static compensator.

Example 4.4

Figures 4.15 and 4.16 of Chapter 4 show the simulation models of the main and control circuits of the SPWM full-bridge three-phase inverter.

Example 4.5

This is a simulation model with a high level of detail, which involved the use of a large number of subsystems. A description of each subsystem is given here.

Figure A.18 shows the general arrangement of the MATLAB–Simulink file used for Example 4.5. The APF block contains the power circuit of the active conditioner, including

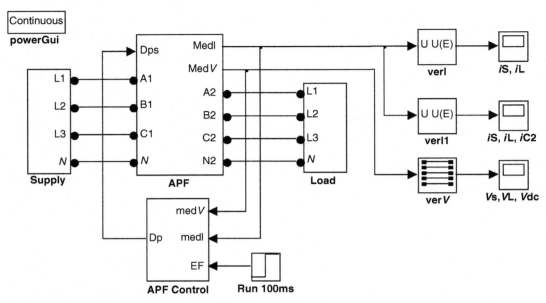

FIGURE A.18 General arrangement of the Simulink file for Example 4.5.

the passive components and the measuring elements (signals grouped in the buses *medV*, *medI*). The block *Supply* contains the voltage sources and the equivalent line impedance of the supply network; and the block *Load* contains the unbalanced nonlinear load. The block *APF Control* calculates the control references and sets the gate signals for the power transistors of the converter (bus *Dp*).

Figure A.19 shows the components that model the supply network, with unbalanced voltages, as well as its equivalent impedances. Figure A.20 shows the load used in Example 4.5.

Figure A.21 presents the power circuit of the APF, with the passive components and coupling transformers, as well as the measuring elements. The simulation model of Example 4.5 has been implemented to model approximately the experimental prototype described in Section 4.6 and Appendix II. The coupling transformers T_P have a turn ratio of 1:2 to adapt the voltage levels between the dc and ac sides of the converter. So, the reference voltage of each dc side capacitor is set to 250 V, and the entire set works with low voltage levels. The model of the transformers includes the short circuit and magnetizing impedances, as well as the L_P inductance and its dc resistance. On the other side, the measurements are grouped in the buses *medI* and *medV*. The bus *medI* contains the supply currents i_{Sabcn}, load currents i_{Labcn} and compensation currents before, i_{CPabc}, and after, i_{C2abc}, the coupling transformers T_P. The bus *medV* contains the supply voltages, v_{Sabc}, and those of the dc capacitors, v_{dc1} and v_{dc2}. The incoming signal bus *Dps* contains the logic switching signals for the six power transistors of the converter.

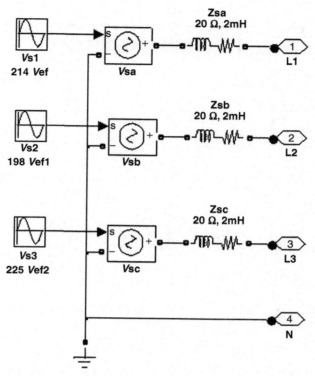

FIGURE A.19 Power supply with unbalance and source impedance.

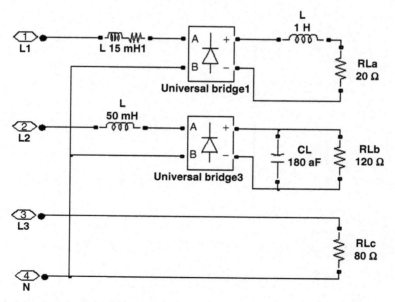

FIGURE A.20 Unbalanced nonlinear load for Example 4.5.

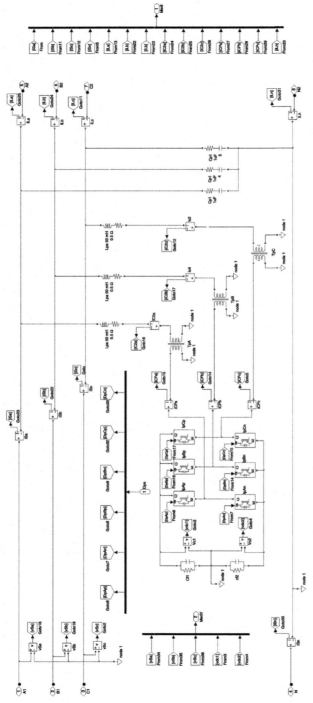

FIGURE A.21 Power circuit of the shunt APF and passive components.

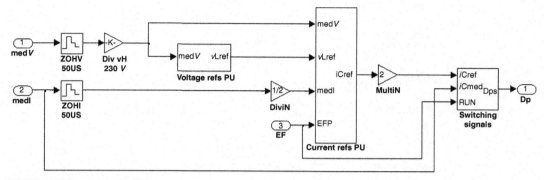

FIGURE A.22 General control block.

Figure A.22 shows the structure of the general control block of the APF. A sample and hold, ZOH, of 50 μs is applied to the input signals to model the effect of the DSP board used in the calculation of this control. Afterwards, the signals are divided by their nominal values to set the later calculations in per unit values. Block *voltage refs PU* calculates the balanced sinusoidal voltage references vLref that are used in Example 4.5b. Block *currents refs PU* calculates the compensating current references, iCref, of the converter; including the regulation of the dc side voltages. Finally, the block *switching signals* sets the gate logic signals for the power transistors, with the periodical sampling, PS, current control described in Section 4.3.2.

Figure A.23 shows the structure of the *voltage refs PU* block. It calculates the balanced fundamental and amplitude regulated component of the supply voltage measurements. In the first stage the signal $v+$, that contains the balanced fundamental component of the voltages, is obtained with the set of phase shift filters. The rest of components are filtered in a second stage with a band pass filter, and its amplitude is regulated to the nominal value. In the third stage the three-phase signal $v1 + PU$ is built, also with the use of a phase shift filter and with null instantaneous zero-sequence component.

Figure A.24 shows the modeling of the trigger circuit of the converter, that in the experimental setup is made with an external circuit that compares directly the measured compensating current, iCmed, with the corresponding reference, iCref, provided by the DSP board; with a previous sample and hold, ZOH, to model the sampling and the corresponding delay, $1/z$, of the digital board. The initial scale factor 2 is due to the turns

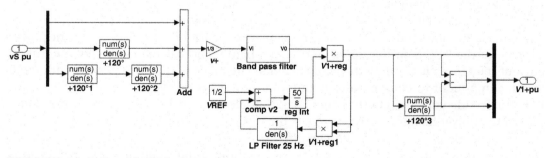

FIGURE A.23 Calculation of the load voltage reference.

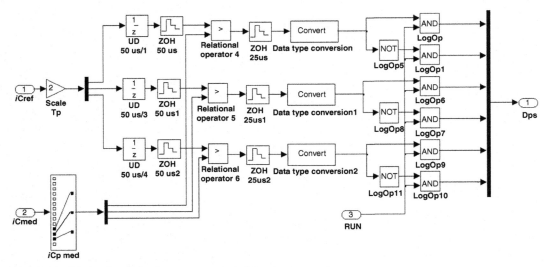

FIGURE A.24 Trigger signals for the current control of the shunt converter.

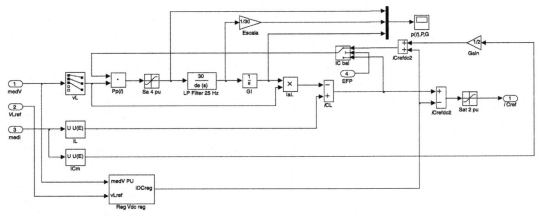

FIGURE A.25 Calculation of the compensating current references for Example 4.5a.

ratio of the coupling transformers T_P. The second ZOH models the PS sampling of the external circuit, and the logic gates set the complementary trigger signals for each branch of the converter, as well as the validation through the *RUN* signal to run/stop the APF.

Figure A.25 shows the control block to calculate the current references *i*Cref. The power balance of the APF is in the upper part of the figure, and calculates the compensating current references *i*CL due to the load currents. The block v_{dcreg} calculates the additional active component to compensate the internal power losses of the conditioner to regulate the dc-side voltages. On the other part, the selector *i*Cbal that appears in the upper part of the power balance allows us to use the measured value of i_C in this balance, when the APF is working. The values of i_C can differ substantially from their reference over relatively long time periods, and this way the estimation of the internal power balanced is

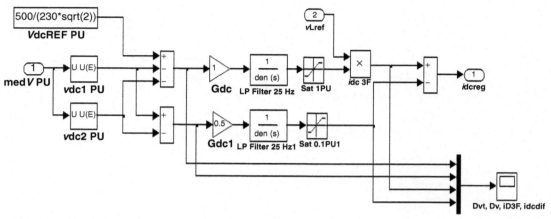

FIGURE A.26 Regulation of the dc side voltages.

enhanced, reducing possible power fluctuations. Thus, the i_{dcreg} component is excluded from this balance, assumed that it has a low dependence on the active/desired component of the load currents. Finally, the saturation block Sat2*PU* limits the compensating currents at the output of the converter, to avoid instantaneous values of these currents higher than those assigned to the physical devices.

Figure A.26 shows the implemented control for the regulation of the dc side voltages. The three-phase signal I_{dc3F} has the same waveform as the voltage reference vLref used, and an amplitude proportional to the difference between the total voltage of both capacitors and its reference V_{dcREF}. A low-pass filter is applied to this signal to allow the oscillating component of the instantaneous power that is being compensated to be provided by the APF, considering that the dc side capacitances had been dimensioned for this task. A small dc component, i_{dcdif}, proportional to the difference between both capacitor voltages, is added to this signal to balance them. When zero-sequence currents are to be compensated, it produces differences between both voltages. The low-pass filter selects the dc component to be compensated, which has an accumulative effect, and so as to maintain this difference in acceptable values.

Finally, Figure A.27 shows the control block *Currents refs PU*, which is used to calculate the current references for the case 4.5b, where the compensated currents are aimed to be proportional to the balanced fundamental component of the voltages. The difference with figure A1.8 is in the construction of the signal i_{aL} with the voltage reference v_{Lref}.

Chapter 5

Examples 5.1, 5.2, and 5.3

In Examples 5.1, 5.2, and 5.3 an active filter with a series connection is used for mitigating the harmonics generated by a HVS-type load. The load consists of a non-controlled three-phase rectifier with a capacitor and a resistor in parallel at the dc side. The system is fed by a three-phase sinusoidal source with a source impedance formed by a resistor in

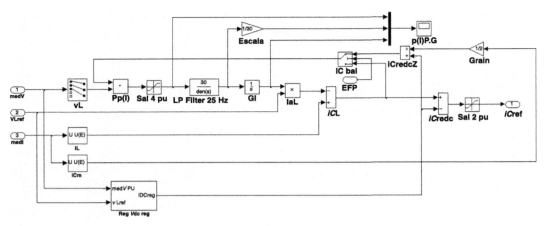

FIGURE A.27 Calculation of the compensating current references for Example 4.5b.

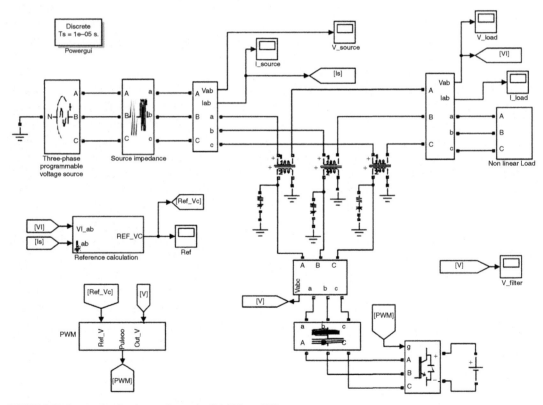

FIGURE A.28 Power circuit scheme, Examples 5.1, 5.2, and 5.3.

series with an inductance. The active filter is connected via three single-phase transformers of 1:1 turn ratio, with a ripple LC filter. The inverter is configured using a three-phase IGBT bridge. Figure A.28 shows the scheme of the power circuit.

The nonlinear load is modeled with the block "Universal bridge" of the library SymPowerSystem from Simulink. Figure A.29 shows the Simulink scheme used.

Furthermore, the block labeled "PWM" generates trigger pulses of the IGBTs from the comparison reference signal and the output signal of the inverter. This includes a D flip-flop, which limits the switching frequency below the clock frequency. Figure A.30 shows the scheme in Simulink in this block.

The "Reference calculation" block has as output signal the reference voltage that the active filter must generate. Their scheme in Simulink is shown in Figure A.31. The fundamental component of the input signal is obtained by the block labeled "Fundamental." This block has been developed using the scheme shown in Figure A.32. The K_v and K values are adjusted depending on the selected control strategy, by detecting the current source, by detection of the load voltage and hybrid strategy.

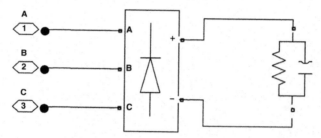

FIGURE A.29 HVS load. Noncontrolled three-phase rectifier with capacitor and resistor connected in parallel at the dc side.

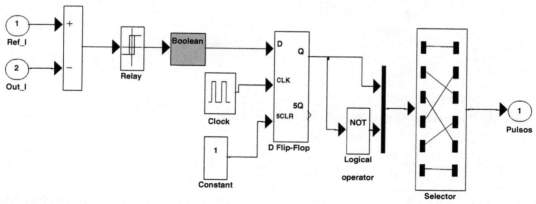

FIGURE A.30 PWM control scheme used in all examples.

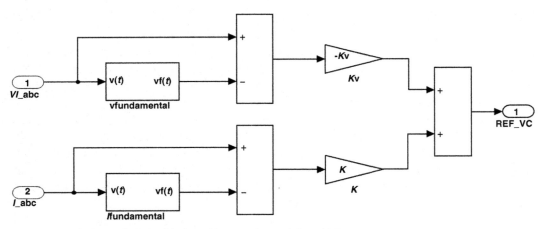

FIGURE A.31 Reference calculation block used in Examples 5.1, 5.2, and 5.3.

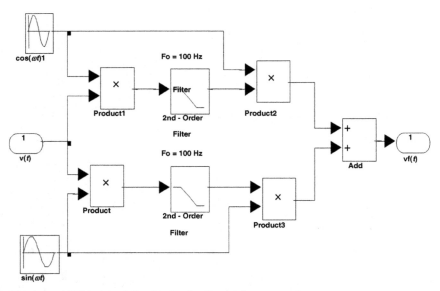

FIGURE A.32 "Fundamental" block to determine the fundamental component.

Example 5.4

In Example 5.4 a case study is analyzed from its state model for the three compensation strategies applied to the series active filter, which compensates a HVS type load. For simulation, a file ".m" from MATLAB is generated with the following script:

```
%Example 5.4.

%Assignment of k and kv value

k=20;

kv=0.95;


%State model

A=[-4089 3446;66.5 -66.5]

B1=[-357;0]

B2=[357 3446;0 -66.5]


% Output signal: source current

C=[1 0]

D=[0]


% System without compensating

Model=ss(A,B2,C,D)

% Control by detection of the source current

K=[k 0]

Ai=A+B1*K


Model_ish=ss(Ai,B2,C,D)


% Control by detection of the load voltage

K1=[-kv*9.65 kv*9.65]

Av=A+B1*K1

K2=[0 kv*9.65]

Bv=B2+B1*K2


Model_vl=ss(Av,Bv,C,D)


% Hybrid control

K11=[k-kv*9.65 kv*9.65]
```

```
Ah=A+B1*K11

K22=[0 kv*9.65]

Bh=B2+B1*K22


Model_h=ss(Ah,Bh,C,D)


%Analysis


ltiview(Model,'r',Model_ish,'b',Model_vl,'g',Model_h,'m')
```

Chapter 6

Example 6.1

In Example 6.1, a system with two nonlinear loads connected to a busbar is simulated. Two loads are connected to a sinusoidal voltage source by means of a line modeled by a RL branch. Subsequently, a passive filter is connected to eliminate the 5th harmonic and to compensate load reactive power, of the load number two. This case allowed illustration of the drawbacks of passive filters connected in parallel to the power system. The circuit scheme in Simulink is shown in Figure A.33.

Examples 6.2, 6.3, and 6.4

Examples 6.2, 6.3, and 6.4 use the same power circuit configuration. This is shown in Figure A.34. Here, a passive filter tuned to the 5th order harmonic is included.

An HCS-type load is used, which is composed of a three-phase noncontrolled rectifier with an inductance and resistor connected in series at the dc side. Figure A.35 shows the scheme of the model used.

The other blocks included in the scheme in Figure A.34 are the same as Examples 5.1, 5.2 and 5.3, according to the compensation strategy applied to the active filter.

Example 6.4

In this Example the hybrid compensator is analyzed for the three strategies. Nonlinear load is HCS type. Analogously to Example 5.4 an ".m" file is written to analyze the results. The script is reproduced below:

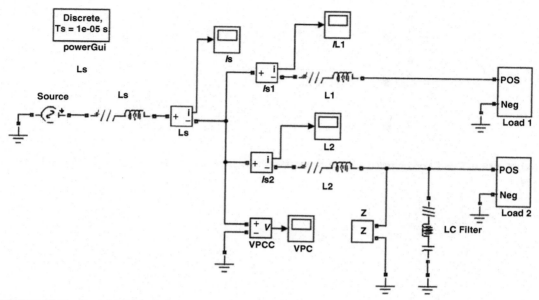

FIGURE A.33 Circuit model, Example 6.1.

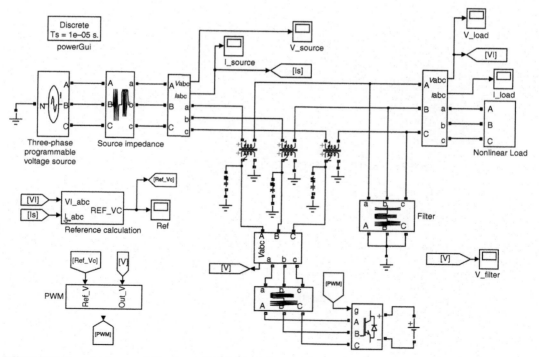

FIGURE A.34 Power system model for Examples 6.2, 6.3, 6.4, and 6.5.

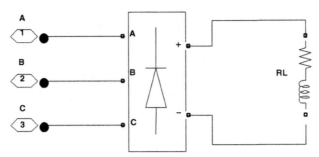

FIGURE A.35 HCS load. Noncontrolled three-phase rectifier with inductance and resistor connected in series at the dc side.

```
% Example 6.4


% Constants

k=40;

kv=-1;


% Space state model of SAPPF filter

% Parameter definition


Rs=1.3; Ls=2.34e-3; L5=13.5e-3;C5=30e-6; R5=2.1; L7=6.75e-3; C7=30e-6;

R7=1.05; RL=14,3; Ll=152e-3;


A =[-(Rs+RL)/Ls RL/Ls RL/Ls RL/Ls 0 0;RL/L5 -(RL+R5)/L5 -RL/L5 -RL/L5 -

1/L5 0;RL/L7 -RL/L7 -(RL+R7)/L7 -RL/L7 0 -1/(L7);

    RL/Ll -RL/Ll -RL/Ll -RL/Ll 0 0;0 1/C5 0 0 0 0; 0 0 1/C7 0 0 0]
```

```
B2=[1/Ls RL/Ls;0 -RL/L5;0 -RL/L7;0 -RL/Ll; 0 0;0 0]

Bl=[-1/Ls; 0; 0; 0; 0; 0]

C=[1 0 0 0 0 0]

D=[0]

%Model without compensating

Model=ss(A,B2,C,D)


% Control by detection of the source current

K=[k 0 0 0 0 0]

Ai=A+B1*K

Model_ish=ss(Ai,B2,C,D)

% Control by detection of the load voltage

K1=[-kv*RL kv*RL kv*RL kv*RL 0 0]

Av=A+B1*K1

K2=[0 kv*RL]

Bv=B2+B1*K2

Model_vl=ss(Av,Bv,C,D)


ltiview(Model_vl)

% Hybrid model

K11=[k-kv*RL kv*RL kv*RL kv*RL 0 0]

Ah=A+B1*K11

K22=[0 kv*RL]

Bh=B2+B1*K22


Model_h=ss(Ah,Bh,C,D)

% Analysis

ltiview(Model,'r',Model_ish,'b',Model_vl,'g',Model_h,'m')
```

Chapter 7

Example 7.1

Figure A.36 shows the general arrangement of the MATLAB–Simulink file used for Example 7.1. The *UPQC* block contains the power circuit of the active conditioner, including the passive components and the measuring elements (signals grouped in the buses *medV*, *medI*). The block *Supply* has the voltage sources and the equivalent line impedance of the supply network; and the block *Load120* contains the unbalanced nonlinear load. The block *Control UPQC* calculates the control references and sets the gate signals for the power transistors of the converter (bus *Dps*).

Figure A.37 shows the components that model the supply network, with unbalanced and harmonic voltages, as well as its equivalent impedances. Figure A.38 shows the load used in Example 7.1, including the ideal switch to produce the load change in the last stage of the simulation.

Figure A.39 presents the power circuit of the UPQC, with the passive components and coupling transformers, as well as the measuring elements. The simulation model of Example 7.1 has been implemented to model approximately the experimental prototype described in Section 7.3 and Appendix II. The shunt coupling transformers T_P have a turn ratio of 1:2 to adapt the voltage levels between the dc and ac sides of the converter. So, the reference voltage of each dc side capacitor is set to 250 V, and the entire set works with low voltage levels. The model of the transformers includes the short circuit and magnetizing

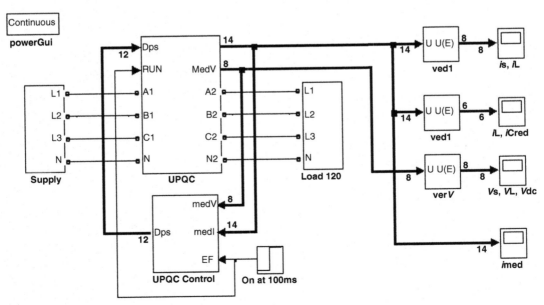

FIGURE A.36 General arrangement of the Simulink file for Example 7.1.

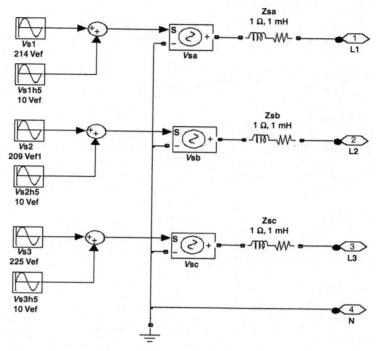

FIGURE A.37 Power supply with unbalance, voltage harmonics, and source impedance.

impedances. The series coupling transformers T_S have a turn ratio of 1:1. The $R_S C_S$ series branch explained in Section 7.2.3 is included as the snubber circuit of the bypass ideal switches below the series transformers. Inductance L_{S2} corresponds to the leakage inductance of these transformers. On the other side, the measurements are grouped in the buses *medI* and *medV*. The bus *medI* contains the supply currents i_{Sabcn}, load currents i_{Labcn} and compensation currents before, i_{CPabc}, and after, i_{C2abc}, the coupling transformers T_P. The bus *medV* contains the supply voltages, v_{Sabc}, load voltages, v_{Labc}, and those of the DC capacitors, v_{dc1} and v_{dc2}. The incoming signal bus *Dps* contains the logic switching signals for the twelve power transistors of the converter.

Figure A.40 shows the structure of the general control block of the UPQC. A sample and hold, ZOH, of 50 μs is applied to the input signals to model the effect of the DSP board used in the calculation of this control. Afterwards, the signals are divided by their nominal values to set the later calculations in per unit values. Block *voltage refs PU* calculates the references associated with the control of the series converter, like the load voltage

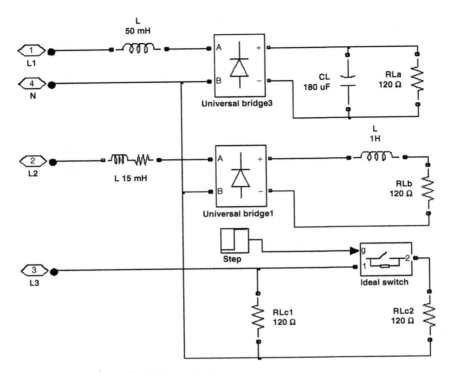

FIGURE A.38 Unbalanced nonlinear load for Example 7.1.

reference vLref and the compensating voltages for the series filter, vCref. These signals are used in the block *Gen_As*, along with the values of the dc side voltages, to calculate the pulse widths, *As*, for the power transistors of the series converter in each sampling period. Block *currents refs PU* calculates the compensating current references, *i*Cref, of the shunt converter; including the regulation of the dc side voltages. It also calculates the corrective signal *REis* (proportional to the tracking error of the supply currents i_S), to be included in the control of the series converter. Finally, the block *switching signals* sets the gate logic signals for the power transistors, modeling the control methods implemented in each converter, as described in Section 7.2.1.

Figure A.41 shows the structure of the *voltage refs PU* block. It calculates the balanced fundamental and amplitude regulated component of the supply voltages v_{Sabc}, to be used as reference for the load voltage vLref (This control block is the same as that used in Example 4.5, (see Figure A.23). With this signal and the prefiltered value of v_S sets the main component of the series compensating voltages, vCref; afterwards, the corrective signal *REis* is added. The saturation block, Sat ± 0.3 pu, limits these compensating voltages to adequate values, according to the turn ratio of the series transformers, the assigned values

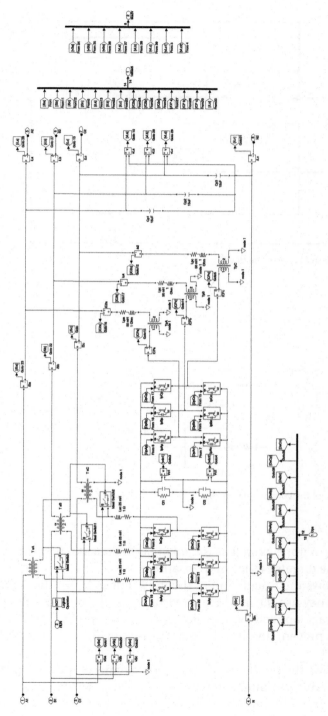

FIGURE A.39 Power circuit of the UPQC and passive components.

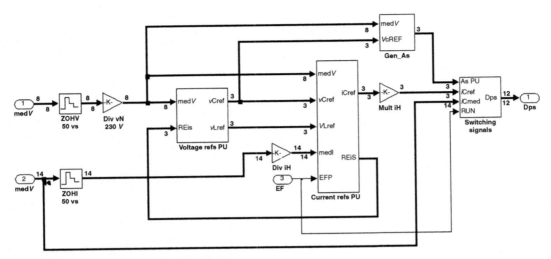

FIGURE A.40 General control block of the UPQC.

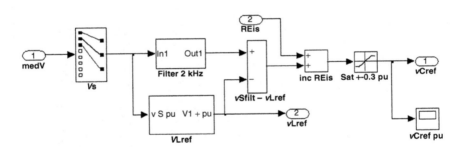

FIGURE A.41 Calculation of the voltage references.

for the dc-side capacitors, as well as the taking into consideration the dimensions of the equipment and the targets of the compensation.

Figure A.42 shows the calculation of the duty ratios, As, for the three branches of the series converter. They use the last measured values of the dc side voltages to set accurate values of the average applied series voltages in each sampling period. Some saturation blocks are used to prevent divisions by zero (Saturation 3), and to limit the output signals As between 0 and 1.

Figure A.43 shows the modeling of the trigger circuits of the converters. The As signals for the series converter are compared with a triangular wave that defines the commutation frequency at 20 kHz, with a previous sample and hold, ZOH, to model the sampling and

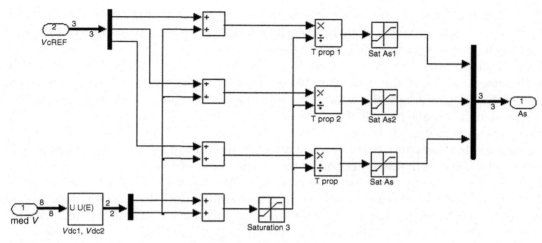

FIGURE A.42 Calculation of the PWM duty cycles, *As*, for the series converter.

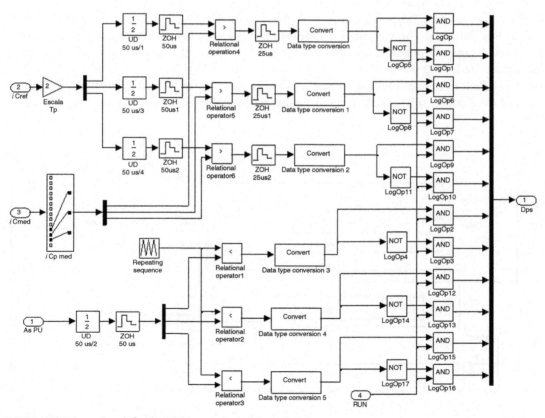

FIGURE A.43 Trigger signals for the UPQC converters of Example 7.1.

the corresponding delay, $1/z$, of the DSP board. The trigger control of the shunt converter is made with an external circuit in the experimental setup, that compares directly the measured compensating currents, i_{Cmed}, with the corresponding references, $iCref$, provided by the DSP board; with the previous sample and hold and corresponding delay. The initial scale factor 2 is due to the turns ratio of the coupling transformers T_P. The second ZOH models the PS sampling of the external circuit, and the logic gates set the complementary trigger signals for each branch of the converter, as well as the validation through the *RUN* signal to run/stop the APF.

Figure A.44 shows the control block to calculate the current references. The power balance of the UPQC is in the upper part of the figure, and calculates the compensating current references iCL due to the load currents and supply voltage distortions. The block $v$dcreg calculates the additional active component to compensate the internal power losses of the conditioner to regulate the dc side voltages. This block $v$dcreg is the same as that implemented for the shunt APF (See Example 4.5, Figure A.26). The *Dev* block calculates the corrective component for damping the load voltage deviations; and the *Dis* block calculates the corrective component for the supply current deviations, that is sent to the control of the series converter. On the other part, the selector iCbal that appears in the upper part of the power balance allows to use the measured value of i_C in this balance, when the APF is working. The values of i_C can differ substantially from their references during relatively long time periods, and this way the estimation of the internal power balanced is enhanced, reducing possible power fluctuations. Thus, the idcreg component is excluded from this balance, assumed that it has a low dependence with the active/desired component of the load currents, and/or a more ideal series compensation. Finally, the saturation

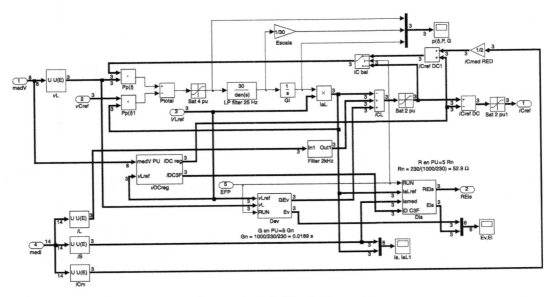

FIGURE A.44 Calculation of the current references. Control of the shunt converter.

block Sat2*PU* limits the compensating currents at the output of the converter, to avoid instantaneous values of these currents higher than those assigned to the physical devices.

Figures A.45 and A.46 show the calculation of the corrective components *GEvL* and *REiS*, as well as their corresponding activation only when the UPQC is working.

Example 7.2

Figure A.47 shows the general arrangement of the MATLAB–Simulink file used for the Example 7.2. The *UPLC* block contains the power circuit of the active conditioner, including the passive components and the measuring elements (signals grouped in the buses *medV, medI*). It has the same power circuit of the UPQC of Example 7.1 (see Figures 7.1 and A.39), and includes the connection of a second supply network between the conditioner and the load, as shown in Figure 7.46. The blocks *Supply1* and *Supply2* have the voltage sources and equivalent line impedances of these networks with the values indicated

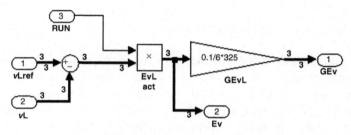

FIGURE A.45 Calculation of the corrective signal **GEvL** for the shunt converter.

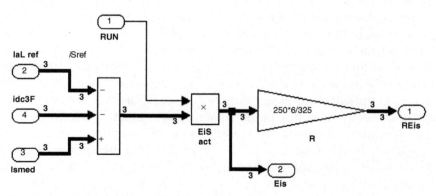

FIGURE A.46 Calculation of the corrective signal **REiS** for the series converter.

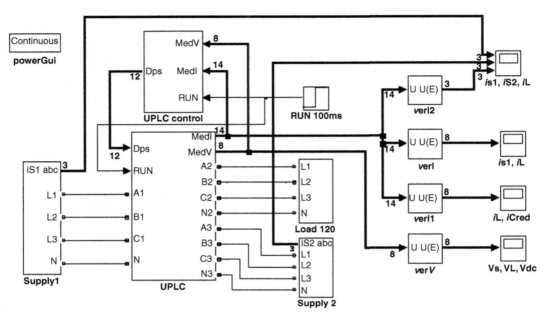

FIGURE A.47 General arrangement of the Simulink file for Example 7.2.

in Example 7.2; and include the output current measurements, i_{S1}, i_{S2}, to see the effects of the UPLC compensation. The block *Load*120 contains the same unbalanced nonlinear load as in Example 7.1 (see Figures 7.1 and A.38) without the ideal switch to produce the load change. The block *UPLC Control* calculates the control references and sets the gate signals for the power transistors of the converter (bus *Dps*).

Figure A.48 shows the structure of the general control block of the UPLC. A sample and hold, ZOH, of 50 μs is applied to the input signals to model the effect of the DSP board used in the calculation of this control. Afterwards, the signals are divided by their nominal values to set the later calculations in per unit values. Block *voltage refs PU* calculates the references associated with the control of the series converter, like the load voltage reference *v*Lref and the compensating voltages for the series filter, *v*Cref. These signals are used in the block *Gen_As*, in the way as in Example 7.1 (see Figures 7.1 and A.42). Block *currents refs PU* calculates the compensating current references, *iC*ref, of the shunt converter; including the regulation of the dc side voltages. Finally, the block *gate signals* sets the gate logic signals for the power transistors, modeling the control methods implemented in each converter, in the same way as in Example 7.1 (see Figures 7.1 and A.43).

Figure A.49 shows the structure of the *Voltage refs PU* block. It calculates the balanced fundamental and amplitude regulated component of the supply voltages v_{Sabc}, to be used as reference for the load voltage *v*Lref. This control block is the same as that used in Example 4.5 (see Figure A.23), but this time including the calculation of the quadrature signal V_q, according to expression (7.36). The block *Fund. Comp.* calculates the fundamental components of the load voltages, *vLf*, to be used in the voltage correcting terms at the

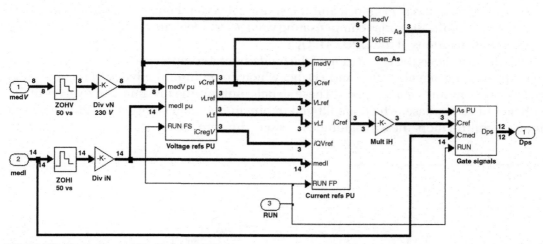

FIGURE A.48 General control block of the UPLC.

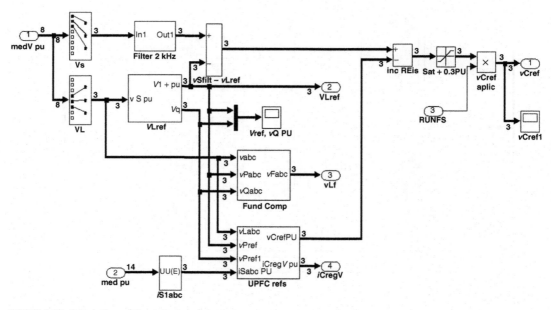

FIGURE A.49 Calculation of the voltage references.

control of the shunt converter. The references for the series compensating voltages, vCref, are built with the difference between the prefiltered value of v_S and the reference for the load voltages, to isolate the load side from the distortions and unbalances of the supply 1 side; as well as the necessary values for the Power Flow Control through the series converter explained in Section 7.4.2 and implemented in the block *UPFC refs*. This block also calculates the current reference, iCregV, for the voltage control in the load side; that is set to the *current refs PU* block and the shunt converter. Finally, the saturation block, Sat ± 0.3 pu, limits the series compensating voltages to adequate values, according to the design considerations of the UPLC/UPQC.

Figure A.50 shows the calculation of the fundamental components of the load side voltages. It uses the signals *v*Pabc and *v*Qabc, which come from the block *v*Lref, to obtain the unitary sine and cosine signals. These signals multiply the instantaneous values of the voltages of each phase, which are filtered to obtain the main component. With a closed integral control loop, these magnitudes are set with zero steady state error and allow to

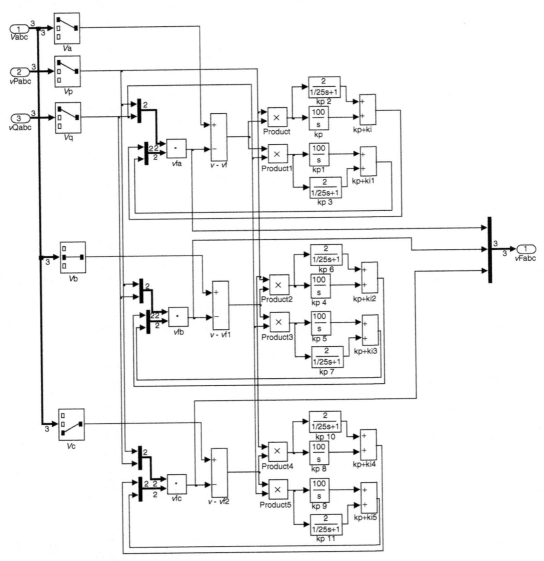

FIGURE A.50 Calculation of the fundamental components of the load voltages.

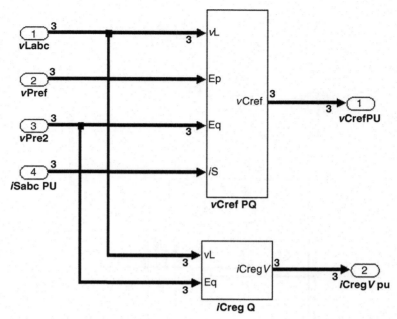

FIGURE A.51 References for the UPFC targets.

build the fundamental component of each phase voltage in approximately two and a half cycles.

Figure A.51 shows the scheme of the *UPFC refs* block. It calculates the references for the series converter in the *vCrefPQ* block, and the references for the shunt converter in the *iCregQ* block.

Figure A.52 shows the calculation of the UPFC references for the series converter, according to the equations (7.37)–(7.44). The step blocks Pref and Qref set the references of the active and reactive powers, to define the current references, *i*REFp and *i*REFq, as well as the voltage signals references *v*Cd and *v*Cq of the power flow controller. The signals E_p, E_q are obtained from the block *v*Lref, and are used to calculate the magnitudes in the d–q frame.

Figure A.53 calculates the additional current reference *i*CregV for the shunt converter. In the first stage it obtains the filtered rms value of the aggregate value of the load voltages, to set the reactive current component according to expression (7.35).

Figure A.54 shows the control block to calculate the current references. It has the same structure as that implemented in Example 7.1, with the addition of the current reference *i*QVref of the UPFC features, and the use of the fundamental components of the load voltages *VLf* to reduce the damping term *GEv* at that frequency in the block *Dev*.

Finally, Figure A.55 shows the calculation of the corrective component *GEv*. This damping term is reduced for the fundamental components of the voltage; and this signal is included in the internal power balance of the conditioner, so it actuates on a net basis only in the transient states.

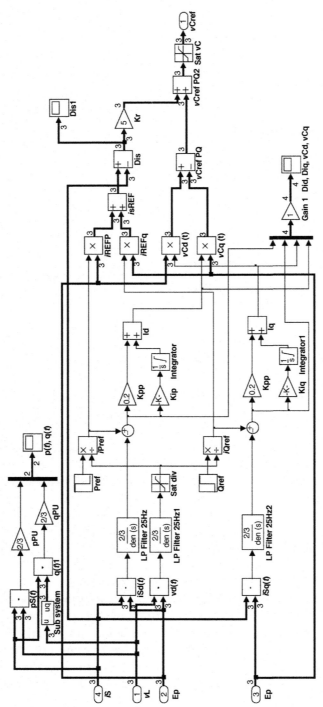

FIGURE A.52 UPFC references for the series converter.

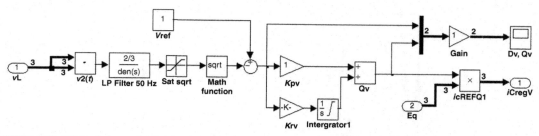

FIGURE A.53 UPFC references for the shunt converter.

Chapter 8

Example 8.1

In Example 8.1, the load flow problem of a power system of four buses is solved. A power source is connected at node 1 and a PV source operating with unity power factor at node 3. Two loads are connected to nodes 2 and 4.

To solve the load flow problem a tool included specifically in the SimPowerSystems library from MATLAB–Simulink has been used. To use this tool the Simulink model of the network shown in Figure A.56 must be built. Each of the nodes must be defined by type. Thus, node 1 will be the swing node, nodes 2 and 4 are PQ type and node 3 is PV type.

Example 8.2

A distributed generation system is connected to a distribution network via a dc/ac converter. The instantaneous unity power factor control strategy is applied to the converter. Its performance is verified in two different situations voltage: sinusoidal, balanced voltage system and nonsinusoidal, unbalanced network voltage. To verify the operation of this strategy, the system is simulated in MATLAB–Simulink. Figure A.57 shows the circuit model.

The distribution grid is modeled by means of a programmable three-phase source of infinite power. This source allows inclusion of imbalances and harmonics in the voltage.

The "PWM" block generates trigger pulses for the IGBT bridge from the reference signal and the output of the inverter dc/ac. This block is the same as the one used in Example 5.2.

The reference signal is calculated by means of the "Reference" block. Since the active power is injected into the grid and the voltage vector at the point of connection, VDG, the reference current signal to be generated by the inverter is determined. Figure A.58 shows control scheme in Simulink. The "u^2" block calculates the instantaneous norm.

Example 8.3

The power circuit of Example 8.3 is the same as in Example 8.2. The difference is in the control strategy converter. In this case, positive-sequence control strategy is applied. The "Reference" block is modified according to the scheme shown in Figure A.59.

The "Direct" block determines the positive-sequence component of the network voltage. Its scheme is shown in Figure A.60; a and a^2 operator are implemented with all-pass filters defined in Simulink using its transfer function.

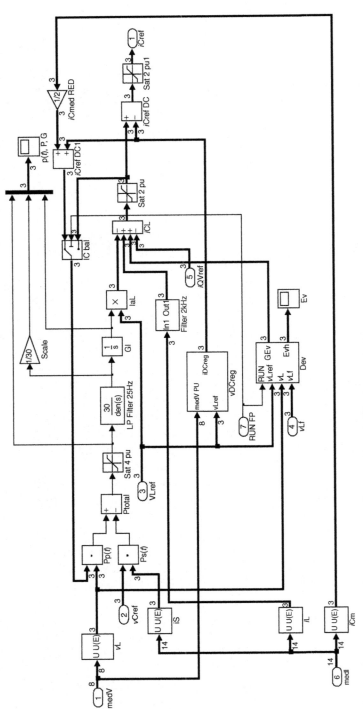

FIGURE A.54 Calculation of the UPLC current references. Control of the shunt converter.

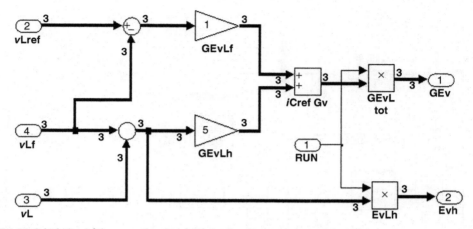

FIGURE A.55 Calculation of the corrective signal *GEv* for the shunt converter of the UPLC.

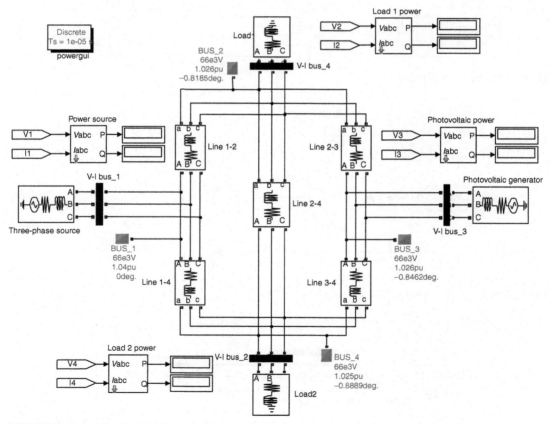

FIGURE A.56 Power circuit model, Example 8.1.

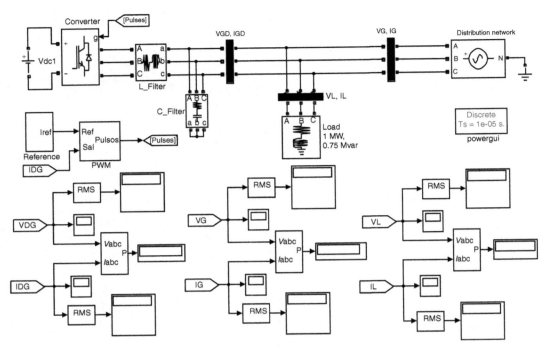

FIGURE A.57 Power circuit model, Example 8.2.

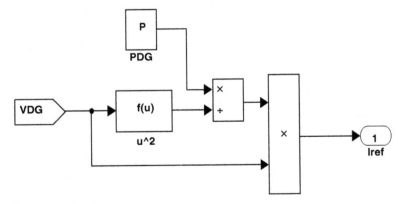

FIGURE A.58 "Reference" block scheme, Example 8.2.

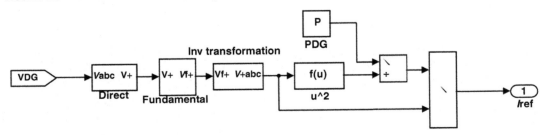

FIGURE A.59 "Reference" block scheme, Example 8.3.

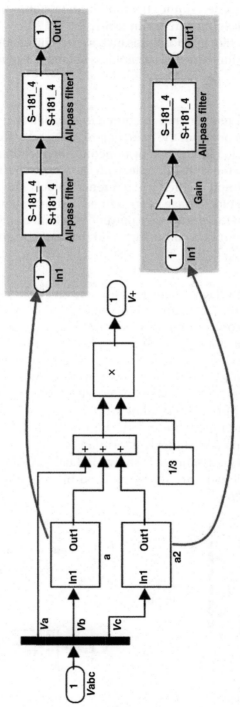

FIGURE A.60 "Direct" block, which determine the direct-sequence component, Example 8.3.

The "Fundamental" block determines the fundamental component of the input signal. Its scheme is the same as described in Example 5.1.

The "Inv Transformation" block performs the inverse transformation, whereby the direct-sequence voltage vector of the fundamental harmonic is obtained. Figure A.61 shows its scheme.

Example 8.4

In Example 8.4 the distributed generation source injects active power and compensates the harmonics and reactive power of the load. Figure A.62 shows the power circuit scheme in Simulink. The nonlinear load is formed by noncontrolled rectifier with a resistor and an inductance connected in series at the dc side, with the same scheme as the Example 6.1.

The control strategy is implemented in "Reference" block. The scheme of this block is shown in Figure A.63. Used blocks have been defined in the previous Examples, except the "Power" block, which calculates the average power of the load from the product of the voltage vector at the point of common connection and the load current vector.

Example 8.5

In Example 8.5, a series interface connected between the utility grid and a microgrid is developed. The interface consists of an inverter connected in series with the system. The objective of the proposed control strategy is to avoid transferring voltage unbalances and current harmonics from the main grid to microgrid or from microgrid to main grid. Thus, the inverter must generate a compensation voltage such that at the network side, voltage unbalances are not transmitted to the microgrid and vice versa. Furthermore, the generated harmonic currents should not be transferred between two networks. Figure A.64 shows the power circuit in MATLAB–Simulink.

For other simulation examples presented, the only block having something new is called "Reference Calculation". This block determines the reference signal to achieve the proposed control objective. The scheme is shown in Figure A.65. In this scheme, "Direct" and "Inv Transformation" blocks are the same used in Example 8.3. Figure A.66 shows "0-Component" block scheme.

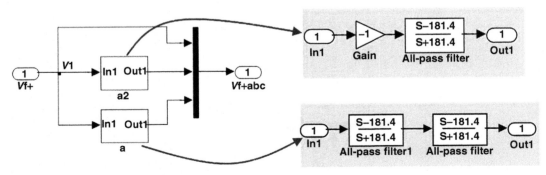

FIGURE A.61 "Inv Transformation" block. It calculates the Fortescue's inverse transformation.

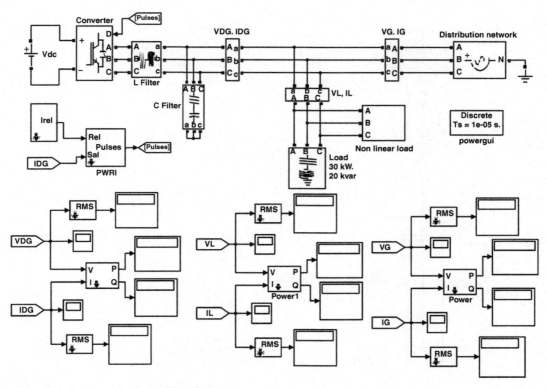

FIGURE A.62 Power circuit scheme, Example 8.4.

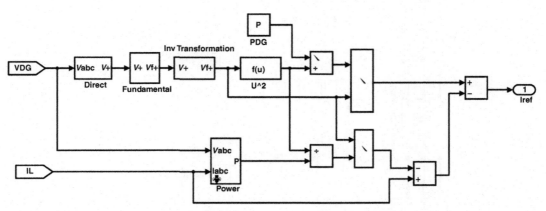

FIGURE A.63 "Reference" block scheme, Example 8.4.

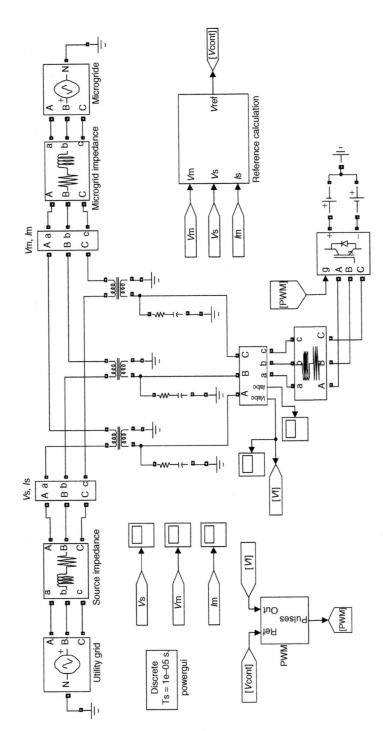

FIGURE A.64 Power circuit scheme, Example 8.5.

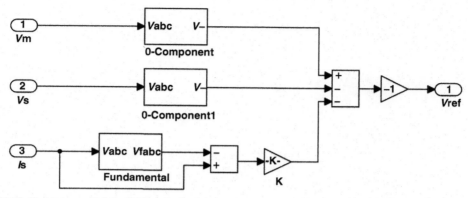

FIGURE A.65 "Reference calculation" block, Example 8.5.

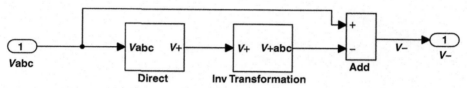

FIGURE A.66 "0-Component" block scheme, Example 8.5.

Appendix
Experimental Implementations

This appendix describes the laboratory setups used to obtain the experimental cases included in this book. Two different platforms have been considered to apply for the various practical cases. The first of these has been used for the development of the series and hybrid examples of Chapters 5 and 6, which has a specific design for these applications. The second one is a general purpose platform that has been selected for the shunt and series-shunt APF cases of Chapters 4 and 7, and is also used for another active power filters configurations and developments of the laboratory. Both platforms share many designs of the individual components and the following exposition will take advantage of these similarities.

B.1 Experimental Platform for Series and Hybrid APF Tests

In Chapter 5 an experimental prototype of series filter is developed and in Chapter 6 a hybrid filter is constructed. In both prototypes the power circuit is very similar. The only difference is due to the parallel connection of a passive filter in the case of the hybrid filter. Also, in both topologies the same control strategies are applied. Therefore, the description of both prototypes will be made jointly because they share the majority of developments.

Figure B.1 shows the general scheme of the laboratory prototype. Each of its components are highlighted. Hereinafter, the main characteristics of the devices used in the prototypes designed for this purpose are presented.

Model 4500-iL from California Instruments has been used as a power supply. It is a programmable three-phase source, which can supply from 4.5 kVA to 10 A in a voltage range from 0 V to 150 V and upto 5 A when the range is selected as 0–300 V. This source allows programming of the voltage signals required for testing Electromagnetic compatibility according to IEC 61000. It has an RS-232 connection that together with the CGUI32 software allows programming (Figure B.2) through a PC. In series with the voltage source is connected an inductance of 2.34 mH with a resistance of 1.3 Ω. These elements allow us to consider the effect of grid impedance on the laboratory prototype.

Two types of nonlinear three-phase loads have been used for different essays. They are based on a noncontrolled three-phase rectifier. The first load at the dc side is connected a high inductance in series with a resistor and the second load a capacitor in parallel with a resistor are connected at its dc side.

A three-phase rectifier configuration is composed by a 36MT60 module from International Rectifier. The maximum permissible current at the dc side of this device is 35 A and the maximum voltage repetitive peak is 600 V. For quick load setup, the rectifier module is

P. Salmerón Revuelta, S.P. Litrán, and J.P. Thomas: Active Power Line Conditioners. http://dx.doi.org/10.1016/B978-0-12-803216-9.00010-9

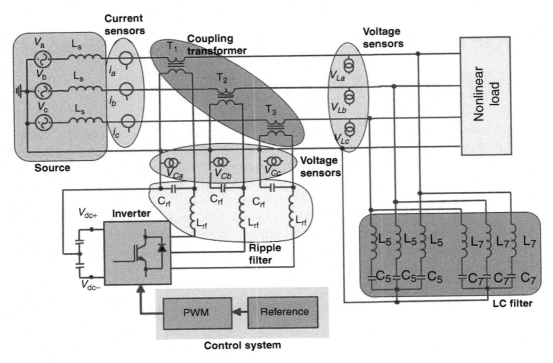

FIGURE B.1 Experimental prototype of hybrid filter.

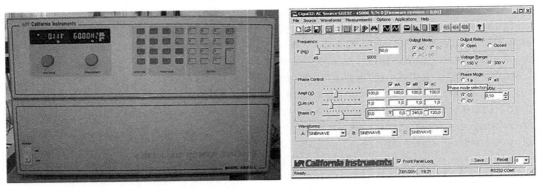

FIGURE B.2 Programmable source, 4500iL model and CGUI32 software from California Instruments.

installed in a box with an inductance of 55 mH and a capacitor of 2200 μF, as it is shown in Figure B.3. A switch allows choice of two load settings, inductance in series or capacitor in parallel. An external variable resistor completes the load.

The second configuration is based on the GBPC2510 rectifier module from International Rectifier. It is a single-phase bridge rectifier which supports a maximum current at the dc side of 25 A and a repetitive peak maximum voltage of 1000 V. Similarly as in the previous configuration was assembled in a box that enables easy setup with an inductance of 55 mH and a capacitor of 2200 μF at the dc side of each rectifier. Figure B.4 shows the load formed by the three single-phase rectifier and the switches that allow selecting energy storage elements at the dc side of each rectifier.

On the other hand, the active filter is composed of three SKM50GB123 IGBT modules from Semikron. Each module is formed by two IGBTs connected with their respective antiparallel diodes as shown in Figure B.5. The main features include the maximum collector-emitter

FIGURE B.3 Three-phase nonlinear load formed by noncontrolled three-phase rectifier.

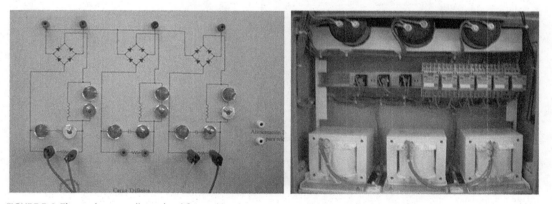

FIGURE B.4 Three-phase nonlinear load formed by three noncontrolled single-phase rectifier.

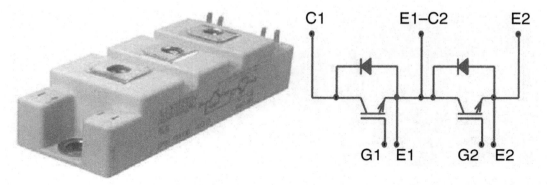

FIGURE B.5 Two IGBTs module, SKM50GB123 model from Semikron.

voltage (V_{CE}) of 1200 V, the collector current (I_C) can be up to 50 A and gate-emitter voltage (V_{GE}) must be between ±15 V. Modules are mounted on a heat sink of anodized aluminum which would have included a fan, allowing cooling of the set.

For gating drivers, three SKHI 22A from Semikron are used. These devices allow control of two IGBTs connected in half-bridge. The setup has additional features such as short-circuit protection, monitoring V_{CE} voltage, transformer isolation between the control circuit and gate, blocking simultaneous firing of the two IGBTs. For gating pulses, an additional circuit is designed to adapt the output 5 V generated by the PWM circuit and input 15 V of the trigger drivers of the transistors. The inverter set is shown in Figure B.6.

At the dc side of the inverter are connected in series two capacitors of 2200 μF and 700 V, allowing, at the midpoint have a fourth conductor for those topologies that require it.

Coupling to the network is performed with three single-phase transformers 1 kVA of 1:1 ratio. Figure B.7 shows an image of the transformer used.

On the other hand, the inverter output has a ripple filter whose function is to remove ripples formed due to high-frequency components, present in the compensation voltage

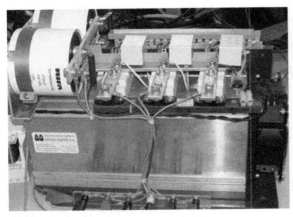
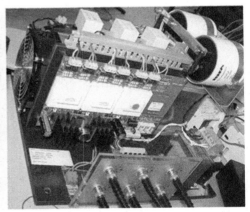

FIGURE B.6 Inverter.

FIGURE B.7 Coupling transformer.

generated by the active filter, Figure B.8. The ripple filter is composed of an inductance of 0.13 mH and capacitor of 50 μF.

In the configuration of hybrid filter, two LC branches tuned at the 5th order harmonic and 7th order harmonic are used. Figure B.9 shows the two passive filters. The branch for the 5th harmonic is formed by an inductance of 13.1 mH and a capacitor of 30 μF. Furthermore, the branch tuned to 7th harmonic is composed of an inductance of 6.5 mH and a capacitor of 30 μF. Two switches allow connection/disconnection of each LC branch.

Different control strategies have been implemented in a modular system that integrates PPC 1005 DS board from dSpace. It has a PowerPC 750GX processor running at 1GHz. Figure B.10 shows this card and Figure B.11 its block diagram. The input-output cards are connected to the DS1005 via a high-speed bus (PHS bus, peripheral high-speed) of 32 bits and a transfer rate of 22 MB/s.

As input, the DS 2004 board is used, Figure B.12. It has 16 differential input channels, each with an A/D with a resolution of 16 bits. The converter performs A/D conversion at 800 ns. Each channel has a buffer of up to 16384 values. To convert the measured data it offers two modes: continuous and events mode. In the first, when the conversion is

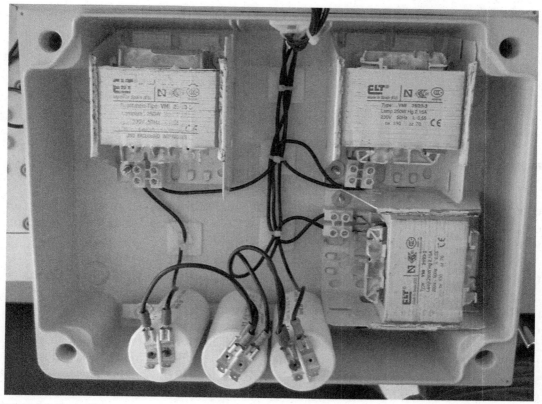

FIGURE B.8 Ripple filter.

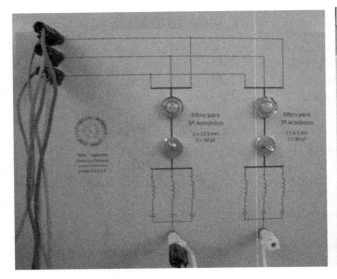

FIGURE B.9 LC filters tuned to 5th and 7th harmonic.

FIGURE B.10 1005 PPC controller board from dSPACE.

complete, it automatically starts the next. In the second, the conversion begins when an event is received (by software or trigger). Figure B.13 shows its block diagram. Connection with DS PPC 1005 is performed by means of the PHS bus.

As output card the DS 5101 DWO board is used, Figure B.12b. This card allows generation of a TTL pulse pattern upto 16 channels. These pulses can be defined by programming. Figure B.14 shows the block diagram.

The set of three cards are connected into an expansion box that includes a communication card for connection to the PC. In this design, the communication is done through an Ethernet card. The computer just shows the user interface: the real-time calculation is performed by the DS 1005 PPC board. There are two connector panels where the input signals from sensors and TTL output signals were connected; the latter are responsible for generating the gating pulse train required for inverter gate. Figure B.15 shows elements of the configuration of the control system.

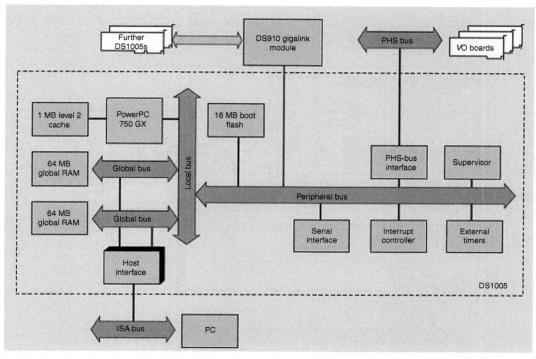

FIGURE B.11 Blocks scheme from DS 1005 PPC board.

(a)   (b)

FIGURE B.12 Input and output board from dSPACE used: (a) DS 2004; (b) DS 5101 DWO.

All cards can be configured and programmed graphically using MATLAB–Simulink. The link between software and hardware is the real-time interface RTI (Real Time Interface) from dSpace. So, when you have a model in MATLAB–Simulink block libraries, RTI allows connection of the cards with Simulink blocks. The RTI block library enables parameterization of the inputs and outputs of the cards. Once connected and configured the inputs and outputs to the model, the C code is generated through the toolbox Real Time Workshop

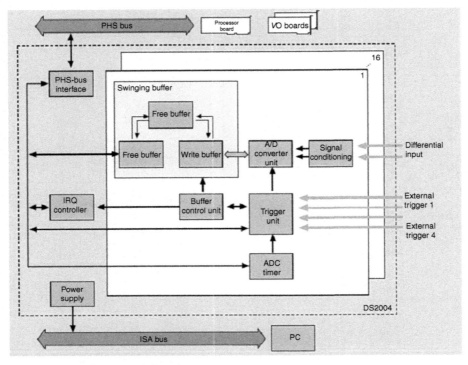

FIGURE B.13 Block scheme from DS 2004 board.

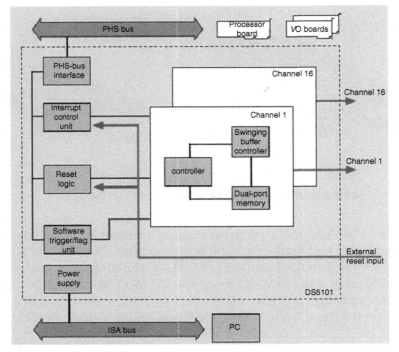

FIGURE B.14 Block scheme from DS 5101 DWO board.

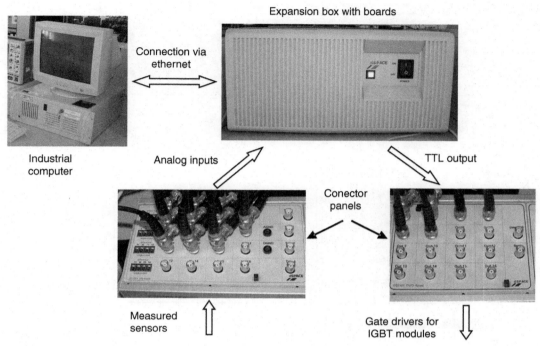

FIGURE B.15 dSpace system configuration.

(RTW) from MATLAB. The real-time model is compiled and read, and starts to run automatically in real-time hardware.

Alternatively, there is another tool called ControlDesk from dSpace that can interact with the system in real time. With this it is possible to monitor the variable set of interest, turning your PC into a virtual instrument that allows the user to control the designed system in real-time. Figure B.16 shows the main screen of the applications that have been done in this work.

The input signals to the control system are voltages and currents. The LV-25-P sensor from LEM is used for the voltage signals, Figure B.17. It is a voltage transducer of the Hall effect. It is a closed loop type, which reduces the effect of magnetic hysteresis, providing an almost linear characteristic with high accuracy. It also provides galvanic isolation between primary and secondary, being able to connect their primary voltage up to 1600 V. The voltage ratio depends on the resistance that is connected to the transducer secondary which should be within a range specified by the manufacturer. In this design, it has chosen a resistance value of 100 Ω, whereby the voltage ratio is 1:240.

Respect to the current sensor has been used the LA35-NP from LEM (Figure B.18). It is a Hall effect sensor of closed loop, with a configurable current ratio. Depending on the type of connection of the pins can achieve until five different relationships. The choice is 1/1000, with a primary nominal current of 35 A.

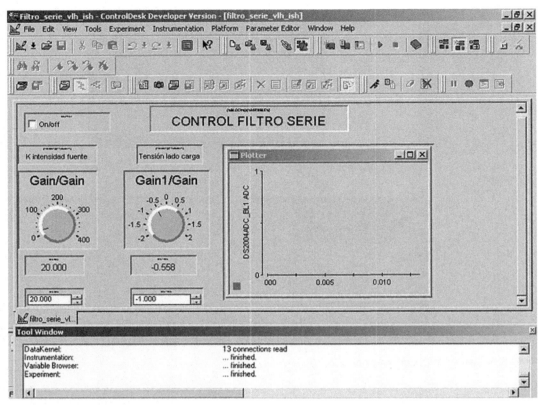

FIGURE B.16 Controldesk.

FIGURE B.17 LV25-P voltage sensor.

The voltage and current sensors are grouped for easy feeding their internal circuitry. Figure B.19 shows a stack composed of 16 current sensor and 9 voltage sensor.

For measures that are included in the different tests that appear in this work has been used a three-phase network analyzer, Fluke 434. This instrument measures virtually every power system parameter, such as voltage, current, power, energy, unbalance, flicker, harmonics, and

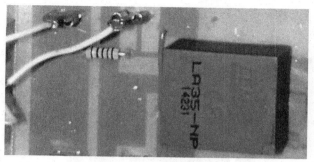

FIGURE B.18 LA35-NP current sensors.

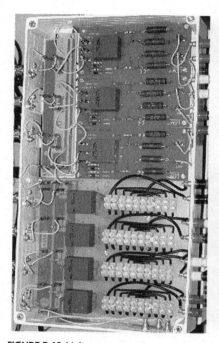

FIGURE B.19 Voltage a current sensor set.

interharmonics. It has four channels in which the voltage and current of all three phases and neutral are measured simultaneously. This device offers a voltage accuracy of 0.1% and meets all the requirements of IEC 61000-4-30 Class A. Figure B.20 shows an image of the measuring equipment.

   On the other hand, a four channel digital oscilloscope is used, exactly the WaveSurfer 424 from Lecroy, Figure B.21. Among its features are a bandwidth of 200 MHz and a sampling rate of 2 GS/s. It has a vertical resolution of 8 bits and a vertical sensitivity of 1 mV/div. This oscilloscope can display and capture instantaneous signals in an easy and practical way, offering different colors for different channels. The waveforms shown as experimental results have been captured with this oscilloscope.

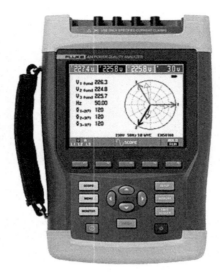

FIGURE B.20 Fluke 434 network analyzer.

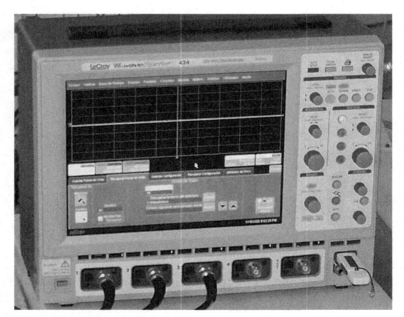

FIGURE B.21 Wavesurfer 424 oscilloscope from Lecroy.

The calculation of the current reference is developed in Simulink. The "Build" tool allows the program to run in the microprocessor built into the DS1005 PPC board from dSpace. Throughout the previous chapters were described the strategies employed, here, Simulink schemes that are designed for different experimental prototypes are described.

In Figure B.22, general scheme in Simulink is shown. The input channels used with DS 2004 board have been included. This scheme has been used for all experimental prototypes developed in Chapters 5 and 6. In some of them has only been necessary measuring the source current or the voltage at the load side, it would be sufficient to cancel in the control scheme the corresponding channels (1–3 for voltage and 5–7 for current). Each input includes a scale factor of 240 to the voltage signals and 1.9644 for the current signals. The main difference between the prototypes to another is the block labeled as "Reference calculation".

Channels 13 and 14 of DS 2004 board are assigned to measure the voltage across the terminals of the capacitors connected to the dc side of the inverter. These values along with the source currents are the inputs of the block that is designated as "secondary control". This block has been implemented control strategy for regulating the voltage on the capacitors. Figure B.23 shows the design of this block, where the reference voltage is 100 V.

Figure B.23 shows the "Direct" block, which determines the positive sequence component of the input vector, in this case the source current vector. Figure B.24 shows the

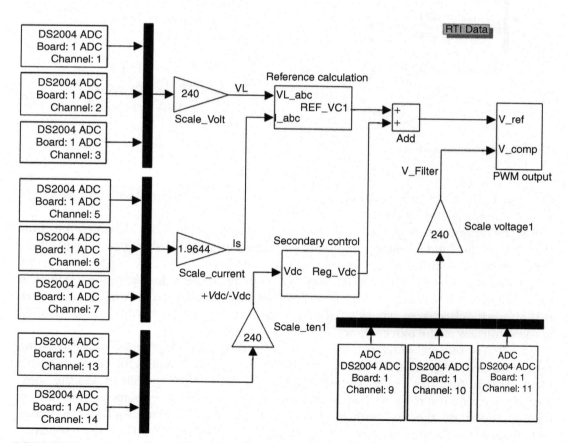

FIGURE B.22 Control scheme from series and hybrid active filters.

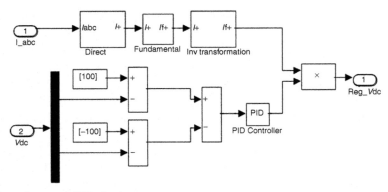

FIGURE B.23 "Secondary control" block scheme.

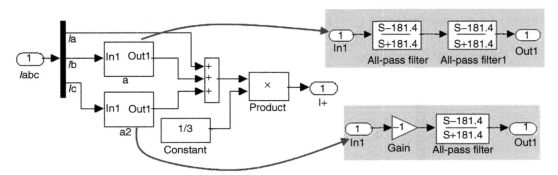

FIGURE B.24 "Direct" block scheme.

structure of the block, where the operator "a" and "a^2" are configured with all-pass filters through their transfer functions.

On the other hand, Figure B.23 includes "Fundamental" block, inside which the computation of the fundamental component of the input signal is developed. Figure B.25 shows its internal structure.

The block labeled "Inv transformation" is the one responsible for obtaining the inverse transformation of Fortescue from the positive sequence fundamental component. Figure B.26 shows block scheme, where "a" and "a^2" operators are included, which are developed in the same way as the block "Direct".

The "PWM output" block is common for all practical developments. In it the gating pulses applied to the different IGBTs are generated. Figure B.27 shows the contents of this block. This in turn has a subsystem designated "PWM", which is what makes the comparison between the reference signal and the output signal of the compensation device. On the other hand, there is another subsystem which will activate or deactivate the active filter from a "stop" signal. Finally, the chosen outputs of DS 5101 DWO board are shown.

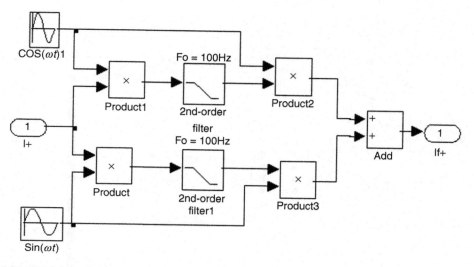

FIGURE B.25 "Fundamental" block scheme.

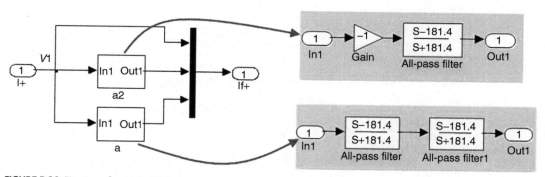

FIGURE B.26 "Inv transformation" block scheme.

In the experimental prototype control strategies have been applied as follows: detection of the source current, detection of the load voltage, and a hybrid strategy, which combines source current and load voltage detection. The block labeled as "Reference calculation" in Figure B.22 includes the scheme shown in Figure B.28. Here, the "Fundamental" block has the same configuration as that shown in Figure B.25. The gain block designated by "K" is used to select the value of the suitable proportionality constant of the source current in the strategies, by detection of the current source and hybrid or cancel source current with $K = 0$ for strategy by detection of the load voltage. Similarly "K_v" is set to "-1" for strategies by detection the load voltage and hybrid. K_v will be "0" for the strategy by the detection of the source current.

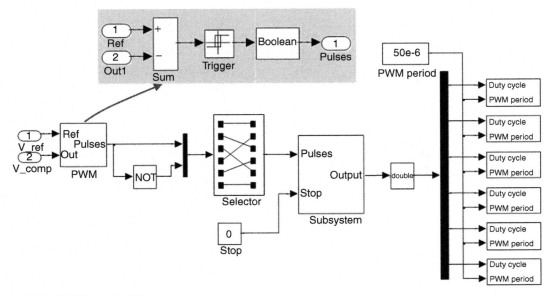

FIGURE B.27 PWM output block.

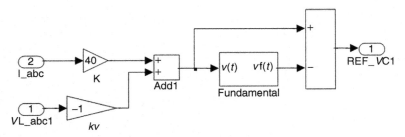

FIGURE B.28 Reference calculation block for control applied to series and hybrid active filter.

B.2 Experimental Platform for Shunt and Series-Shunt APF Tests

This section describes the experimental setup used for the realization of the experimental cases explained in Chapters 4 and 7, with shunt and series-shunt active power filters. A modular experimental platform has been used for this purpose, which is prepared to test different configurations of APFs as well as their respective measurements and passive components. Figure B.29 shows the configuration applied for the UPQC experimental test, where both – series and shunt converters – have been connected to the power circuit, with their corresponding passive components. For the shunt APF experimental case, the series converter is bypassed with the panel selector and a different passive branch $R_P C_P$ is

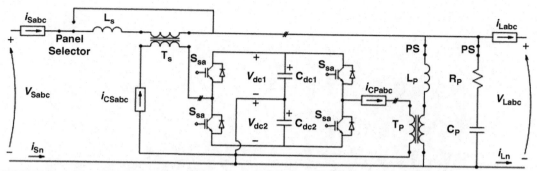

FIGURE B.29 Configuration of the experimental platform for the UPQC case.

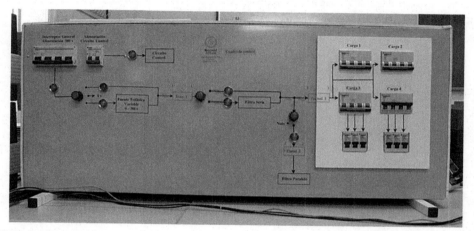

FIGURE B.30 Configuration panel.

connected. In both cases the converters will be controlled through the DSP board dSPACE-DS1103, with the acquisition of the corresponding voltage and current measurements.

The configuration of the experimental platform is made with a connector panel, whose front view is shown in Figure B.30. The series and shunt inverters can be included with panel selectors, between the supply side and load side measurements; and provides for up to four three-phase or single-phase loads to be preconnected. The panel also allows selection between two power sources: directly with the laboratory network or through a variable autotransformer. Figure B.31 shows the back view of the configuration panel with the wiring to the different components of the experimental platform.

The power supply used for the experimental cases is a three-phase variable autotransformer, composed by three single-phase units of 0–230 V, 50 Hz, 25 A, with star connection. The fundamental phase to neutral voltages can be set independently, and the harmonic distortion introduced is the existing in the supply network of the laboratory. Figure B.32 shows the autotransformer set, with the manual cursors and voltage indicators for each phase.

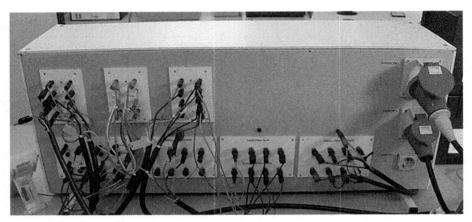

FIGURE B.31 Back front of the configuration panel. Circuit connections.

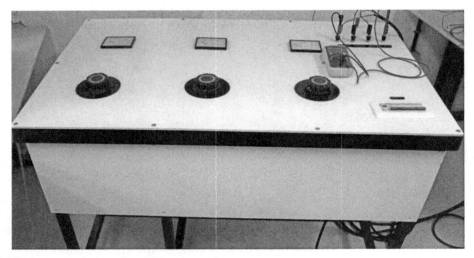

FIGURE B.32 Power supply through variable autotransformer.

The three-phase unbalanced nonlinear load is similar to those applied in the simulation cases (see Figure 4.30): a single-phase bridge diode rectifier with inductive load on the dc side, DLR load, for phase 1; a single-phase bridge diode rectifier with capacitive load on the dc side, DCR load, in phase 2; and a linear resistor in phase 3. Figure B.33 shows the general arrangement of this load, and the right side picture of the figure shows in detail the connection of the diode bridges, smoothing inductances, and the snubber circuit of the DLR load. The diode bridges are made with a GBPC2510 rectifier module which supports a maximum current at the dc side of 25 A and a maximum repetitive peak voltage of 1000 V. The main linear components have been taken from the selectable impedance boxes shown in the figure. The resistances box has a nominal power of 1200 W per phase,

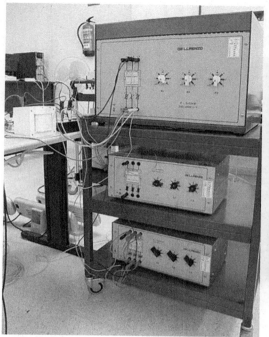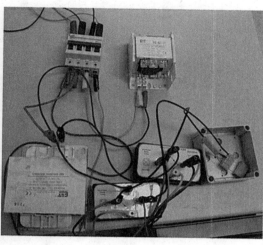

FIGURE B.33 Variable three-phase four-wire unbalanced nonlinear load.

and 900 VAr for the inductors and the capacitors box. This modular set allows to easily configure different load compositions, to produce load changes, etc.

The experimental platform has two three-phase IGBT converters (Semikron SKM50G-B123D) that are connected back to back, with a common dc link composed by two electrolytic capacitors, C_{dc+} and C_{dc-} of 2200 µF, 400 V each one. The middle point of the dc link is connected with the neutral wire of the three-phase line. Figure B.34 shows the arrangement of both inverters. The shunt converter is at the left side of the picture and provides the dc side capacitors; and the series converter is at the right side. The IGBTs of the converters are commanded by SKHI 22A gate drivers that works with input signals of 0–15 V. The series converter has an additional circuit to adapt the 5 V pulses of the DSP board to the voltages levels of the gate drivers, including a validation input to run/stop the gate signals from the control program. The voltage level adaptation for the shunt converter is made in the external circuit that provides the PS current control (see Figure B.45).

The coupling of the shunt converter to the network is made with three single-phase transformers of 4 kVA, with a turn relation of 1:2; to set the reference values of the dc side capacitors at 250 V each. Figure B.35 shows the arrangement of the coupling transformers.

Figure B.36 show the inductors L_P (50 mH) used at the output of the shunt converter. The smaller inductors that appear in the lower part of the picture (5 mH) form part of the flexible configuration system of the general platform to test different compositions for

FIGURE B.34 Shunt and series IGBT converters.

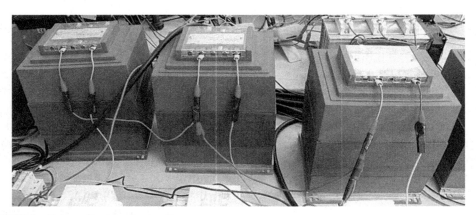

FIGURE B.35 Parallel coupling transformers T_P.

the passive components of shunt APFs; as well as the manual switches that appear at the left side.

Figure B.37 shows the $R_P C_P$ branch used for the shunt APF experimental case, with capacitors of 1 μF, 250 Vac; and resistances of 22 Ω. Figure B.38 shows the $R_P C_P$ branch for the UPQC case, with capacitors of 15 μF, 300 Vac; and resistances of 2.5 Ω.

Figures B.39 and B.40 show the series coupling transformers T_S and smoothing inductors L_S applied at the output of the series converter. The coupling transformers have a nominal power of 4 kVA and a turn relation 1:1; and the series inductors are of 25 mH.

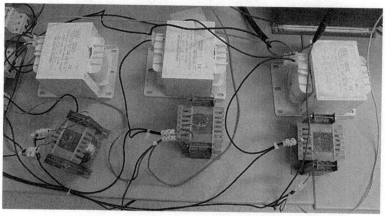

FIGURE B.36 Shunt inductors L_p.

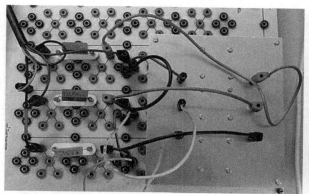

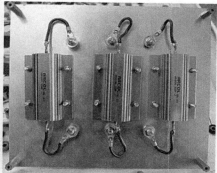

FIGURE B.37 $R_p C_p$ branch for the shunt APF.

The voltage and current sensors used in the platform are the same as those described in Section B.1 (see Figures B.17 and B.18); but the current transducers are configured for a range of 14 A and therefore the ratio between the measured current and the voltage output signal is 2 A/V. Figure B.41 shows the arrangement of the sensors for the whole platform, in a stack of three trays with four current and three voltage measurements each.

Both converters are controlled with the DSP board DS1103 from dSPACE. It has similar features to the DS1005 system explained in Section B.1, but it works with a single board design (see Figure B.42). The DS1103 board has a main processor, PowerPC 604e that runs at 400 MHz and executes the real-time applications. The board is connected with the host PC via an ISA bus and a host interface communications protocol, where the host PC acts only as a user interface; although some control variables can be set from the ControlDesk applications. The Master PPC provides up to 20 analog input channels and 8 analog output channels with a voltage range of ± 10V. Four analog input channels have their own A/D

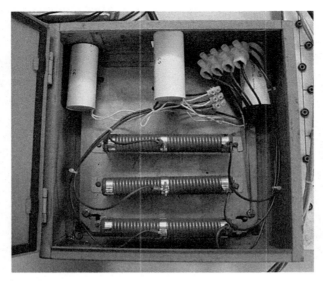

FIGURE B.38 $R_P C_P$ branch for the UPQC.

FIGURE B.39 Series coupling transformers T_S.

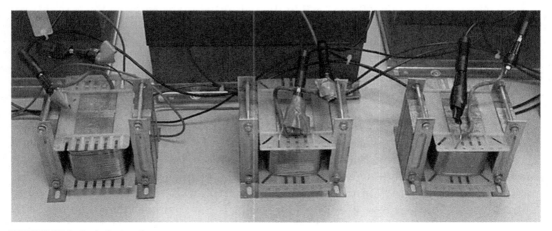

FIGURE B.40 Series inductors L_S.

FIGURE B.41 General stack of voltage and current LEM sensors.

converter and a resolution of 12 bits, and the remaining 16 inputs have a resolution of 16 bits and share the A/D converters in groups of four multiplexed channels; while the analog outputs have a resolution of 14 bits. It also has several configurable I/O bit channels with TTL voltage range, which in this case are used as status signals from external circuits. The board also has a slave DSP (TMS320F240) that runs at 20 MHz, with its own inputs and outputs, focused in PWM pattern generation.

The DS1103 board is connected in a 19″-rack industrial PC, see Figure B.43, where the MATLAB–Simulink and dSPACE–ControlDesk applications are installed to design and develop the control and monitoring programs. The input/output signals are plugged through the connector panel CP1103 shown in Figure B.44. The BNC connectors of the left side correspond to the analog input and output signals of the main processor, while at the right side there are various pin connectors for digital input/output, or the concentrated signals of the slave processor that goes to the series converter.

Figure B.45 shows the external circuit for the realization of the Periodic Sampling current control for the shunt converter. It receives the signals from the i_{CPabc} LEM current sensors and their references generated by the DS1103 board through BNC cables. In the first stage these signals are compared with fast voltage comparators (LM311) to generate a trigger signal per phase. These signals are retained in "D"-type flip-flop circuits (CD4013) that are activated with the rising edge of a clock signal that defines the commutation frequency (20 kHz). These circuits also generate the direct and complementary signals for the gates of each phase branch. In the final stage these six signals are validated, through AND

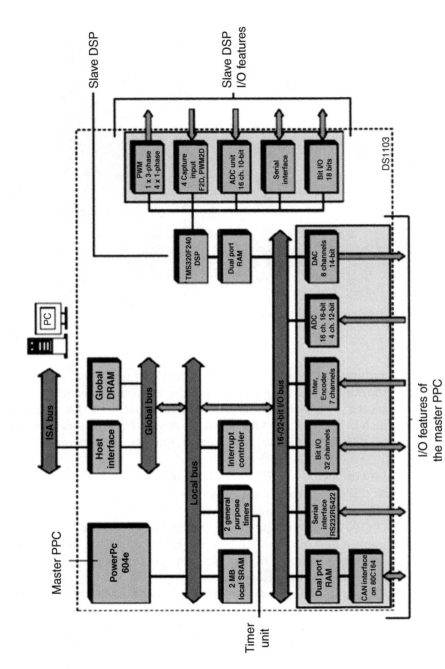

FIGURE B.42 Functional structure of the dSPACE DS1103 board.

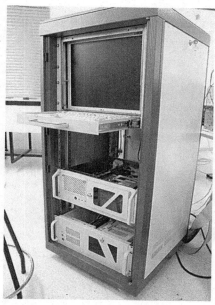

FIGURE B.43 Industrial PC with the dSPACE DS1103 DSP board.

FIGURE B.44 Connector panel CP1103 and signal wiring.

logic gates (SN74LS08N) with an additional digital output generated by the DS1103 board to start/stop the shunt converter.

Figure B.46 shows the external circuit for checking the series converter currents $i_{CS\text{-}abc}$. It checks that these currents are under admissible values for the physical components (±14 A). The analog signals provided by the LEM transducers for these currents come

FIGURE B.45 Implementation of the PS current control for the shunt converter.

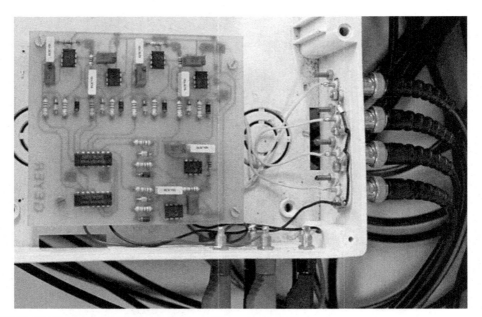

FIGURE B.46 Checking circuit for the currents, i_{Csabc} of the series converter.

through three BNC connectors. In a first stage, these three voltage signals are compared with preset references (±7 V) to check that they are within the acceptable range. This stage is made with differential comparators (LM741) and in the second stage the voltage levels are adapted to the TTL standards. Finally, with AND logic gates (SN74LS08N) a final status signal is derived, with 0 V level as alarm condition. This output signal goes through the fourth BNC connector to a digital input of the connector panel CP1103 (see Figure B.36), where it acts as a validation signal for the activation of the power converters of the UPQC.

Finally, as mentioned in the description of the experimental platform, the control programs can be developed in the Simulink environment; where they are compiled and sent to the DSP board for real-time execution. On the following, the programs used in the experimental cases of this section will be described.

Figure B.47 shows the general arrangement of the MATLAB–Simulink file used to obtain the experimental results of the UPQC case. The block "Medidas" takes the signals of the voltage and current LEM sensors, and scale them to their real values. The block "Control" calculates the internal references of the conditioner in the way exposed in Chapter 7, and determines the values of the compensating currents i_{CPref} of the shunt converter and the duty cycles As of the series converter. The block "Trigger signals" uses the calculated references to set the outputs of the DSP board. The block "Safety" checks that the internal

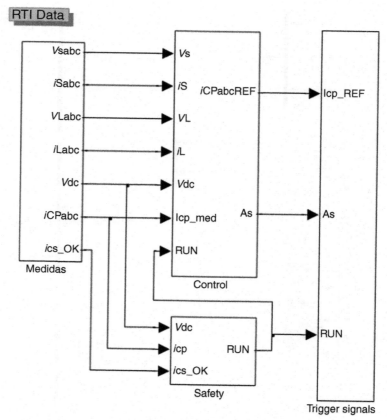

FIGURE B.47 General arrangement of the Simulink file for the UPQC experimental results.

variables of the active conditioner (v_{dc}, i_{CP}, i_{CS}) are under the limits admitted by their physical components, and sets the variable *RUN* to allow the activation of the converters. This variable also activates some parts of the calculation of the control references.

Figure B.48 shows the components of the block "Medidas", with the acquisition, scaling and conditioning of the input signals. The analog signals of voltages and currents of the supply and load sides are acquired through the multiplexed A/D converters CON1 to CON3. Each multiplexer is devoted to the measurements of one phase. After the corresponding scaling, they are rearranged in three-phase buses by magnitude. The dSPACE DS1103 board has four input channels with their own A/D converter, that is used for the measurement of the shunt converter currents, i_{CPabc}, and for the total voltage of the dc side capacitors, V_{dctot}. One channel of the fourth multiplexed A/D converter is used for the measurement of the voltage of the upper dc side capacitor, V_{dc1}. Finally, the currents that flow through the series converter, i_{CSabc}, are also measured with LEM transducers and the results are checked in an external circuit to be under limits. The digital input *Ics_ok* is the output of this circuit and is used to validate the activation of the converters (see Figure B.46).

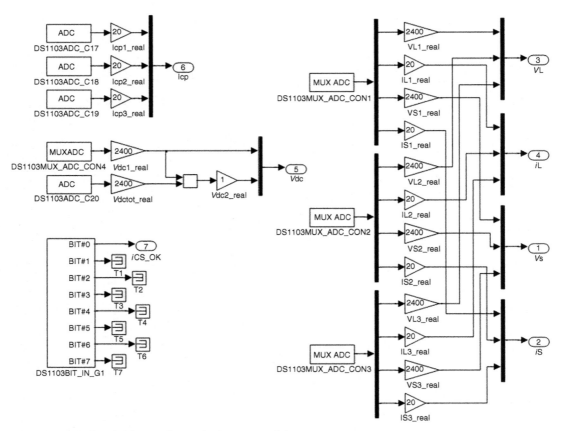

FIGURE B.48 Block "Medidas". Scaling and arrangement of the measurements.

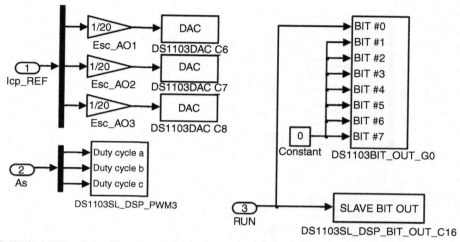

FIGURE B.49 Block "Trigger signals". Scaling and arrangement of the DSP board outputs.

The outputs of the control program are located in the block "Trigger signals", shown in Figure B.49. The current references for the shunt converter are scaled and sent to their respective D/A converters, for the external circuit that makes the PS current control (see Figure B.45). The duty cycles A_S are sent to the slave processor of the DSP board, where it generates the PWM patterns with a switching frequency defined in this control block (20 kHz). The signal RUN is sent as a digital output to validate the gate signals of the power transistors. The signal sent with the main processor is the validation signal for the PS current control circuit (see Figure B.45), and the signal sent with the slave processor goes to the circuit that adapts the TTL voltage levels of the DSP outputs to the CMOS levels needed by the IGBT drivers of the series converter (see Figure B.34).

Figures B.50–B.54 show the checks made in the *Safety* block. The blocks "Cond_Icp" check that the shunt converter currents are below 14 A, in absolute value. When this condition is not satisfied the block "Memory" retains this situation and inhibits the activation of the RUN signal. This condition can be monitorized in the ControlDesk application, and it must be reset manually to activate the power converters again. The block "Cond_Ics_ok" acts only as an interlock for the corresponding signal. The minimum and maximum conditions for the voltage of the dc side capacitors are checked separately to monitorize the condition that has inhibited the gate signals. Finally, the constant *Manual_RUN* is used to activate manually the start/stop of the UPQC, through the ControlDesk application.

Figure B.55 shows the structure of the block "Control". The internal power balance is in the upper part of the figure, with a similar implementation to that explained in the simulation models (see Figure A.9), and sets the main shunt converter current reference *IcpL*. The block "Vs_fund" obtains the balanced, sinusoidal, regulated load voltage reference *VL_REF*. The block "Comp_FS" calculates the reference for the series compensating voltage, and the block "Refs_FS" calculates the duty ratios for the three branches of the series converter, considering the actual values of the dc side voltages. The block "Reg_Vdc" obtains

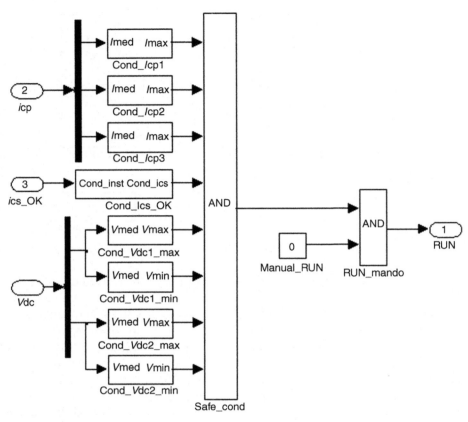

FIGURE B.50 Safety block. Internal checks.

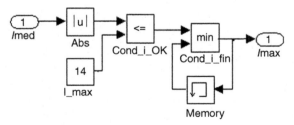

FIGURE B.51 Block "Cond_Icp1".

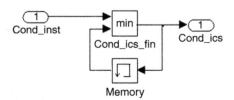

FIGURE B.52 Block "Cond_Ics_ok".

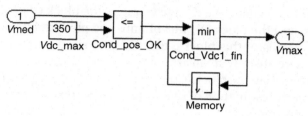

FIGURE B.53 Block "Cond_Vdc1_max".

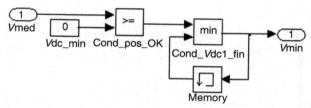

FIGURE B.54 Block "Cond_Vdc1_min".

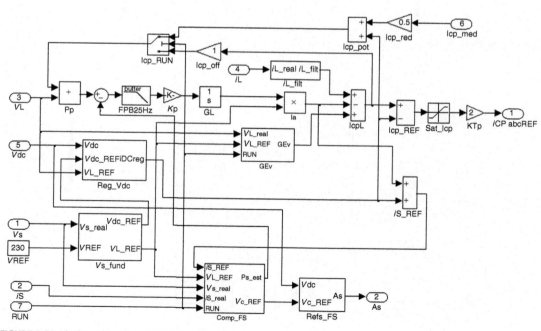

FIGURE B.55 Block "Control". Calculation of the references.

the additional current to regulate the dc side voltages, the block *GEv* calculates the voltage error damping term, and the block "IL_filt" obtains the pre filtered load currents. Figures B.56–B.61 show the details of this blocks, that are very similar to those explained in the simulation model. There are only small arranging differences, like the concentration of the series converter references in the block "Comp_FS", including the calculation of the current damping term *REis* or the estimation of the instantaneous power *Ps* of the series converter for the internal power balance.

Everything stated so far corresponds to the file used for the UPQC experimental case. For the experimental case with the shunt APF included in Chapter 4, a simplified version of this file has been used, eliminating the unnecessary components. The block "Medidas"

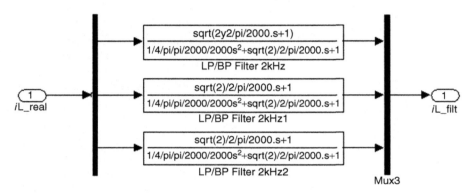

FIGURE B.56 Block "IL_filt".

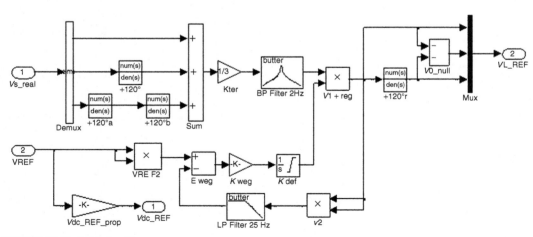

FIGURE B.57 Block "Vs_fund".

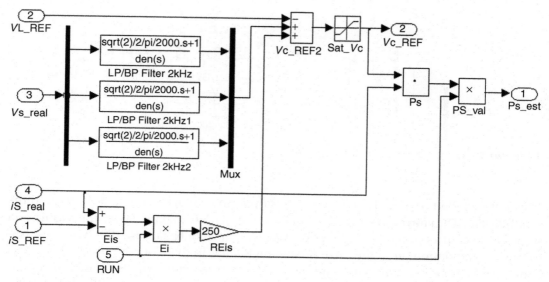

FIGURE B.58 Block "Comp_FS".

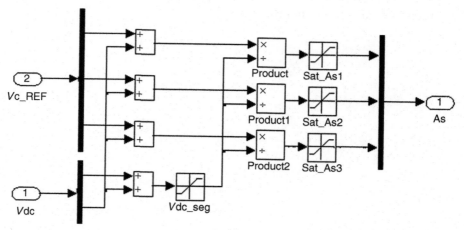

FIGURE B.59 Block "Refs_FS".

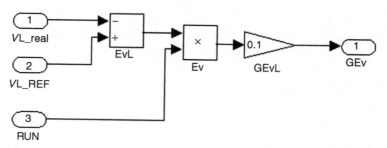

FIGURE B.60 Block "GEv".

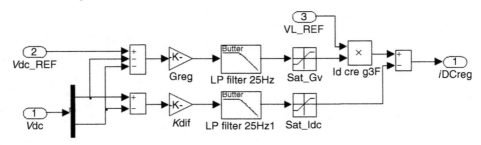

FIGURE B.61 Block "Reg_Vdc".

does not include the measurements of the supply side voltages V_S and currents I_S in the multiplexed ADC converters, nor the validation bit *ics_ok*; making faster the acquisition stage of the control program. The block "Trigger signals" does not use the PWM3 generator nor the validation bit of the slave processor. Finally, in the block "Control", the blocks "Comp_FS", "Refs_FS," and "GEv" are removed. And the load currents iL are directly included in the power balance, without the previous filtering. These simplifications give a reduced real time control program for the DSP board, that allows to set a sampling period of 50 μs for the shunt APF experimental case.

Index